Decoding the Stars

Jesuit Studies

MODERNITY THROUGH THE PRISM OF JESUIT HISTORY

VOLUME 16

The titles published in this series are listed at *brill.com/js*

Decoding the Stars

A Biography of Angelo Secchi, Jesuit and Scientist

By

Ileana Chinnici

BRILL

LEIDEN | BOSTON

Cover illustration: portrait of Angelo Secchi. Courtesy of INAF-OAR.

Library of Congress Cataloging-in-Publication Data

Names: Chinnici, Ileana, author.
Title: Decoding the stars : a biography of Angelo Secchi, Jesuit and scientist / by Ileana Chinnici.
Description: Leiden ; Boston : Brill, [2019] | Series: Jesuit studies: modernity through the prism of Jesuit history, ISSN 2214-3289 ; volume 16 | Includes bibliographical references and index.
Identifiers: LCCN 2018051687 (print) | LCCN 2018053239 (ebook) | ISBN 9789004387331 (ebook) | ISBN 9789004387294 (hardback : alk. paper)
Subjects: LCSH: Secchi, Angelo, 1818-1878. | Astronomers–Italy–Biography. | Jesuits–Italy–Biography. | Astronomy–History–19th century. | Astrophysics–History–19th century.
Classification: LCC QB36.S4 (ebook) | LCC QB36.S4 C45 2019 (print) | DDC 520.92 [B]–dc23
LC record available at https://lccn.loc.gov/2018051687

Typeface for the Latin, Greek, and Cyrillic scripts: "Brill". See and download: brill.com/brill-typeface.

ISSN 2214-3289
ISBN 978-90-04-38729-4 (hardback)
ISBN 978-90-04-38733-1 (e-book)

Contents

Foreword

Br. Guy Consolmagno, S.J.
Director, Vatican Observatory

Angelo Secchi might be best known today for the two instruments that bear his name. The Secchi Disk he apparently made one afternoon, as a favor for a friend looking for a repeatable way to measure the clarity of littoral water; the other is an instrument package on the NASA STEREO spacecraft monitoring solar activity, the *Sun–Earth Connection Coronal and Heliosphere Investigation* (SECCHI) package. Perhaps one cannot ask for more lasting fame nowadays than to be turned into a NASA acronym. Unlike the Secchi Disk, this instrument does at least actually relate directly to a field of science that Secchi pioneered.

It is a shame that Secchi is not better known. His life and personality was as rich and colorful as Galileo Galilei's. In fact, this comparison may be unfair—to Secchi.

Galileo, for all his famous problems with church authorities, was never forced into exile from his homeland. Secchi was driven out of Italy once, and nearly had to flee a second time, due to the agitations of his enemies in the highest levels of the anti-clerical Italian regime.

Galileo was supported by two of the wealthiest and most powerful city-states of his era, first Venice and then Florence, and he operated at the center of the scientific world of his era. Secchi had to work with the limited budget of the Roman College, in competition with the national observatories of France and England and all their resources. And he mostly worked and wrote in a language that many of his contemporaries did not read. Indeed, his rivals in the UK (including the editor of *Nature*, Norman Lockyer) made sure that Secchi's most important works were never translated into English.

Galileo, after his first discoveries with the telescope, spent the rest of his career writing works of philosophy and popularization. Secchi was also well published in popularization and philosophy, but he continued to produce profound advances in science, in half a dozen widely divergent fields: meteorology, time-keeping, surveying and geodesy, planetary geology, spectroscopy of the stars and the planets, stellar classification, solar physics, and the above-mentioned "Sun–Earth connection" between solar activity and the Earth's magnetosphere.

Before Secchi, astronomy was only *astrometry* and *celestial mechanics*: the study of the positions of the stars and planets and the calculation of their orbits. This meant merely carrying on the same program that Tycho Brahe and

Johannes Kepler had been doing back in the sixteenth and seventeenth centuries (and indeed pioneered by the ancients, Hipparchus and Ptolemy). Astrometry had a very practical purpose: the ancients wanted to be able to predict planetary positions to improve their astrology, while the national observatories of Secchi's time wanted better star positions for commercial navigation. But Secchi was the first person to think about the lights in the sky as places with their own compositions and histories. He was curious about them not for commercial reasons, but for their own sake. He, and those he encouraged, invented modern *astrophysics*—and planetary sciences.

Others had noted that the Sun had a unique spectrum; Secchi used spectroscopy to answer specific questions about the solar constitution. Others had noted that one could even see spectra in bright stars; only Secchi planned to measure the spectra of hundreds of stars, looking for patterns and ways to classify them. Furthermore, he also took the spectra of planets, trying to identify their atmospheres.

Larger telescopes than Secchi's had existed since the end of the eighteenth century, and Mars makes an especially close approach to Earth every twenty-two years or so. But no one before Secchi had bothered to actually try to inspect the surface of Mars and make sense of its surface markings in terms of geological processes. Secchi's work inspired the maps of Giovanni Schiaparelli at the next favorable conjunction, twenty-two years later. Schiaparelli and Percival Lowell, infamously, borrowed Secchi's term "canale" to describe features that turned out to be optical illusions. By contrast, the features that Secchi described as *canale* are real.

My comparison of Secchi and Galileo can also extend to their combative personalities, which reacted to the turbulent currents of the politics of their times. One result was that Secchi was subjected to no shortage of slander, and as mentioned above, much of his most important work was suppressed by English scientists who perhaps did not want to reveal how much of his work anticipated their own efforts.

For that very reason, the publication of this book, in conjunction with the celebrations of Secchi's bicentennial of birth in 2018, offers a wonderful opportunity to rectify that slight. It is also a welcome chance to introduce the world to a fascinating and colorful character.

One final comparison: Galileo died quietly in his home in Arcetri, at the age of seventy-eight. Sadly, Secchi died at the much younger age of fifty-nine, his health worn down by the stresses of his travels and his political fights. (Of course, if Galileo had died at fifty-nine, he'd never have had to endure his infamous trial ...)

Foreword

Nichi D'Amico
President of the National Institute for Astrophysics (INAF)

Angelo Secchi is one of the most complex and interesting characters of nineteenth-century astronomy. Although a physicist by training, he became a director of an astronomical observatory, where he applied his knowledge of physics to the field of astronomy, testing new techniques such as photography and spectroscopy. As a result of this work, Secchi is now recognized as one of the founders of modern astrophysics: his early classification work has provided the basis for the current spectral classification of stars, and modern astronomical research has confirmed many of his intuitions about the Sun and the Sun–Earth connection, as well as the existence of dark matter and exoplanets. As a physicist, he was interested in exploring all natural phenomena, looking to their connections. Accordingly, he extended his studies from astronomy and astrophysics to meteorology, geomagnetism, and geophysics. Despite his eclecticism, reminiscent of the natural philosophers of the seventeenth and eighteenth centuries, Secchi was in many ways a very modern scientist. Indeed, his multidisciplinary approach to astronomical phenomena heralded the more recent development of multi-messenger astronomy, related to the detection of gravitational waves.

The following pages outline not only Secchi the scientist but also Secchi the man: a man who appreciated life, loved to travel, and enjoyed meeting people; someone who dared to challenge the ideologies of his epoch and who believed that every social class should have access to a scientific education and knowledge of the latest scientific discoveries. As well as providing an invaluable account of Secchi's life and work—something that has been sorely lacking in the English-language scholarship—this biography will be especially stimulating for those interested in the evolution of astronomy as a discipline from the nineteenth century onward. The book also serves to confirm the importance of preserving the heritage connected with nineteenth-century astronomy, including the books and archives without which this biography could never have been written.

Preface

Ileana Chinnici

One of the main difficulties in writing the biography of a figure like Angelo Secchi, Jesuit and pioneer of astrophysics, is to distinguish between history and rhetoric, namely to ensure that the biography does not become a hagiography. The circumstances of Secchi's life and the historical period in which he lived would make a fine plot for a novel, and it is difficult to avoid the risk of making this scientist a martyr or a romantic hero. Moreover, Secchi's Jesuits companions at the Collegio Romano, who were among his first biographers, were often biased and used contradictory sources, so that reliable biographies are often hard to find.

To avoid this risk, this book is conceived as a sort of "autobiography" by Secchi himself, in the sense that it is a biography that is mostly written in his own words. The continual references to quotations of his texts, publications, letters, or diaries aim to give the reader a well-rounded portrait of the Jesuit scientist, letting him tell the story of his life himself. His beautiful prose allows the biographer to have just to provide a frame, to set the scientific and historical context of the time.

Although this book examines Secchi's main scientific contributions, especially to astrophysics, the reader will not find detailed scientific content or complicated technical language: the book is deliberately written for a non-specialist audience, one having only a basic background in physics and astronomy.

Finally, it must be stated that this biography does not purport to be comprehensive, as it is beyond the scope of this work to provide an encyclopedic survey of all the questions in which Secchi was involved. The complexity of the figure of Secchi would require considerably more time and expertise to be analyzed in such depth. There are many aspects of his life and scientific work that remain unexplored, while others are only mentioned briefly here. There are plenty of documents and sources (manuscripts, publications, and bibliographies, especially those written by Jesuits) that have not yet been fully examined. The extended archives on Secchi at the Gregorian University, which have been the starting point for this biography, are still largely unexplored, especially the immense corpus of his correspondence, which would deserve to be examined in detail. Secchi's relationships with physicists and meteorologists are mostly uninvestigated, as well as some aspects related to his religious life and his influence on Catholicism. Hence the intention of this book is to stimulate scholars to study the figure of Secchi in a wider context. The bicentennial of his birth represents a good opportunity to highlight lesser-known interdisciplinary aspects of this protagonist of nineteenth-century science.

Acknowledgments

A wide range of people and institutions have contributed to the production of this book which took many years. First and foremost, I am extremely grateful to the Specola Vaticana and the National Institute for Astrophysics (INAF) for generously providing the finance to cover the costs of copy-editing, and I would also like to thank Guy Consolmagno and Nichi D'Amico for writing the prefaces. I am especially indebted to Tim Page, whose excellent and accurate copy-editing work greatly improved the quality of this book.

Many other people have also contributed to this work, providing their assistance in finding documents, publications, and pictures. I would like to thank Aldo Altamore (Roma Tre University, Rome), Renzo Lay (Roccadipapa, Rome, Italy), Marco Faccini, and Marco Ferrucci (INAF-Osservatorio Astronomico di Roma, Italy) for helping me retrieve materials related to Angelo Secchi in Rome and for sending me relevant photographs. I am also grateful to the director of the Pontifical Gregorian University Archives in Rome, Martín M. Morales, as well as the staff at the archives, Cristina Berna, Lorenzo Mancini, and Irene Pedretti. Mention must also be made of Fr. Massimo Nevola, S.J. (rector of St. Ignatius Church, Rome) and Maria Macchi (Historical Archive of the Euro-Mediterranean Province of the Society of Jesus, Rome), who provided invaluable assistance in finding materials related to Secchi's entrance into the Society of Jesus.

Thanks are also due to David Knight (Stonyhurst College, United Kingdom) for his help in my archival research into Secchi's sojourn at Stonyhurst. Credit should also be given to Antonio Garrido (INAF-Osservatorio Astronomico di Palermo, Italy) for helping me contact Ignacio Husillos Tamarit O.C.D. (Archivo Desierto de Las Palmas, Benicàssim, Castellón de la Plana, Spain), whom I wish to thank for the information he provided about Secchi's expedition to Spain. In addition, I am grateful to William Sheehan and Richard McKim (British Astronomical Association) for providing information about Secchi's colored drawings of Mars, which have been retrieved at Rome Observatory.

A considerable amount of assistance was also provided by Donata Randazzo (INAF-Osservatorio Astronomico di Palermo, Italy), Antonella Gasperini, Simone Bianchi (INAF-Osservatorio Astronomico di Arcetri, Florence, Italy), and Paolo Brenni (CNR, Fondazione Scienza e Tecnica, Florence, Italy), who provided me with documents and publications related to Secchi's scientific work. Similarly, I wish to thank Giuseppe Adriano Rossi (Deputazione di Storia Patria per le Antiche Province Modenesi—Sezione di Reggio Emilia, Italy) and Simonetta Secchi (Reggio Emilia, Italy) for their precious help in cross-checking biografical information on Secchi's childhood and family as well as Don Augusto Gambarelli (Archivio Diocesano di Reggio Emilia) for the identification of Secchi's teachers. Thanks also to Roberto Passanisi (Rotary Club Syracuse, Italy), who kindly provided the photograph of the sundial Secchi built at Augusta (Syracuse, Italy), and to Oleh Petruk (Institute for Applied Problems in Mechanics and Mathematics, Lviv, Ukraine), who translated Arseny Tarkovsky's poem on Secchi into English.

Special mention is due to Louis Caruana (Pontifical Gregorian University, Specola Vaticana) for suggesting that I write this book during a conversation about Secchi. He gave me numerous suggestions and helped me greatly during the difficulties I encountered while writing the biography—without his encouragement, this book would never have been written.

Above all, I would like to thank all the staff of the Specola Vaticana (Vatican City), and especially Sabino Maffeo, S.J., Guy Consolmagno, S.J., Paul Mueller, S.J., and Giuseppe Koch, S.J. Their encouragement, support, and assistance played a crucial part in this book: many of its pages were born in the quiet silent rooms of the Specola, in the library and the guest-room, often after sharing prayers and meals with the Jesuit community.

Illustrations

Abbreviations

APUG	Archivio della Pontificia Università Gregoriana (Archive of the Gregorian Pontifical University), Rome
ASPEM	Archivio Storico della Provincia Euro-Mediterranea dei Gesuiti (Historical Archive of the Euro-Mediterranean Jesuit Province), Rome
ACREA	Archivio del Consiglio per la ricerca in agricoltura e l'analisi dell'economia agraria (Council for Agricultural Research and Economics, Historical Archive), Rome
AOAB	Archive of INAF-Osservatorio Astronomico di Brera (Brera Astronomical Observatory), Milan
AOAPa	Archive of INAF-Osservatorio Astronomico di Palermo (Palermo Astronomical Observatory)
AOAR	Archive of INAF-Osservatorio Astronomico di Roma (Rome Astronomical Observatory), Monte Porzio Catone (Rome).
CNR	Consiglio Nazionale delle Ricerche (National Research Council)
FS	Fonds "Angelo Secchi"
INAF	Istituto Nazionale di Astrofisica (National Institute for Astrophysics)
OAA	Osservatorio Astrofisico di Arcetri (Arcetri Astrophysical Observatory), Florence
OAC	Osservatorio Astronomico di Capodimonte (Capodimonte Astronomical Observatory), Naples
OAPa	Osservatorio Astronomico di Palermo (Palermo Astronomical Observatory)
OAR	Osservatorio Astronomico di Roma (Rome Astronomical Observatory)
SV	Specola Vaticana (Vatican Observatory), Vatican City
UEHA	University of Exeter, Historical Archives, Fonds "Lockyer"

INTRODUCTION

A Century in Ferment

In the history of science, the nineteenth century was marked by increasing specialization and the transformation of entire scientific disciplines due to the introduction of new scientific techniques and theories. Before proceeding further, it is important to take note of the historical context in which these changes took place in order to understand the broader context in which Secchi lived and the events and ideas that informed his work. As we will see, Secchi was very much a man of his time, his life and career having been strongly influenced by the cultural and political changes that took place throughout this highly turbulent period.

Politically and militarily, the early nineteenth century was dominated by a number of upheavals, beginning with the Napoleonic Wars (1803–15), followed by an attempt to restore a stable order on the continent with the Congress of Vienna (1814–15). In the second half of the century, a number of additional adjustments took place as a result of the Crimean War (1853–55), the Austro-Prussian War (1866), and the Franco-Prussian War (1870–71). These conflicts in turn led to the consolidation of the German Empire, the political unification of Italy, and the establishment of the French Republic, resulting from the policies of Otto von Bismarck (1815–98, in office 1871–90), Camillo Benso, count of Cavour (1810–61, in office 1848–61), and Adolphe Thiers (1797–1877, in office 1871–72) respectively.

In Eastern Europe, meanwhile, the Russian Empire had begun to totter in the face of attempts at reform, an upsurge of nationalism, and as a result of its expansion into Central Asia, as well as uprisings in Poland, attacks on the tsar, and war with Turkey (1877–78). Across the Channel, the economic and commercial expansion of Victorian Britain, with its many colonies, was proceeding apace, while, overseas, the United States was mired in the Civil War (1861–65).

The nineteenth century was a period of remarkable transformations in science and technology. In physics and chemistry, these changes included the development of the concept of the atom, the kinetic theory of gases, and electromagnetism, as well as advances in organic chemistry and the compilation of Dmitri Mendeleev's (1834–1907) periodic table. In biology, the discovery of Gregor Mendel's (1822–84) laws and the controversy over the origin of species between Darwinists and creationists were major topics of debate. In geology, the debate over the age of the Earth and the dispute between uniformitarianism and catastrophism (theories that interpreted the origin of the Earth

 | DOI:10.1163/9789004387331_002

as the result of slow geologic and geomorphic processes and of supernatural and catastrophic events) continued for many decades. The nineteenth century was also a period of numerous inventions, including the steam locomotive, the telegraph, the telephone, the electric motor, and the combustion engine, which led to profound changes in transportation and communication, with remarkable economic and social consequences. There were also improvements in medicine and hygiene, with vaccination, sanitization, and disease prevention policies. Science began to specialize, and new approaches to the development of data collection emerged with the aim of cataloging natural objects and studying their "evolution"—a concept that passed from the purely biological to the general context, including the field of astronomy.

1 Astronomy in the Nineteenth Century

In the field of astronomy, the nineteenth century opened with the discovery of the first minor (or dwarf) planet, Ceres, in 1801, followed by Pallas (1802), Juno (1804), and Vesta (1807); the publication of the final volumes of the *Mécanique céleste* (Celestial mechanics), the epoch-making treatise by Pierre-Simon de Laplace (1749–1827), in 1799–1825; and the development of powerful mathematical tools, like the theory of least squares, by Carl Friedrich Gauss (1777–1855) for orbit calculations in special cases, as formulated in the treatise *Theoria motus corporum coelestium* (Theory of the motion of the heavenly bodies [1809]). Throughout the century, celestial mechanics triumphed: the main result was the discovery of Neptune (1846), identified by making calculations of the perturbations of Uranus and pointing a telescope at the calculated position. Improvements in optical instrumentation also allowed for the discovery of many asteroids and comets and the successful application of the laws of celestial mechanics to calculate their orbital parameters. The compilation of extended catalogs of stars and other celestial objects, which was one of the main projects of eighteenth-century astronomy, also continued: the star catalog by Jérôme de Lalande (1732–1807) appeared in 1801; the catalog by Giuseppe Piazzi (1746–1826), in two editions, in 1803 and 1814; the *Bonner Durchmusterung* (Bonn systematic survey) by Friedrich Argelander (1799–1875) was compiled in 1859–63; and the *Uranometria Argentina* was published in 1879 by Eduard Schönfeld (1828–91) and Benjamin A. Gould (1824–96). A catalog of double and multiple stars by Friedrich Wilhelm von Struve (1793–1864) appeared in 1837, while a catalog of nebulae, completed in 1802 by William Herschel (1738–1822), was expanded in 1864 by his son, John (1792–1871).

In the second half of the century, there was a major breakthrough in the discipline of astronomy: the application of spectroscopy[1] led to the birth of a new scientific discipline, called "physical astronomy" or astrophysics.[2] The laws of radiation, published in 1859 by Gustav Kirchhoff (1824–87),[3] supplied the theoretical foundation for a correct interpretation of spectral lines, first observed by William Wollaston (1766–1828) in 1802 and classified by Joseph von Fraunhofer (1787–1826) in 1817. Thus it became possible to study the physical constitution of heavenly bodies—a result hitherto considered unattainable.[4] Initially, only a few astronomers—and Secchi was among them—recognized the promise of astrophysics and its advantages in achieving important results with only modest technical equipment. These "innovators," ready to bet on new astronomy, were mainly amateur astronomers with a professional background in other fields, as well as astronomers who had a background in physics, chemistry, engineering—hence free from the weight of a "classical" tradition.

The new astronomy—so called in order to distinguish it from traditional astronomy, which was mainly based on celestial mechanics and positional astronomy—was slow to gain acceptance from the "classical" astronomers, who did not recognize it as a scientific discipline, or at least failed to do so initially. Most of these astronomers viewed astrophysics with suspicion and held to a traditional or "purist" view of astronomy: "What astronomy must do has always been clear—it must lay down the rules for determining the motions of the heavenly bodies as they appear to us from the Earth. Everything else that can be learned about the heavenly bodies [...] is not properly of astronomical interest."[5]

1 On the history of spectroscopy, its techniques, and applications, see Charlotte Bigg and Klaus Staubermann, eds., "Spectroscope Histories," *Nuncius* 17, no. 2 (2002): 583–689; 18, no. 2 (2003): 735–852.

2 On the early applications of spectroscopy to astronomy, see John B. Hearnshaw, *The Analysis of Starlight* (Cambridge: Cambridge University Press, 1986); and Arthur J. Meadows, *Early Solar Physics* (Oxford: Pergamon Press, 1970). Astrophysics obtained the status of a scientific discipline at an earlier date than spectroscopy; see Klaus Hentschel, *Mapping the Spectrum* (Oxford: Oxford University Press, 2002).

3 See Gustav R. Kirchhoff, "Über den Zusammenhang zwischen Emission und Absorption von Licht und Wärme," *Monatsberichte Berliner Akademie* (1859): 783–787.

4 See, e.g., Auguste Comte, *Cours de philosophie positive*, tome 2 (Paris: Bachelier, 1835), 8–9.

5 Friedrich Wilhelm Bessel (1784–1846) quoted in Karl Hufbauer, *Exploring the Sun* (Baltimore: Johns Hopkins University Press, 1993), 43.

For a long time, this attitude led many astronomers to view astrophysics as a spurious scientific discipline, "a subject both physically and conceptually messy":[6]

> As yet, astrophysical investigations are far from the standard of scientific accuracy possessed by classical astronomy, which [...] rightfully occupies the premier place among the experimental sciences. God forbid that astronomy should be carried away by fascination with novelty and diverge from [its] essential basis, which has been sanctified for centuries [...].[7]

Hence many "classical" astronomers kept their distance from so-called physical astronomy; although they did not openly oppose the new science, they would often admit their unease at its development:

> This new branch of the old astronomy has a completely new and distinctive character. Here, geometry and mechanics are replaced by physics and chemistry. Everything, [...] instruments and staff, had to take a separate path. We find physicists, such as Huggins,[8] [...] or Lockyer,[9] associated with chemists such as Frankland.[10] [...] This new branch has taken root and is developing very quickly. For us, old-fashioned astronomers, it is very difficult to recognize, for its ideas, its methods, its goals, and even the spirit that reigns, are very different from ours. [...]
>
> It is the case of applying an important law [...] that of division of work. [...] The two sciences [physical astronomy and classical astronomy] will develop in parallel, without getting in each other's way, but using different aptitudes.[11]

---

6 See Arthur J. Meadows, "The Origin of Astrophysics," in *The General History of Astronomy, 4A: Astrophysics and Twentieth-Century Astronomy to 1950*, ed. Owen Gingerich, part A (Cambridge: Cambridge University Press, 1984), 3–15, here 14. See also the quotation from Admiral William Henry Smyth (1788–1865).

7 Text by Otto Wilhelm Struve (see Chapter 4) quoted in Arthur J. Meadows, "The New Astronomy," in Gingerich, *General History of Astronomy*, 59–72, here 61.

8 William Huggins was a pioneer in the spectroscopic investigations of stars and nebulae (see below); on his life and work, see Barbara J. Becker, *Unravelling Starlight: William and Margaret Huggins and the Rise of the New Astronomy* (Cambridge: Cambridge University Press, 2011).

9 Norman Lockyer discovered helium and was one of the most important contributors to the development of solar physics (see below); on his life and work, see, e.g., Arthur J. Meadows, *Science and Controversy: A Biography of Sir Norman Lockyer* (Cambridge, MA: MIT Press, 1972).

10 Edward Frankland (1825–99) was one of Lockyer's most important collaborators.

11 *[L'on] voudra bien considérer le caractère tout nouveau que revêt cette branche nouvelle de la vieille Astronomie. Ce ne sont plus ici la géométrie et le Mécanique qui dominent: c'est*

Consequently, astrophysics mainly developed in new institutions, separate from the traditional observatories. Within a few years, between 1874 and 1880, thanks to astrophysicists such as Jules C. Janssen (1824–1907), Hermann C. Vogel (1841–1907), and J. Norman Lockyer (1836–1920), the astrophysical observatories of Potsdam (Germany), Meudon (France), and South Kensington (England) had been established as standalone scientific institutions with the aim of developing the field of new astronomy without interfering with classical astronomy.[12]

Secchi was a forerunner in this field. Though initially a "traditional" observatory, the Collegio Romano Observatory, which Secchi directed for almost thirty years, soon came to specialize in physical astronomy and, in a few years, was directed toward the new frontiers of astrophysics. Other astronomers later tried to introduce spectroscopic studies into "traditional" observatories, and the establishment of the "Società degli Spettroscopisti Italiani" (see Chapter 8) is the remarkable result of this cooperative effort, inspired by Secchi, which played an important role in the gradual recognition of astrophysics as a new scientific discipline.[13]

This is one of the myriad reasons why Secchi stands out among the protagonists of nineteenth-century astronomy. In some of his autobiographical notes, Secchi presented himself as someone who "brought, in a manner of speaking, the flavor of physics into astronomy and can be considered one of the most active and happy founders of this branch of modern astronomy."[14]

---

la Physique et la Chimie. Tout, [...] *instruments et personnel, tout a dû prendre un tour spécial. Nous y trouvons des physiciens, comme Huggins* [...] *ou Lockyer, associés à des chimistes, comme Frankland.* [...] *La branche nouvelle a pris racine dans un sol à elle, et s'y développe rapidement. Nous autres, astronomes anciens, nous avons peine à nous y reconnaître, tant les idées, les méthodes, les objets que l'on a en vue, et jusqu'à l'esprit qui y règne, diffèrent des nôtres.* [...] *C'est ici le cas d'invoquer une grande loi* [...] *celle de la division du travail.* [...] *Les deux sciences se développeront ainsi parallèlement sans se gêner, en utilisant des aptitudes différentes*; Hervé A. Faye, "Rapport de la Commission nommée le 17 août pour préparer une réponse à la lettre adressée par M. Le Ministre de l'Instruction Publique, au sujet de l'opportunité de la création d'un Observatoire d'Astronomie Physique aux environs de Paris," Académie des Sciences, séance du 2 novembre 1874, Paris Observatory Archives, Meudon Section, Dossier Janssen.

12 On the formation and consolidation of the solar physics community, see, e.g., Hufbauer, *Exploring the Sun*, 59–67.

13 See Ileana Chinnici, "The Società degli Spettroscopisti Italiani: Birth and Evolution," *Annals of Science* 65, no. 3 (2008): 393–438.

14 Archives of the Pontifical Gregorian University, Fonds "Angelo Secchi" (hereinafter APUG, FS) 23.I.7.

2 The Nineteenth Century in the Catholic Church

This century of crucial changes could not but affect the Catholic Church. The key figure of this era was Pius IX (1792–1878, r.1846–78), who initially attempted a renewal of the Papal States, turning them into a "modern" state in line with the movement of reform granted by other sovereigns in the Italian Peninsula in 1848, but who then, worried by the spread of anti-clericalism, locked himself into a defense of the establishment. In 1848, Pius first suggested that the Jesuits leave Italy in order to keep them from the anti-clerical persecutions, and then, because of republican insurrections, left Rome himself and retired into voluntary exile at Gaeta, in the Kingdom of the Two Sicilies. He returned to Rome seventeen months later, after the failure of the Roman Republic (1849), at which time the Jesuits were also able to return from their diaspora. However, in the 1860s, the Papal States lost part of their territories when they were annexed by the Kingdom of Italy, which suppressed the religious orders and confiscated church properties. Pius reacted by excommunicating the Savoy Government and, in the face of the progress of liberalism, atheism, and socialism, issued the Syllabus of Errors (1864) condemning those ideologies. In 1870, when French troops who were defending Rome left the city to fight on the front lines of Prussia, the Kingdom of Italy launched its final attack and conquered Rome. The eternal city became the capital of the new Italian state, and the pope remained confined within the Vatican palaces, declaring himself a political prisoner. Difficult negotiations were opened with the Italian government, which recognized some of the pope's rights, including the inviolability of his person and the extraterritoriality of the Vatican buildings he inhabited (Law of Guarantees [1871]). But in 1873, the confiscation of church properties, undertaken in the territories of the former Papal States, was extended to Rome. In the face of this unilateral act, the pope replied with the *Non expedit* decree of 1874, which declared that it was no longer acceptable for Catholics to participate in Italian politics. The century ended with this conflict not yet healed: the relationship between the church and the Italian state would only be regularized around fifty years later with the Lateran Treaty, signed by both sides in 1929.

Yet this was a period in which there was turmoil not only outside but also *within* the church. Politically, many Catholics favored liberal ideas; the nineteenth century thus saw a flowering of liberal Catholicism, which attempted to combine patriotism and religion, whose exponents supported freedom of conscience, a free press, and free association, as well as the separation of church and state. On the doctrinal level, some texts "demythologizing" religion also circulated in this period, one of the most popular being Ernest Renan's (1823–92) *Vie de Jésus* (Life of Jesus [1863]), which accepted the historical existence of Christ but denied his divinity and acknowledged only the "exemplary value" of his teaching. This and other "new heresies," as well as the role of the pope and the relationship between church and state, were among the topics discussed in the First Vatican Council (1869–1870), summoned by Pius IX in 1868. The council defined the dogma of *ex-cathedra* papal infallibility and the relationship between faith and reason as separate orders of knowledge, not in contradiction. The latter pronouncement in turn served to legitimize the scientific interests of many clergymen and members of religious orders who, over the centuries, had made significant contributions to the development of disciplines such as biology, meteorology, physics, seismology, and astronomy.[15]

As a Jesuit, all of these events had crucial repercussions for Secchi's life and work. He could count on Pius's protection and financial support, but he suffered many difficulties and criticisms because of his allegiance to the papacy and the Society of Jesus in the critical years in which the papacy lost its temporal power and the church was shaken by external and internal conflicts.

To complete the setting, it is worth mentioning that there were also a number of "social saints," such as Giovanni Bosco (1815–88) and Leonardo Murialdo (1828–1900), who were concerned by the Industrial Revolution's social consequences. These individuals sought to promote education and professional training for young people from the most disadvantaged social classes by promoting social organizations and creating educational facilities. This concern with the disadvantaged and the dispossessed was ultimately reflected in the Catholic Church with the encyclical *Rerum novarum* (1891) by Leo XIII (1810–1903, r.1878–1903), which laid the foundations for the church's social doctrine, acknowledging the challenges presented by modernity for social justice and workers' rights.

In the same year that the encyclical *Rerum novarum* was issued, the pope also authorized the establishment of the Vatican Observatory with the motu proprio *Ut mysticam*, relaunching a dialogue between faith and science and providing the Vatican with a modern center of astronomical research, located in the Vatican buildings. It should also be noted that a decree issued by Pius VII (1742–1823, r.1800–23) in 1820 had approved mentioning the Copernican theory in the writings of Catholic authors, thus formally concluding the Galilean affair. In the seventy years that elapsed between these two documents being issued, there had been a fierce dispute between opponents and supporters of the dialogue between science and faith inside and outside the church, until the unexpected return of astronomy inside Saint Peter's walls. Secchi, however,

---

15 On science and religion, see "Documentazione interdisciplinare"; www.disf.org (in Italian) (accessed March 22, 2018).

could not see these changes, as his untimely death prevented him from participating in this "new course" of Vatican astronomy.

3 Focusing on Italy: Astronomy in the Years of Unification

On both the pre-Napoleonic political map and that redrawn by the 1815 Congress of Vienna, Italy was fragmented into several states, each of which had at least one astronomical observatory. After the unification process, which concluded in 1870, the Italian government "inherited" as many as ten different astronomical observatories located in different regions, from north to south. All the observatories were provisionally connected to the local universities.[16] As such, they were under the authority of the Ministry of Public Education, which provided financial support for their maintenance. The young nation had to reorganize its scientific resources as well as its educational system. Literacy was a priority in many regions, especially in southern Italy, and financial resources were primarily employed to fund educational institutions. In such a context, scientific research—including astronomy—suffered due to funding restrictions, which hampered Italian astronomy from becoming competitive on the world stage as there was a lack of adequate financial support for new equipment, additional staff, and other essentials. Astronomical research funding could only rely on ad hoc grants, as in the case of the scientific expeditions to observe the total solar eclipse of 1870 and the transit of Venus in 1874. Even these occasional resources, inadequate as they were, were often obtained thanks to personal and political connections. And though many Italian scientists played a role in politics during and after political unification, with a number of them occupying important offices in the post-unification government or becoming influential advisors,[17] Italy continued to lack a clear policy about science[18] until the creation of the Consiglio Nazionale delle Ricerche (CNR) in 1923.

16 A reform project for the Italian observatories, drafted in 1874, became law in 1876 but was never implemented; see Francesco Poppi, Fabrizio Bònoli, and Ileana Chinnici, "Il progetto Tacchini e la riforma degli Osservatori italiani," in *Cento anni di astronomia in Italia 1860–1960*, Atti dei Convegni Lincei 217 (Rome: Bardi Editore, 2005), 123–71.

17 Schiaparelli (see below) was the only astronomer who received sufficient government funding to acquire a large instrument—a fifty-centimeter Merz equatorial refractor—for the Milan (Brera) Observatory.

18 That the Italian political elite had a limited interest in astronomy is apparent from a remark quoted by Tacchini (see below) in a report to a government minister: "To obtain the allocation of some funding, not even large, for one of our observatories, a minister said at

As a young nation, the Italian state was hardly in a position to support all of the observatories that had been established on its territory. At the same time, however, the government wanted to avoid clashes resulting from various local sensitivities. The government thus decided to maintain the *status quo*, which had the unintended consequence of hindering progress in astronomical research. In the words of the famous astronomer Giovanni Virginio Schiaparelli (1835–1910):[19] "Here [i.e., in Italy] you can never get anything, because there are too many observatories and the government spends a considerable sum without anyone of them [...] able to do some important work [...] and the government promises, promises, but then we don't get anything."[20] An attempt at reform was made in 1876 with the Bonghi decree,[21] but the decree was never actually implemented, and Italian astronomy continued to operate with woefully inadequate funding.

The Italian astronomical community was largely unable to form a common front to determine the post-unification policy for astronomical research. Marked by fragmentation and individualism, this community appeared weak and lacking in innovative ideas and initiatives.

One exception was Pietro Tacchini (1838–1905), a pragmatic, creative, and active "dreamer" who tried to stimulate his colleagues by promoting scientific expeditions, proposing and drafting reforms, and establishing societies and scientific institutions.[22] Hence it is far from surprising that Tacchini was Secchi's main scientific partner: despite their differences in age, convictions, and lifestyle, they shared the same passion for the development of science (see

---

the House: vote it and then we need not care any further about astronomy" (*Un ministro alla Camera per far votare una somma, e non grande, per uno dei nostri Osservatorii, disse: votatela, perché fatto ciò non avremo più altro da pensare per l'astronomia*); Pietro Tacchini, *Eclissi totali di Sole* (Rome: Tipografia Eredi Botta, 1888), 113.

19 Giovanni Virginio Schiaparelli, director of the Brera Observatory from 1862 to 1900, was one of the most influential scientists in determining the Italian government's policy toward astronomy.

20 *Qui non si può mai ottener nulla, e la causa sono i troppi osservatorii che abbiamo, per i quali il Governo spende una somma ragguardevole, senza che perciò in nessuno* [...] *si possa far qualche lavoro importante* [...], *ed il governo promette, promette, ma poi non si ottiene nulla.* Schiaparelli to Secchi, Milan, June 28, 1868, in Letizia Buffoni, Alessandro Manara, and Pasquale Tucci, eds., *G.V. Schiaparelli, A. Secchi: Corrispondenza (1861–1878)* (Milan: Edi. Artes, 1991), 175.

21 Ruggiero Bonghi was minister of education from 1874 to 1876 (see Chapter 10). On the Bonghi decree, see Poppi, Bònoli, and Chinnici, "Il progetto Tacchini e la riforma degli Osservatori italiani."

22 For a short biography of Tacchini, astronomer at Palermo Observatory, see Ileana Chinnici, "Pietro Tacchini (1838–1905): A Key Figure in the Post-unitarian Italian Astronomy," *Memorie della Società Astronomica Italiana: Supplementi* 9 (2006): 28–34.

Chapter 8). It was mostly thanks to Tacchini's mediation that Secchi was always involved in initiatives relating to Italian astronomy, whether it was the expedition in Sicily to observe the eclipse of 1870 or the discussion of reform during the congress of Italian scientists in Palermo in 1875.

In many ways, Secchi was a peculiar "Italian" astronomer in the sense that he was obedient to the pope, not to the king—but his scientific authority was unquestioned. For this reason, Secchi was long courted by the Italian government, which offered him attractive assignments—albeit without success, because of Secchi's difficulties in reconciling them with his religious commitments.

Hence it is clear that, in the "constellation" of Italian astronomy of that time, Secchi was the most brilliant "star." He was mentioned and quoted far more frequently than his contemporary Italian colleagues, whose names only occasionally emerge in noteworthy contemporary popular books such as Simon Newcomb's (1835–1909) *Popular Astronomy* (1878) and Agnes M. Clerke's (1842–1907) *The System of the Stars* (1890). These publications helped to enhance the Jesuit's popularity, especially outside of Italy, after his death.

CHAPTER 1

Secchi's Early Life and Education: From Childhood to the Priesthood

Biographical information on Secchi's childhood is scarce, and what little that does exist is often incorrect. In some "hagiographic" obituaries of Secchi, for example, unfounded information is provided that more recent biographical research has revealed to be inaccurate.[1] Accordingly, what follows seeks to provide a more accurate reconstruction of Secchi's early life on the basis of what is currently considered to be reliable, if admittedly incomplete, archival information.

1 Secchi's Family

Angelo Secchi was born into a middle-class family in Reggio, a small town in the Emilia region, in the heart of the Po Valley, on June 28, 1818, at 19:45.[2] He was the son of Giovanni Antonio Secchi (1758–*c.*1839) and Luigia Belgieri (1777–1867). Secchi's modest house, located in the San Lorenzo district, at via Porta Brennone no. 14, is today identified by a plaque.[3]

1 See Ileana Chinnici, "Il profilo scientifico e umano di Angelo Secchi," in *Angelo Secchi: L'avventura scientifica del Collegio Romano*, ed. Aldo Altamore and Sabino Maffeo (Foligno: Edizioni Quater, 2012), 43–64.

2 Some biographers – and Secchi himself (see "Punti d'informazione che i candidati danno di sé in probazione," Archivio Storico della Provincia Euro-Mediterranea [Jesuit Euro-Mediterranean Province Historical Archive] – hereinafter ASPEM – Fondo Provincia Romana 589, 12 [530–31]) – report June 29, but this is the date he was baptized, as confirmed by the baptismal register preserved at the diocese archive of Reggio Emilia. On June 29, 1818, the register (in Latin) reports Secchi's full name (*Angelus Franciscus Ignatius Balthasar*) and birthdate (*heri Vesperi hora prima cum tribus quadrantibus* [yesterday, 19:45]), as also reported in the certificates that are kept in the historical archives of the Jesuit Roman province (see fig. 1.2) and INAF-OAR at Monte Porzio Catone. The *Biographical Encyclopedia of Astronomers* (BEA) (New York: Springer, 2014), reports June 18, but this evidently comes from an incorrect source. Moreover, the name (Pietro Angelo) is also incorrect (see note 4).

3 Some sources report that he was the last son of a large family and that he became fatherless at a young age, but this is incorrect. The signature of his father, in fact, appears in a document dated 1833 granting Secchi permission to enter the novitiate (see fig. 1.3).

 | DOI:10.1163/9789004387331_003

FIGURE 1.1 Angelo Secchi's birthplace. On the left: the house at the beginning of twentieth century; in the middle: the house today; on the right: detail of the memorial plaque. Courtesy of Giuseppe Adriano Rossi.

Secchi was baptized the day after his birth in the church of St. John the Baptist, better known as the Baptistery of Reggio Emilia, with the name Angelo Francesco Ignazio Baldassarre.[4]

His grandfather, Genesio (1708–87), worked as a farmer in the nearby Augustinian nunnery of St. Ilarius, while his father was a cupboard-maker who had a small carpentry shop. After the death of his first wife, Teresa Poli, Angelo's father married Luigia Belgieri, a seamstress, almost twenty-years younger than him. Angelo was the last-born son, after Tommaso (1791–1844), born from his father's first marriage, and Anna (1802–?).[5] As often happened in modest families, only the eldest son could be supported in his studies: Tommaso became a lawyer, while the younger son, Angelo, was sent to become a priest. Angelo's father was certainly still alive in 1833, when both parents signed a document granting permission for Angelo to enter the novitiate of the Society of Jesus.

4 Many biographers report the name of Pietro Angelo—see, for instance, Juan Casanovas, "Secchi, Angelo (Pietro Angelo)," in *Diccionario Histórico de la Compañia de Jesús*, ed. Charles E. O'Neill and Joaquín M. Domínguez (Rome: Institutum Historicum; Madrid: Universidad Pontificia Comillas, 2001), 4: 3542–43—but it is possible that this comes from some confusion with his cousin Giovanni Pietro (or Giovampietro, see below). About Angelo's godfather, presently only his name is known, Giuseppe Baracchi, and that of his father, Claudio.

5 From a family tree, in possess of some descendants of a brother of Genesio, it appears that, after Anna, four other babies were born in 1805, 1807, 1809 and 1815, but they did not survive. Angelo was therefore a long-desired son.

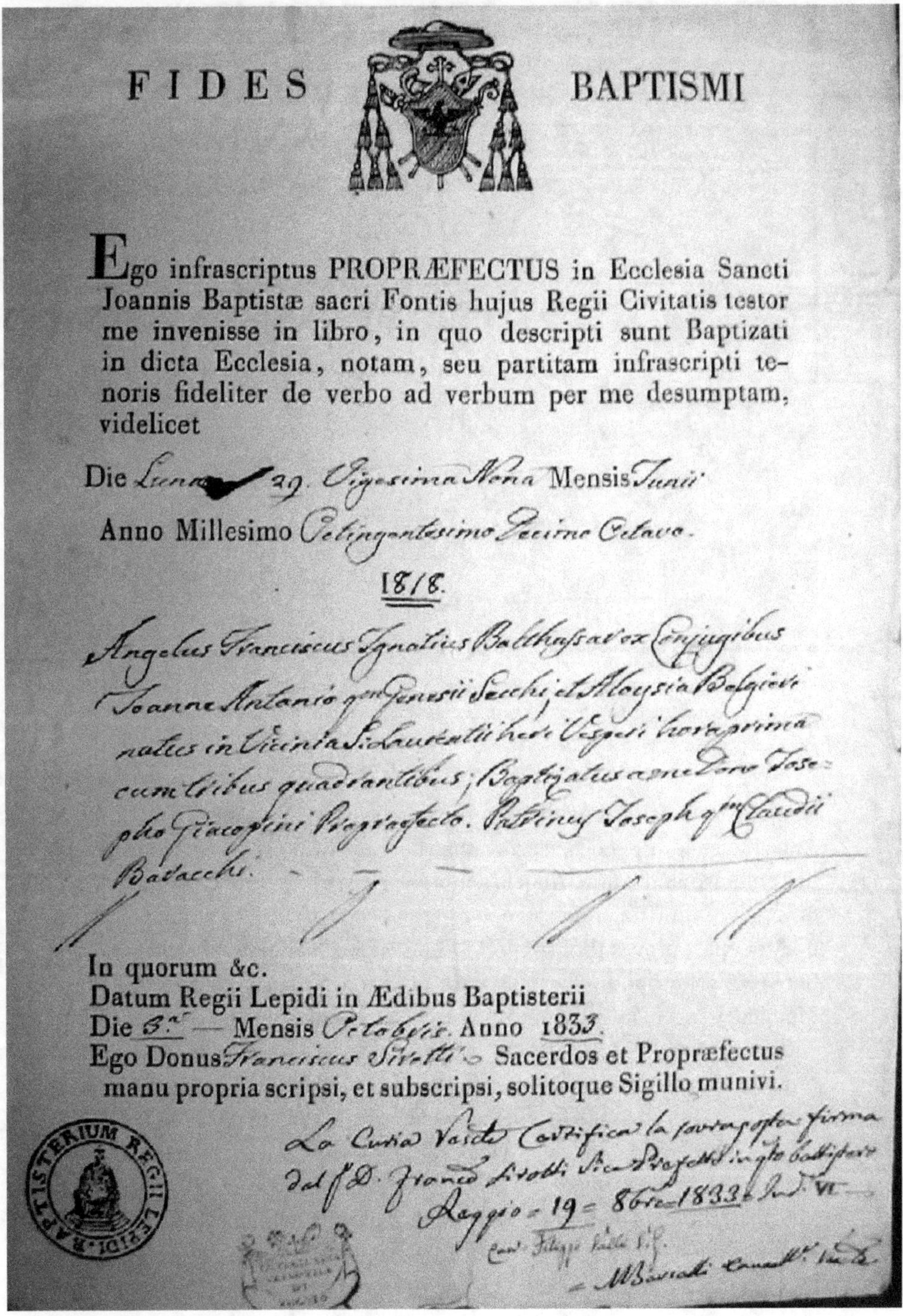

FIDES BAPTISMI

Ego infrascriptus PROPRÆFECTUS in Ecclesia Sancti Joannis Baptistæ sacri Fontis hujus Regii Civitatis testor me invenisse in libro, in quo descripti sunt Baptizati in dicta Ecclesia, notam, seu partitam infrascripti tenoris fideliter de verbo ad verbum per me desumptam, videlicet

Die Lunæ 29. Vigesima Nona Mensis Junii

Anno Millesimo Octingentesimo Decimo Octavo.

1818.

Angelus Franciscus Ignatius Balthassar ex Conjugibus Joanne Antonio qm Genesii Secchi, et Aloysia Belgieri natus in Vicinia S. Laurentii heri Vesperi hora prima cum tribus quadrantibus; Baptizatus a me Domno Joseph Giacopini Proprafecto. Patrinus Joseph q. Claudii Batacchi.

In quorum &c.
Datum Regii Lepidi in Ædibus Baptisterii
Die 8.a — Mensis Octobris. Anno 1833.
Ego Donus Franciscus Sivotti Sacerdos et Propræfectus manu propria scripsi, et subscripsi, solitoque Sigillo munivi.

La Curia Vescle Certifica la surrogata firma dal S.D. Franco Sivotti Vice Prefetto in qto battistero
Reggio = 19 = 8bre 1833 Ind. VI.

BAPTISTERIUM REGII LEPIDI

FIGURE 1.2 Angelo Secchi's baptismal certificate. Courtesy of ASPEM.

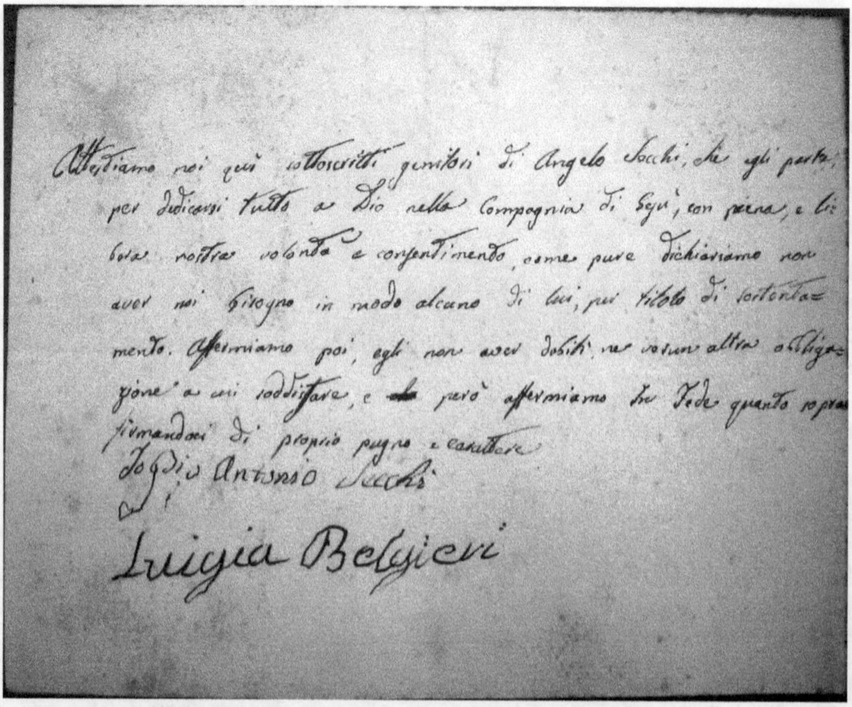
Attestiamo noi qui sottoscritti genitori di Angelo Secchi, che egli parte, per dedicarsi tutto a Dio nella Compagnia di Gesù, con piena, e libera nostra volontà e consentimento, come pure dichiariamo non aver noi bisogno in modo alcuno di lui, per titolo di sostentamento. Affermiamo poi, egli non aver debiti, nè nessun altra obbligazione a cui soddisfare, e però affermiamo in fede quanto sopra firmandoci di proprio pugno e carattere
Io Gio Antonio Secchi
Luigia Belgieri

FIGURE 1.3 Document granting permission from Angelo's parents for him to enter the Jesuit novitiate. Courtesy of ASPEM.

Angelo's father presumably died in 1839, because, in November of that year, Angelo signed a deed declaring that he had formally given his inheritance to Tommaso and Anna. Angelo's mother remarried around 1838; she had no other children. The address appearing in Secchi's letters ("at Cervi's pharmacy, Modena") proves that she left Reggio (Cervi was probably the surname of her second husband, who may have been a pharmacist with a drugstore in Modena). Angelo loved his mother very much and exchanged touching letters with her throughout his life; according to a few biographical notes, written by brother Marchetti (see Chapter 3), she said of Angelo that she took comfort and solace "not so much for the knowledge he has acquired but for the religion to which he was dedicated."[6] This seems to confirm that Secchi was encouraged by his family to become a clergyman. In his letters, Secchi kept his mother informed about all his steps in the religious life and in his early scientific career, also sending her some of his publications and small gifts; in 1853, for instance, he wrote to tell her that "I shall send you a small rosary blessed by the pope;

6 See APUG, FS 23.III.19.

I'm sorry that the short time did not allowed me to find a better blessed one; but at the next audience I will present him a better one to be blessed for you."[7]

Secchi's half-brother, Tommaso, twenty-eight years older than Angelo, was his confirmation witness in 1827; he married Antonia Mattioli (d.1876), the daughter of a lawyer, Fortunato, and had two sons: Giuseppe (1825–67) and Alessandro (1829–92), who became a priest.[8]

Secchi's sister Anna, seventeen years older than Angelo, married Gaetano Zanardi and had two children, Barbara and Massimiliano; she was still alive in 1858. It is quite likely that Anna was Angelo's first tutor. At home, he learned not only how to read and write but also some skills typically associated with women, such as knitting and sewing, which he was probably taught by his mother; even after he had become a renowned scientist, Secchi did not refrain from revealing these details to some of his Jesuit confrères.[9] It could be speculated that these skills may have given him a propensity for the repetitive and precision work that would later prove valuable in his astronomical research. Secchi's niece Barbara married Giuseppe Zoboli and had four children: the second-born daughter was proudly named Angelina, after her famous uncle, according to the will of great-grandmother Luigia. Massimiliano Zanardi (d.1872) married Anna Paglia; they were childless, and Anna assisted her great-grandmother Luigia in her last years.[10]

Though working in places that were far away from his family members, Angelo always remained very attached to them and never missed an opportunity to visit his relatives. He was in touch with his nieces and nephews as well as his cousins, who continuously gave him news about his family. Sometimes, they asked for his help or advice when dealing with delicate family matters,

7 *Le spedirò una coronella benedetta dal Papa; mi dispiace che la strettezza del tempo non mi ha permesso di trovarne una migliore benedetta ma alla prima udienza che avrò gliene presenterò una migliore da benedire per Lei.* Secchi to his mother, Luigia Belgieri, Rome, May 28, 1853; APUG, FS 9.V-bis. Secchi asked for the rosary beads of the Sorrowful Mother as a token after his mother's death, which were then returned back to Anna Paglia, Secchi's grand niece, after his death (see Anna Paglia to Br. Francesco Marchetti, Mantua, March 30, 1878; APUG, FS 21).

8 Giuseppe married Teresa Montalto (1835–72) and had two children, Gabriella (1859–91) and Riccardo (1863–?). After the premature death of their father, they seem to have been largely abandoned by their mother, who started a liaison with a certain Magionato Casali and married him at the point of death. The children were raised by their grandmother Antonia; after Antonia was struck by paralysis in 1871, they were cared for by their uncle priest, Alessandro. Secchi was highly attached to his nephew Alessandro and his grandniece Gabriella, who were the only relatives mentioned in his will (see APUG, FS 23.V).

9 See APUG, FS 23.III.19 and 23.IV.

10 In 1885, Fr. Stanislao Ferrari (see below) acquired from Anna Paglia nine letters written by Secchi to his mother, which are an important source of information about his early years as a Jesuit (see Anna Paglia to Fr. Stanislao Ferrari, S.J., January 17, 1885; APUG, FS 9.Vbis).

trusting in his judgment and occasionally seeking to capitalize on his influence once he had gained in stature as an astronomer. They all felt a sense of pride at being related to such a famous personality. As one of his cousins wrote to him in 1871: "I am proud of the distinctions that are afforded to me precisely because I am your relative, as people gathered in small groups at crossroads will repeat when they see me: look at him, he is the cousin of the great astronomer […]."[11]

2 Early Studies and Novitiate

During the Napoleonic age, the city of Reggio was part of the territory of the short-lived Cispadane and Cisalpine Republics, but in 1815 the Congress of Vienna re-established the previous Duchy of Modena and Reggio, ruled by Francesco IV, archduke of Austria-Este (1779–1846, r.1814–46), whose government was often judged dictatorial and sanguinary by contemporary sources.[12]

In the seventeenth century, the Jesuits had established a college in Reggio, near St. George's church, where the City Library is located today.

Angelo entered the college at the age of nine years. He stood out from the other students due to his intellectual qualities, his reliability, and his love of studying, especially the classics—he obtained a prize in the Greek language—as well as his "serious and sensible" personality.[13] Angelo's teachers would later recall that "he applied himself very much to each new subject that was taught, without caring about the other ones, as if he was looking for a pasture appropriate for his talent."[14]

On November 3, 1833, Secchi was admitted to the novitiate of the Society of Jesus at St. Andrew at Quirinale in Rome.

According to the remarks made by his doctor on the certificate required for admission:

> Mr. Angelo Secchi is endowed with a hot-blooded temperament, […] no alteration is found in his physique, and […] he was never affected by eyes

11 *Vado poi superbo delle distinzioni che mi vengono usate appunto perché son Vostro parente, ove van ripetendo nei crocchi, allorché mi vedono: Ecco là il cugino del grande astronomo.* Pietro Secchi to Angelo Secchi, Gorizia, May 23, 1871; APUG, FS 19. Secchi's relatives often asked him for help, especially after 1870, when he was not removed from the directorship of his observatory, as they inferred that he could count upon important political protectors.

12 See, e.g., Taddeo Grandi, *Ciro Menotti e i suoi compagni* (Modena: Tipografia Azzoguidi, 1880).

13 See APUG, FS 23.III.19.

14 *Ad ogni nuova materia che si prendeva ad insegnarsi vi si applicava con grandissimo impegno senza curarsi più del rimanente come se cercasse un pascolo adattato alla sua inclinazione.* APUG, FS 23.III.

FIGURE 1.4 The Jesuit College at Reggio Emilia, currently the City Library. Courtesy of Giuseppe Adriano Rossi.

FIGURE 1.5 St. Andrew at Quirinale, church of the Jesuit novitiate in Rome. Courtesy of Aldo Altamore.

or head diseases. The stomach is intact from any alteration so that digestion is done normally, as well as respiration. Every part of his body is free from chronic diseases and also from genetic disorders, his parents having

> a strong and robust constitution. He has always been in good health since his birth, so that he was never affected by epilepsy [...] and never was subjected to hernia, and after a scrupulous examination, I have found that smallpox was inoculated in him.[15]

During his probation—a preliminary formation period—Secchi wrote the following about himself in the required introductory notes: "I feel inclined to the study of classics as well as to preaching [...]. I have a healthy and good memory; I am hot-blooded and solidly built, with a small stature."[16] Angelo took the first vows on November 4, 1835. After the novitiate, his superiors decided to foster his talent for classical studies, and he stayed at the college for a further two years to study rhetoric; he obtained excellent results, mainly in the study of the Greek classics, taught by the Jesuit philologist and epigraphist Antonio Angelini (1809–92),[17] who had already been his teacher in Reggio. Thanks to his studies in rhetoric, Secchi developed an extraordinary talent for writing and public speaking, with an elegant, clear, and fluent style, later becoming the author of a large number of successful scientific treatises and a brilliant public speaker.

3 Scholastic at Collegio Romano

In 1837, Angelo began his study of philosophy at the Collegio Romano.
From its establishment in the sixteenth century, this prestigious institution had developed an important tradition in teaching sciences,[18] especially thanks to

---

15 *Il Sig. Angelo Secchi è dotato di temperamento sanguigno,* [...] *nessuna alterazione riscontrasi nel di lui lato fisico e che non mai andò soggetto a malattia di occhi, né di capo. Lo stomaco trovasi illeso da qualunque affezione per cui la digestione viene compita normalmente come pure anche le funzioni della respirazione. Qualunque punto del suo corpo è privo da malattia cronica, ed anche gentilizia, tantoché i di lui genitori sono di complessione forte e robusta. Esso ha sempre goduto di un'ottima salute, fino dalla sua nascita, cosicché non fu mai molestato da Epilessia* [...] *né mai soggetto andò ad ernie, ed avendolo assoggettato ad una scrupolosa visita ho riscontrato in esso essergli stato inoculato il vaiolo.* Reggio, October 20, 1833, certificate by Orazio Mattioli; Archives of the Roman Jesuit province, Rome.

16 *Sentomi inclinato tanto allo studio di belle lettere, come ancora alla Predicazione.* [...] *Ho la memoria vegeta, e buona, sono di temperamento sanguigno e di complessione robusta, e una statura piccola* "Punti d'informazione che i candidati danno di sé in probazione," ASPEM. In this interesting document, recently retrieved, Secchi also briefly describes his family and his early studies at Reggio; he also mentions his teachers: the priests Don Domenico Canali (1788–1863), Don Rivi – probably Giuseppe (1779–1847) – and Don Giovanni Battista Tarasconi (1802–1882) and the Jesuits Ignazio Gurri (1805–1879) and Antonio Angelini.

17 See APUG, FS 23.III and 23.IV.

18 It was the first—and for many years the only—college where mathematics was regularly taught. See Sabino Maffeo "Il Collegio Romano e l'insegnamento delle scienze," in Altamore and Maffeo, *Angelo Secchi*, 16–22.

FIGURE 1.6 Engraving of the Collegio Romano by Giuseppe Vasi (1710–82). From Giuseppe Vasi, *Delle magnificenze di Roma antica e moderna, libro nono* (Rome: Nella Stamperia di Niccolò e Marco Pagliarini, 1759), 162, pl. 9.

the work of Christopher Clavius (1538–1612), who trained an entire generation of Jesuit mathematicians who would later disseminate their astronomical and mathematical knowledge all over the world.

In the years he spent at the Collegio Romano, Secchi excelled in mathematics and physics, subjects taught by Francesco de Vico, S.J. (1805–48), who was also director of the Collegio Romano Observatory, and Giovan Battista Pianciani, S.J. (1784–1862). From then on, he never ceased his study of these two disciplines, and even when he was sent to teach literature at the Collegio dei Nobili in Rome (today known as the Palazzo Gabrielli-Borromeo),[19] he continued to study mathematics by annotating and commenting on the works of Andrea Caraffa, S.J. (1789–1845).[20] However, even though Secchi was able to carry out

19 Some sources suggest that his superiors, having noticed his abilities in science, tasked Secchi with giving private tutoring to some of the students of the Collegio de' Nobili (see APUG, FS 23.III), but this probably confuses his work at this time with what he would actually do later.

20 Caraffa, professor of mathematical analysis and mathematical physics at Collegio Romano, was the author of the handbooks *Elementorum matheseos* [Elements of mathematics (1835)] and *Elementorum physicae-mathematicae* [Elements of physical mathematics (1840)].

mathematical calculations effectively, in a letter written to renowned astronomer Virginio Schiaparelli in 1873, Secchi acknowledged his own "inability for mathematics."[21] This is not just an excess of modesty but a reflection of his awareness that he had a different mindset, one less theoretical and more empirical.

The young Jesuit showed a clear preference for physics. Pianciani noticed this and strongly encouraged his pupil. He used to have long conversations with Secchi and would often send him important publications, such as the letters on the "radiant caloric" (heat radiated by a body) by the famous physicist Macedonio Melloni (1798–1854). Pianciani probably saw Secchi as his future assistant and requested to his superiors that Secchi, as a scholastic, be sent to teach physics at the Collegio Illirico[22] in Loreto, on the Adriatic coast of Italy.

4 Teacher at Loreto College

Secchi spent four years at the Collegio Illirico, from 1841 to 1844.[23] During this period, he continued his scientific studies and read the texts of mathematician and physicist Gabrio Piola (1794–1850), the *Traité de mécanique* (Treatise on mechanics) by Siméon-Denis Poisson (1781–1840), and the works of the famous Swedish chemist Jacob Berzelius (1779–1848).

However, despite his initial good health, the young Secchi became sickly, probably because of overwork. Indeed, studying and teaching seemed to be his favorite activities: "In addition to the daily lessons, on Sunday he taught catechism to the class with uncommon clearness and eloquence; the rest of the day, he stayed closed in his room, from which he went out only for spending a few hours in the physics cabinet he had created."[24]

21 *La mia incapacità per la matematica.* Secchi to Schiaparelli, Rome, January 30, 1873, quoted in Buffoni, Manara, and Tucci, *G.V. Schiaparelli, A. Secchi*, 219.

22 About the history of the Collegio Illirico, see the related webpage by Maria Macchi at the virtual exhibition on Secchi "Tra cielo e terra/From heavens to earth" (https://tracieloeterra.bicentenarioangelosecchi.it/lavita/the-society-of-jesus/illyrian-college-lauretano-2/?lang=en).

23 The only source containing any information about Secchi's stay in Loreto is a biographical note by his confrère Ugo Molza (see below) later rector of the Gregorian University, who met him there for the first time and became his friend (see APUG, FS 23.IV).

24 *Alle due scuole che faceva ogni giorno, aggiungeva nella domenica il catechismo alla scolaresca, che spiegava con chiarezza e facondia non comune, tutto il resto del giorno se ne stava chiuso in camera, onde usciva solo per passar qualche ora nel gabinetto fisico da lui formato.* APUG, FS 23.IV.

FIGURE 1.7 The Sanctuary of the Holy House at Loreto; on the right, the buildings of the Collegio Illirico. Courtesy of Specola Vaticana.

Thus Secchi was able to establish a physics laboratory in the college (a "cabinet" in nineteenth-century parlance); though he had only pneumatic and electrical devices, he made other instruments by himself with the help of some local artisans:

> It was really admirable to see how, in his hands, every nail that he found thrown away at home, became a physics instrument [...], and how able he was in instructing those skillful artisans, who were inexpert in these kinds of works, so that they made with great exactness the machines that he needed.[25]

25 *Era veramente mirabile come in sua mano, ogni chiodo che trovava gettato per casa, diventava uno strumento di fisica* [...] *e come egli sapeva dirigere quei bravi artisti, di tali lavori al tutto inesperti, sicché facessero con grande esattezza le macchine ond'egli abbisognava.* APUG, FS 23.IV.

This practical skill would prove useful in Secchi's future research work, as he was already accustomed to designing, constructing, and adjusting various scientific instruments.

The young professor would conclude the school year with a public demonstration, prepared by his pupils, of experiments to test the physical laws they had learned: "As a teacher, he was [...] very successful, and this was proved by the public demonstrations annually made by his classes."[26]

5 Back to Rome: Assistant to Pianciani

In October 1844, Secchi returned to Rome to start his theology studies, eventually leading to ordination. At that time, the course was taught and administered by renowned teachers such as Carlo Passaglia (1812–87), Giovanni Perrone (1794–1876), Francesco Saverio Patrizi (1797–1881), and Antonio Ballerini (1805–81), most of whom were active in political, social, and religious life.[27] The experience Secchi gained at Loreto served to prepare him to become an assistant to Pianciani at the Collegio Romano. The close collaboration with his former professor was excellent training for Secchi. According to Brother Marchetti: "He hung on that learned professor's words like a child on his own mother's [and] he dedicated to the study of physics all the time that he could spend, often abnegating walking to stay longer in the cabinet."[28] Pianciani's worldview, which was open to new discoveries, was based on the conviction that scientific results could be consistent with the metaphysical order as well as with the Bible.[29] Secchi was clearly influenced by his theories, which are recognizable

26 *Nell'insegnamento, fin da allora era felicissimo, come dimostravano i saggi pubblici a cui, al cadere di ogni anno, esponeva alcuni suoi scolari.* APUG, FS 23.IV.

27 See Stefan Hughes, *Catchers of the Light. The Forgotten Lives of the Men and Women Who First Photographed the Heavens* (n.p.: ArtDeCiel Publishing, 2012), 768. Passaglia was a supporter of Italy's political unification and left the order to become an activist of liberal Catholicism; Perrone wrote important theological treatises, while Ballerini was an opponent of Rosminian ideas: they were both involved in the preparation work for Vatican Council I.

28 *Dalla bocca di quel dotto professore pendea siccome fanciullo alla madre sua e da lui può dirsi ebbe i primi raggi di luce di quella celebre opera nata dalla sua penna, l'Unità delle Forze fisiche.* APUG, FS 23.III.

29 On Pianciani, see Roberto Mantovani, "Un fisico dimenticato: Il gesuita G.B. Pianciani"; http://www.uniurb.it/PhysLab/immagini/mantovani_sisfa2002.pdf (accessed May 29, 2018).

in Secchi's later treatise *Sull'unità delle forze fisiche* (On the unity of physical forces [1864]) (see Chapter 6).[30]

At the same time, Secchi gave review sessions of physics for the students of the Collegio de' Nobili. At Secchi's instigation and under his supervision, his students would prepare a public demonstration, given at the end of the school year, which "was very appreciated [as] for the first time in Rome, experiments with the quadrant telegraph were carried out."[31] His interest in telegraphic systems and his efforts to popularize scientific work would be a constant trait in his future work.

Secchi does not appear to have shown any particular interest in astronomy at that time. De Vico was a reputed astronomer in the field of cometary observations—he discovered six comets in the years 1843–47—but he taught mathematics, and in his teaching plan, astronomy may simply have been a field for applying the exact science of mathematics, a pretext for calculation exercises. This may well have been less attractive to Secchi's bright mind, which was more attuned to practical sciences. Thus, rather than focusing on astronomy, Secchi's early publications, dating back to the years 1846–47, contain descriptions of various inventions, such as two prototypes of magnetic clocks and a new telegraphic apparatus.[32] Secchi also began working on some studies of electrical resistivity[33] during this period that would later be developed and published in the United States (see Chapter 2) as a result of his collaboration with Pianciani.

6 Priestly Ordination and the Revolutions of 1848

On September 8, 1847, Secchi was ordained a deacon, and a few days later, on September 12, he received priestly ordination. He had announced these plans for ordination to his mother a few months earlier, at Easter:

30 It is no accident that the future director of the Collegio Romano Observatory would be the best biographer of his teacher; see Angelo Secchi, "Intorno alla vita e alle opere del P. Giambattista Pianciani," *Giornale Arcadico di Scienze: Lettere ed arti*, n.s., 62 (1862): 1–48.

31 See *Che fu ammiratissimo, nel quale per la prima volta a Roma si fecero gli esperimenti del telegrafo a quadrante.* APUG, FS 23.IV.

32 See Angelo Secchi, "Sugli orologi elettromagnetici," *Raccolta scientifica di Roma*, anno secondo (November 15, 1846): 353–58; Secchi, "Descrizione di un nuovo apparato per trasmettere i segni nei telegrafi elettrici," *Raccolta scientifica di Roma*, anno terzo (February 1, 1847): 40–45.

33 See Angelo Secchi, "Sopra alcune esperienze di reometria elettrica," *Gazzetta scientifica* 3 (1847): 385–94.

> Pray the Lord, because such a tremendous act could be made by me [...] as little unworthily as possible. [...] So far, we had some hope to meet again, but now [...] perhaps this hope is lost [...]: anyway, your sacrifice and mine must be perfect, and we hope to meet again in heaven, if I'll be worthy.[34]

Secchi was concerned for his mother because in 1846 the northern part of the Roman province of the Society of Jesus, including his native territories, had been taken over by the Venetian province; it was unlikely that a Jesuit would be assigned to teach in a college of another province and, consequently, the number of opportunities for him to visit his mother was drastically reduced. He could not imagine that upcoming political events would increase their distance even more and, in a few months, lead to the voluntary exile of the Jesuits.

Early in 1848, the first revolutionary riots broke out and spread throughout Italy from south to north, leading to the first Italian War of Independence (1848–49). In the wake of these events, a popular protest against the Jesuits spread throughout Italy, leading to their expulsion from Genoa, Turin, Naples, and even some cities of the Papal States, such as Loreto, Spoleto, and Ferrara; according to the ideas of Vincenzo Gioberti (1801–1852), the patriot-priest who advocated the political unification of Italy, the Jesuits were enemies of Italian unification. Overall, the animosity was the result of a range of motives and prejudices.[35] The Jesuits were widely resented, partly due to their perceived reputation as power-hungry double-crossers, partly because they were identified as the main obstacle to the diffusion of modern anti-clerical ideologies. In any case, the situation was dangerous for the Jesuits and thus, on the pope's urging, Superior General Jan Roothaan (1785–1853, in office 1829–53) ordered them to leave Italy and go abroad (March 28–29, 1848). Secchi's priestly ordination should have been followed by the solemn profession of the perpetual vows, but these events hampered the last step of his religious formation, which had to be deferred. After a protracted period of political crisis, the pope himself left Rome in November 1848 to seek refuge under the protection of Ferdinand II (1810–1859), king of the Two Sicilies. In February 1849, an ad hoc national assembly proclaimed the Roman Republic, while Italy as a whole experienced great political instability, beset by revolutions and repression.

---

34 *Ella preghi il Signore perché un tale atto così tremendo venga fatto da me* [...] *il meno indegnamente possibile* [...] *Finora potevamo avere qualche speranza di rivederci, ma adesso che è stata divisa la provincia Romana e gli stati Estensi coi collegi suoi non appartengono più ad essa ma alla Veneta, forse questa è perduta,* [...]*: ad ogni modo il Suo sacrifizio ed il mio deve essere perfetto, e speriamo di rivederci in Paradiso, se io ne sarò degno.* Secchi to his mother, Luigia Belgieri, Rome, April 4, 1847; APUG, FS 9.Vbis.

35 See Stefano Tomassini, *Storia avventurosa della rivoluzione romana: Repubblicani, liberali e papalini nella Roma del '48* (Milan: Il Saggiatore, 2008), 89.

CHAPTER 2

The Experience of Exile: A Tragic Opportunity

After receiving the superior general's order, the Jesuits left Rome for other colleges in Europe and the United States. Passing through France, Secchi went to Great Britain where, together with other exiled companions, he found hospitality at the Jesuit College of Stonyhurst, in Lancashire.

He arrived on April 26, 1848, as recorded in the *Minister's Journal*:

> This afternoon two Italian Fathers Perrone[1] & Secchi with a junior Melandri arrived here.[2] The accounts they brought from Rome regarding religion etc. were of the most wicked description. Irreligion and misrule seem to be the order of the day.[3]

At last, far from Roman political turmoil, Secchi and his companions were safe.

1 A Refugee at Stonyhurst College

Secchi's stay in England, though short (about seven months), was nonetheless important. The few letters he wrote to his mother contain interesting first-hand information about this period. In one of the letters, probably anticipating her disapproval that he had moved to such a distant location, Secchi explained why he had left Rome to go to England:

> The sacrifice we made to God, me by consecrating myself and you by letting me enter the Society of Jesus, must be complete [...]. [In the present situation], whoever intended to continue in the divine service [...] had to accept to do it under foreign skies: [...] I believed that the best thing to do was to seek refuge in England, ready to return when the situation turns better. [...] My condition [i.e., his belonging to the Jesuit order] rendered my person hateful and suspicious for the people, and therefore condemned me, at least for a time, to complete inaction in any place of

1 Giovanni Perrone (see below) was a well-known theologian, later rector of the Collegio Romano.

2 Giuseppe Melandri (1826–1882) S.J. was not yet ordained priest in 1848; see ASPEM, Fondo Provincia Romana, "Fascicoli personali," 1814.

3 Fintan O'Reilly, "Stonyhurst and Angelo Secchi," *Stonyhurst Magazine* (1999): 28–34, here 28.

 | DOI:10.1163/9789004387331_004

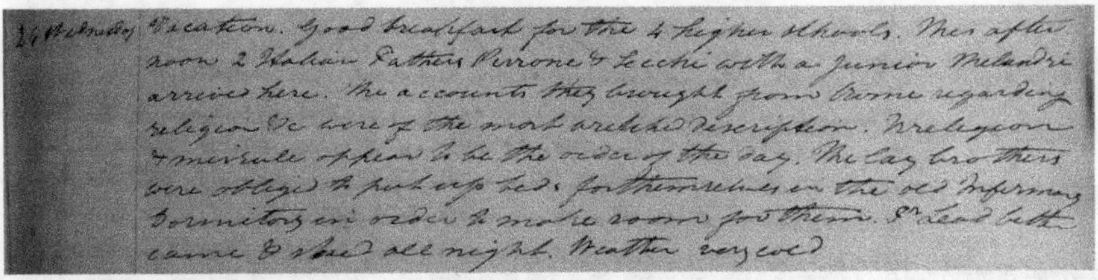
[illegible] Vacation. Good breakfast for the 4 higher schools. This afternoon 2 Italian Fathers Perrone & Secchi with a junior Melandri arrived here. The accounts they brought from Rome regarding religion &c were of the most wretched description. Irreligion & misrule appear to be the order of the day. The lay brothers were obliged to put up beds for themselves in the old Infirmary Dormitory in order to make room for them. [illegible] came & slept all night. Weather very cold

FIGURE 2.1 Detail of the minister's journal reporting the arrival of Secchi. Stonyhurst College Archives, by kind permission of the Stonyhurst governors.

Italy: in other times, I say honestly, the skills that I acquired through my studies, would have been helpful for me, but now, nothing at all; to wait inert in England or in Italy was the same thing; indeed this stay is highly instructive, and I hope that, once I come back, it will be even more advantageous: here, among other acquisitions, I learn a new language, new customs, and I can gain much more knowledge in the sciences of my competence, that is, natural sciences. […] Here we are as safe as in Italy, even safer, among good people and, after some time, if things in Italy will be re-established, we can return more apt to work than before, if not, our standard of living will depend on the circumstances: I'm sure that at any time I knock at the door of your house, I would find it open. I say this because, […] it would not be bad to have always ready the sum needed to return to Italy in case of risk, […] the trip costs at least ninety Roman scuda, almost five hundred francs: but I don't have a penny, in fact, as I said, I was sent here on charity. […]

Please, note that, for me, there was another reason to accept the journey, namely that I was troubled for six months from intermittent fevers that gave me an assault almost every month. The air of England is eminently fever reducing, therefore I couldn't get a better way to heal. It is true that the winter here is quite frigid, and could let me fall ill to another extent, but we'll see. […] I'm fine now and, since two months, it seems that the fever has gone; anyway, this is a great place; solitary, it's true, and tending to make one a little melancholy, but we'll try to overcome it by being busy, which is soon done. Now I study English, but it is a very difficult language, and it is very rare that a foreigner can speak it well: therefore, for a long time I'll be useless except to myself.[4]

4 *Il sacrifizio che noi due facemmo a Dio, io consecrandomi ed Ella lasciandomi andare alla compagnia, dovea essere completo* […]. [*Nelle attuali circostanze*] *a qualunque fosse stato proposto di seguitare nell'intrapreso divino servizio, unito agli altri, anche sotto cielo straniero dovea accettarlo* […] *Credetti miglior bene il riparare in Inghilterra pronto a ritornare quando*

FIGURE 2.2 "Stonyhurst [College], 1799". Watercolor by William Turner (1775–1851); Stonyhurst College Archives, by kind permission of the Stonyhurst governors.

---

questo mi paresse meglio. [...] *La mia condizione rendeva la mia persona odiosa, sospetta, e quindi condannata almeno per qualche tempo ad una totale inazione in qualunque sito di Italia: in altri tempi lo dico sinceramente le abilità acquistate collo studio mi sarebbero state utili, ma ora niente affatto; l'aspettare inerte in Inghilterra o in Italia era tutt'uno; anzi questo viaggio mi è stato sommamente istruttivo, e spero che il ritorno lo sarà ancor di vantaggio: qui acquisto tra le altre cognizioni una nuova lingua, imparo nuovi costumi e posso acquistare molte cognizioni di più nelle scienze di mia professione, cioè le naturali.* [...]. *Qui intanto si sta sicuro e tanto bene quanto in Italia se pure non meglio, fra ottima gente e dopo passato qualche tempo se le cose nostre in Italia si ristabiliranno potremo ritornare più abili di prima a faticare, se no dipenderà dalle circostanze il tenore di nostra vita: del resto sono sicuro che in qualunque ora io arrivassi alla porta della sua casa, la troverei aperta. Dico questo perché* [...]*non sarebbe male se io avessi sempre pronto la somma necessaria per tornare in Italia ad ogni rischio, perché* [...] *il viaggio costa almeno 90 scudi romani pressocchè 500 franchi: ma io qui non ho un soldo anzi come dico sono stato mandato qua per carità.* [...] *Però osservi che per me vi fu un'altra ragione per accettare il viaggio, cioè che era già da sei mesi travagliato da febbri intermittenti che mi davano un assalto quasi ogni mese. L'aria d'Inghilterra è eminentemente febbrifuga, quindi non poteva darmisi miglior mezzo di guarire. E' vero che l'inverno qui è alquanto fiero e mi potrebbe guastare per un altro verso, ma staremo a vedere. Frattanto si accomoderanno le cose se piacerà a Dio; se no vedremo quello che bisognerà fare. Io sto bene e da due mesi pare che la febbre se ne sia ita, ad ogni modo questo è un ottimo luogo; solitario è vero e tendente a mettere un po' di malinconia ma procureremo di vincerla stando occupati, il che presto si trova come fare. Ora studio l'inglese, ma è una lingua molto difficile ed è assai raro che uno straniero riesca a parlarla bene: quindi sarò per molto tempo inutile fuorché a me stesso.* Secchi to his mother, Luigia Belgieri, Stonyhurst College, June 18, 1848; APUG, FS 9.Vbis.

He was, however, optimistic about the future:

> The end of the war will decide many things uncertain so far, and if the reports are true it seems that soon the Italians should remain winners. After all, the rational Italians will have understood that there was no relationship between us[5] and those against whom they are fighting, and after this rage, they will know the truth and let religion be free even in its special regular corporations. Please, pray the Lord in order to give me patience and constancy in every event […].[6]

It is unclear if, during his inactive stay, Secchi studied mathematics at St. Mary's Hall, the Jesuit philosophate located at Stonyhurst.[7] Besides studying English, he certainly taught mathematics and physics to two refugee Italian scholastics.[8] The college also contained a well-equipped astronomical and meteorological observatory, built in 1838.[9] At that time, the observatory was directed by Alfred Weld, S.J. (1823–90), who observed many astronomical phenomena during Secchi's stay.[10] Weld is reported to have been the person "who introduced Angelo Secchi to the work of an astronomical observatory."[11] This would appear to be confirmed by a manuscript datable to around 1880, in which the director of the Stonyhurst observatory, Stephen J. Perry, S.J. (1833–89), states that "the meteorological and astronomical work at the observatory soon

5 Underlined in the original.

6 *La fine della guerra deciderà di molte cose finora incerte, e se sono vere le notizie dei fogli pare che presto gli Italiani debbano restare vincitori. Del resto gli Italiani di giudizio avranno ben capito non vi era relazione tra noi e quelli che essi combattono colle armi, e dopo passata questa febbre, conosceranno la verità e lascieranno la religione libera anche nelle sue particolari corporazioni regolari. Mi raccomandi al Signore che mi dia costanza e pazienza in ogni evento* […] Secchi to his mother, Luigia Belgieri, Stonyhurst College, June 18, 1848; APUG, FS 9.Vbis.

7 See O'Reilly, "Stonyhurst and Angelo Secchi," 29.

8 See Secchi to father provincial, October 20, 1850; ASPEM, Fondo Provincia Romana, "Fascicoli personali," 2604. The names of the two scholastics are not mentioned, but Melandri could be one of them (see note 2).

9 The equipment included a four-inch equatorial telescope, a meridian circle, and many meteorological instruments. See Agustín Udías, *Searching the Heavens and the Earth: The History of Jesuit Observatories* (New York: Kluwer Academic, 2003), 67–68, 185–189.

10 An accurate list of the observations carried out by Weld in those months is reported in O'Reilly, "Stonyhurst and Angelo Secchi," 30–32.

11 O'Reilly, "Stonyhurst and Angelo Secchi," 29.

attracted [Secchi's] attention and exercised a fascinating effect on his active mind."[12] However, despite these statements, Weld never mentioned Secchi in his reports, and it is difficult to prove that the two collaborated on any astronomical work.[13]

In June, Secchi and his companions received a short visit from Fr. De Vico, who stayed just two days at Stonyhurst: he "made his appearance […] not to stop as he had made arrangements to go and send his things to America."[14]

A new opportunity for the Italian Jesuits was in fact arising in the United States, where the local Jesuit superiors were ready to open the doors of their colleges to the highly respected teachers of the Collegio Romano and their confrères. Consequently, De Vico went to the United States in order to explore this possibility and find appropriate positions for his companions. De Vico's visit was highly successful. His scientific reputation was well known, and in July, when he arrived at Georgetown, his presence was considered a political achievement for the whole nation.[15] De Vico wrote to a friend:

> I pass over telling you about the meetings that I had with several authorities and senators: but I will note that all, without exception, at our first meeting, incessantly rejoiced to be able to welcome an Italian scientist with every demonstration of esteem, and this (they said) to show actually how different is real freedom in America from the ridiculous freedom of those European countries where illustrious persons are expelled or, in order to have a quiet life, must become expatriates.[16]

After receiving such an encouraging welcome, De Vico went back to Stonyhurst on September 21 and persuaded Secchi and his other confrères to follow him to Georgetown College in Washington. He left the day after and started to

12 Agustín Udías, private communication, March 2015. This idea is also reported in O'Reilly, "Stonyhurst and Angelo Secchi."

13 A few days after Secchi's arrival, on April 30, a nova, visible to the naked eye, appeared in the constellation of Ophiuchus; though the event attracted a considerable amount of attention, there is no record of Secchi's involvement in this observation.

14 O'Reilly, "Stonyhurst and Angelo Secchi," 33.

15 See Robert Emmett Curran, *The Bicentennial History of Georgetown University: From Academy to University, 1789–1889* (Washington, DC: Georgetown University Press, 1993), 1:143.

16 *Lascio di contarvi gli abboccamenti che ho avuti con parecchi deputati e Senatori: ma non lascerò di notare che tutti senza eccezione, al primo vedermi non finivano di congratularsi seco medesimi di potere accogliere con ogni dimostrazione di stima uno scienziato italiano, e ciò (dicevano essi) per mostrare col fatto quanto sia diversa la vera libertà americana dalla ridicola libertà di quei paesi Europei, donde persone illustri o vengono espulse, o per viver quieti debbono spatriare.* De Vico, quoted in Angelo Angeletti, "Padre Francesco De Vico, astronomo maceratese un ricordo nel bicentenario della nascita," *Astronomia* 1 (2006): 4–10.

FIGURE 2.3 Portrait of father Francesco De Vico. Courtesy of INAF-OAR.

arrange their transfer. However, in the meantime, in September 1848, a cholera epidemic erupted in England. The increasing danger accelerated their preparations and, on October 21, Secchi and the other Italian Jesuits left Stonyhurst. Their departure was preceded by the spectacular display of a brilliant aurora borealis on 17–18 October, recorded in the college's *Meteorological Journal*: the luminous phenomenon inevitably stimulated Secchi's scientific curiosity, and he observed and monitored its evolution while simultaneously trying "to explore the electricity of the atmosphere with a common atmospheric electrometer"[17] This was probably the last memory that Secchi took away from Stonyhurst, one that clearly made a deep impression, as we shall see.[18] Before leaving, Secchi wrote another long letter to his mother in which he announced his departure for the United States:

17 Angelo Secchi, "Aurora borealis," *Proceedings of the American Association for the Advancement of Science* 11 (1849): 2–9, here 5.

18 See O'Reilly, "Stonyhurst and Angelo Secchi," 32. Secchi again visited Stonyhurst ten years later (see Chapter 5).

In the last letter, I announced my arrival in England as well as my temporary stay in this College of Stonyhurst: now I have to inform you about my departure for America, that is, at least twice as far as I am now. Six months have elapsed since I arrived in this lucky island, where the disorder that ravages the European continent has not yet penetrated, but there is incumbent another danger, not minor perhaps, and it is cholera, already arisen in many sites. Since the Lord is pleased to put us on the road of sacrifices, we must go forth merrily, being sure that he will give us aid and grace to take advantage from them for our spiritual good. The day fixed for the departure from the port of Liverpool is at present the twenty-third, or a few days later, and when you read these lines I shall just be navigating the Atlantic Sea, headed to the United States. I shall disembark there, in New York or Boston (I don't know well yet), and I will reach the capital Washington and then the nearby Georgetown College where I shall be professor of physics or mathematics. Some years ago, I was asked to go there, and I would have gone, if I was allowed; now that there is nothing to hope for the present in Italy, providence reopens the doors of America. Many Catholics and Protestants have joined to pay the travel expenses for a number of my companions to go to work there, and since I was explicitly required in these last days, I will go with them. The American people love freedom and science and do not believe that our presence there can destroy the first, but rather introduce the second. [...] The bishops call us there and desire to have us for spreading the Catholic religion and supporting many missions. All these things encourage us very much and we hope that Our Lord will use the unworthy tools we are for the propagation of his glory. [...] During my stay in England, I was fine, I even grew in health, which I needed, due to the enormous stresses, fears, and travails suffered in Italy. I regret not being able to entertain you with long stories about the things I knew and viewed on my travels, some of which are very interesting; in general, however, I say that it has vanished from my head the prejudice that you can feel good only in Italy. There are not here, it is true, those great memories of antiquity who swell the Italians, or the many monuments erected in the past times of power; but, instead, there is a life, an activity, a currently flourishing wealth that are certainly unseen in Italy. Railways, steam navigation, ports full of ships, and immense manufactures are things that contribute to the welfare of a nation far more than temples, libraries, museums, and palaces, however not missing here at all. But Italy will never reach this condition, because in its inhabitants there is not that spirit of application to work and industriousness animating the industrious Englishmen. The climate itself, being colder and less delightful, hardens the men who find pleasure in being closed indoors at home to work, while in Italy it invites one to have fun and to enjoy the beautiful nature. However, in my opinion, I think that the Italians, taken individually, even in low classes, are much happier than their counterparts in England, where the life that is led by so many of them, from birth to death, is non-stop working, like a machine. The natural [character] of the Englishmen, although it is considered strange and melancholic, is not recognized in this manner by those who deal with them in their home […]; they (with appropriate exceptions, of course) are loyal, honest, and religious, and many Protestants, perhaps indeed the majority of them, would shame many Catholics in the point of their natural honesty. Education and practice of the Catholic religion is perfectly free here, and Catholicism spreads quickly, although perhaps it will never become the state religion, which is best, in order to avoid those pressures that governments ordinarily put on the church. Because of these virtues perhaps they deserve the prosperity which they enjoy […]. Now I'm going to find an even newer people, descended from this one: great things are said of their greatness, and after I have seen them, I will certainly not fail to write you about them, if before arriving I do not became lunch for the fishes.[19]

---

19 *Nell'ultima mia le annunziava il mio arrivo in Inghilterra e insieme il mio stato di provvisoria permanenza in questo collegio di Stonyhurst: ora sono a certificarle la mia partenza per l'America cioè almeno due volte più lontano che non sono adesso. Sono appunto scorsi sei mesi dacchè sto in quest'isola fortunata ove non è ancora penetrato il disordine che devasta il continente Europeo me se ne prepara un altro forse non minore, nel colera morbus che già manifestasi in molti siti. Giacché il Signore si è compiaciuto di metterci su la strada dei sacrifizi, dobbiamo andare innanzi allegramente sicuri che ci darà aiuti e grazia onde trarne il dovuto vantaggio per il nostro bene spirituale. Il giorno fissato per la partenza dal porto di Liverpool è il 23 corrente o pochi giorni dopo, e mentre Ella leggerà la presente io appunto starò traversando il Mare Atlantico diretto agli Stati Uniti. Ivi sbarcherò a New York o a Boston (che bene nol so ancora) e di là passerò alla capitale Washington e quindi al vicino collegio di Georgetown dove sarò professore di fisica o matematica. Già alcuni anni fa io fui richiesto di andare colà, e vi sarei andato se mi fosse stato concesso; ora che non vi è nulla da sperare pel presente in Italia la provvidenza riapre l'America. Molti cattolici e protestanti si sono uniti per pagare il viaggio ad un certo numero de' miei compagni per andare a faticare colà e io che era stato nominatamente richiesto anche in questi ultimi giorni andrò con loro. Il popolo americano ama la libertà e la scienza e non crede che la nostra presenza colà possa distruggere la prima, ma anzi introdurre la seconda.* […] *i Vescovi colà ci chiamano e desiderano per averci a propagare la religione cristiana cattolica, e coltivarvi le numerose missioni. Queste cose ci incoraggiano assai e speriamo che il Sig.*[nore] *N.*[ostro] *vorrà servirsi di questi indegni strumenti quali noi siamo alla propagazione della sua gloria.* […] *Durante la mia dimora in Inghilterra sono stato benissimo anzi mi sono rimesso bene in*

2 From Stonyhurst to Georgetown

On October 24, Secchi, with his other exiled companions, sailed from Liverpool to New York on board the packet ship *Patrick Henry*. Secchi was already considered a leader by his confrères; one of them, Ugo Molza (1821–91), recalled: "We were a caravan of twenty-one Jesuits whose nominal head was Fr. Pianciani, but the real head was Fr. Secchi who already knew enough English."[20] In the caravan, together with Secchi and Pianciani, who was sent to Georgetown to teach dogmatic theology, were the above-mentioned Molza, who later became rector of the Collegio Romano; Paolo Rosa Antonisi (1825–74), future assistant to Secchi at the Collegio Romano Observatory; Michele Saverio Tomei

salute, del che abbisognava non poco attesi gli enormi strapazzi, paure e travagli sofferti in Italia. Mi dispiace di non poterla trattenere con lunghi racconti sulle cose vedute e conosciute in questi miei viaggi, alcune delle quali sono assai interessanti; in generale però dico che mi è svanito dal capo il pregiudizio che si possa star bene solo in Italia. Non si trovan qui è vero quelle grandi memorie dell'antichità che tanto gonfiano gli Italiani, né i molti monumenti eretti in tempi di passata potenza; ma invece si trova una vita, una attività, una ricchezza attualmente fiorente, che certo non si vede in Italia. Le strade ferrate, la navigazione a vapore, i porti pieni di navi, le immense manifatture sono cose che contribuiscono ben più al ben essere di una nazione che i templi, le biblioteche, i musei o i palazzi, i quali pure tuttavia qui non mancano affatto. Ma a questo stato non arriverà mai l'Italia, perché nei suoi abitanti non è quello spirito di applicazione al lavoro, di attenzione e di fatica che anima l'industre inglese. Il clima stesso coll'essere più freddo e men delizioso indura l'uomo e gli fa trovare diletto nello stare chiuso in casa a lavorare, mentre in Italia lo invita a divertirsi e a godere la bella natura. Tuttavia per me io credo essere assai più felici gli Italiani presi individualmente anche nella classe bassa che i loro simili in Inghilterra, ove la vita menata da tanti e tanti dal nascere al morire consiste in lavorare sempre come una macchina. Il naturale [*carattere*] *degli Inglesi benché sia stimato strano e malinconico pure non si trova tale da chi li tratta in casa loro, e i viaggiatori di questa nazione possono molto incompletamente rappresentarne il carattere, il che dipende dallo stare essi estremamente riservati con quelli che non conoscono, e dall'essere ordinariamente occupati in affari. Essi* (*colle debite eccezioni però come già s'intende*) *sono probi, leali e religiosi, e molti protestanti, anzi forse la massima parte di loro, farebbero vergogna a tanti e tanti cattolici in punto di onestà naturale. L'educazione e l'esercizio della religione cattolica è liberissimo qui, e il cattolicismo rapidamente si propaga, benché forse non arriverà mai ad essere religione di stato, il che sarà meglio per evitare le oppressioni che i governi fanno d'ordinario alla chiesa. Queste loro virtù forse meritano loro dal cielo la prosperità temporale di cui godono* [...]. *Adesso io vado a trovare un popolo più nuovo ancora e discendente da questo, grandi cose si dicono delle sue grandezze e io dopo che le avrò vedute non mancherò di scrivergliene, se pure prima di arrivare non vado ad essere colazione di pesci.* Secchi to his mother Luigia Belgieri, Stonyhurst College, October 19, 1848; APUG, FS 9.Vbis.

20 *Partimmo insieme da Liverpool ai 24 di ottobre: eravamo una carovana di 21 gesuiti de' quali il capo nominale era il P. Pianciani, il reale p. Secchi che già conosceva sufficientemente l'inglese* [...] APUG, FS 23.IV.

(1792–1850), sent to teach moral theology; Giuseppe Bixio (1819–89),[21] who would contribute to the establishment of the Catholic Church in California; theologian Giovanni Perrone, who had arrived at Stonyhurst with Secchi; mathematician Francesco Saverio Provenzali (1815–94); Antonio Maraschi (1820–97), who was later considered the founder of the University of San Francisco; Torquato Armellini (1823–1901); historian Giuseppe Brunengo (1821–91); and philosopher Salvatore Tongiorgi (1820–65). Molza also recalled that, during the crossing, Secchi had the opportunity to have frequent conversations with the captain's wife, who was a Quaker. He "tried to let her think about the truths of religion and left her with the hope that the seed he had planted took well in her heart and was ready to germinate."[22] This statement reveals that Secchi did not hold back from having serious conversations with women,[23] which was far from typical for those times. Many Jesuits avoided conversing with women, because, in their opinion, they were "generally speaking, inconstant in their resolutions, and talk so much, that a great deal of time is wasted with them, and very little lasting fruit comes from it."[24] Apparently, Secchi did not share such sexist prejudices and, on the contrary, was quite in favor of women's rights (see Chapter 4), appreciated female beauty,[25] and was sympathetic to women—all attitudes that may have derived from a healthy childhood relationship with his mother and sister.

The ship *Patrick Henry* was renowned for its fast Atlantic crossing, and it reached its destination on November 20, less than a month after it had departed.[26] Jesuit refugees were an unexpected resource of highly qualified men for the United States. The arrival of this "extraordinary band of émigré scholars"[27] was seen by the local superiors in the United States as an opportunity to expand the Jesuit presence in the country and to establish schools in many dioceses. The provincial superior, Ignatius Brocard (1793–1852), offered them asylum in Maryland province: "[He] saw this turmoil as the opportunity to provide an

21 He was brother of Nino (1821–73), a general in Giuseppe Garibaldi's (1807–82) army.

22 *Si adoperò in modo da farla pensare alla verità della religione e la lasciò con la speranza che il seme da lui gettato avesse preso nel suo cuore e fosse per germogliare.* APUG, FS 23.IV.

23 In 1867, Secchi met Empress Eugénie of France (1826–1920) at the Universal Exhibition in Paris and remarked that "she thoroughly wished to speak and to remain and […] would have liked to stay longer" (*avea gran voglia di parlare e di trattenersi, e mi accorsi che avrebbe voluto tirare assai a lungo*). APUG, FS 23.II.A.

24 Gerald McKevitt, *Brokers of Culture: Italian Jesuits in the American West 1848–1919* (Stanford: Stanford University Press, 2007), 282.

25 In one of his diaries, Secchi states that he had traveled in the company of a Florentine gentleman "with a very beautiful daughter" (*con una bellissima sua figlia*). APUG, FS 23.II.D.

26 See Hughes, *Catchers of the Light*, 770.

27 Curran, *Bicentennial History of Georgetown University*, 1:133.

excellent theological and philosophical faculty at Georgetown, as well as a chance to strengthen the collegiate curriculum there, especially in mathematics and natural sciences, including astronomy."[28] This "infusion" was expected to allow Georgetown University "to rival the best European Universities."[29]

Preceded by so many expectations, Secchi and his companions arrived at the college on November 22. They found "an enthusiastic reception" at Georgetown. "I congratulate our country upon your arrival"—said an American Jesuit in his welcome speech—"and am tempted to cry out, O felix culpa! O happy frenzy of deluded Italy that has borne such precious fruit for America."[30]

The situation at Georgetown, however, was decidedly unconventional for Italian Jesuits, who were accustomed to the hierarchy and the regularity of Rome. They were surprised to see that the order's traditional protocol—namely to travel with a companion, to avoid late night visits outside the residence, and so on—was casually disregarded. For example, Molza was disturbed to find that, during the laying of a cornerstone for a new church, students of the college and women guests shared seats on the stage with the clergy.[31] The Italian Jesuits were also dismayed by "the frequency with which Americans socialized with women." Indeed, Fr. Benedetto Sestini (1816–90), assistant to De Vico, was scandalized by the behavior of some of the Georgetown Jesuits, including the provincial, who had attended a wedding reception and stayed until midnight "in the midst of a crowd of women [...] naked from the head to halfway down the chest and on the shoulders." He noted: "I confess, I would rather die than find myself in such an ugly situation."[32] In the midst of all this, Secchi prepared for his serious examination "ad gradum" (equivalent to the examination for the doctorate in theology), while substitute teaching mathematics, physics, and astronomy for almost three hours per day. He did not pass his examination, a failure he attributed to his teaching duties.[33]

Secchi's stay at Georgetown, albeit short, played an important role in his scientific training. He had the opportunity to keep abreast of the latest theories in physics and meteorology by visiting the US Naval Observatory and other academic institutions in Washington, and he also became acquainted with many US scientists, especially Commodore Matthew F. Maury (1806–73), one of the

28 Curran, *Bicentennial History of Georgetown University*, 1:132. See also George A. Fargis, "Georgetown College Observatory, 1843–1893," *Publications of the Astronomical Society of the Pacific* 7, no. 41 (1895): 89–97, here 89–91; Udías, *Searching the Heavens and the Earth*, 104.

29 See Curran, *Bicentennial History of Georgetown University*, 1:133.

30 McKevitt, *Brokers of Culture*, 45.

31 See McKevitt, *Brokers of Culture*, 68.

32 McKevitt, *Brokers of Culture*, 68.

33 See Secchi to father provincial, October 20, 1850; Archive of the Roman Jesuit province.

founders of dynamic meteorology, and physicist Joseph Henry (1797–1878), Secretary of the Smithsonian Institution. James Curley, S.J. (1796–1889), the resourceful professor of sciences at Georgetown who had been able to build an astronomical observatory in the college, may have played an important role in introducing Secchi to local scientists and institutions.

Curley was one of the two reviewers for Secchi's first significant article, published in the *Smithsonian Contributions to Knowledge*, a periodical issued by the Smithsonian Institution, which had been established just two years earlier. Secchi's paper was probably published at Curley's suggestion: it reviewed the research on electricity that he had carried out in Rome and dealt with a special case of attraction of a magnetic pole on circular electric currents. The article contained both a theoretical analysis and an experimental test.

The apparatus for the experiment was made by Secchi himself in "a room in the infirmary [that] was fitted up for his especial use to make these instruments."[34]

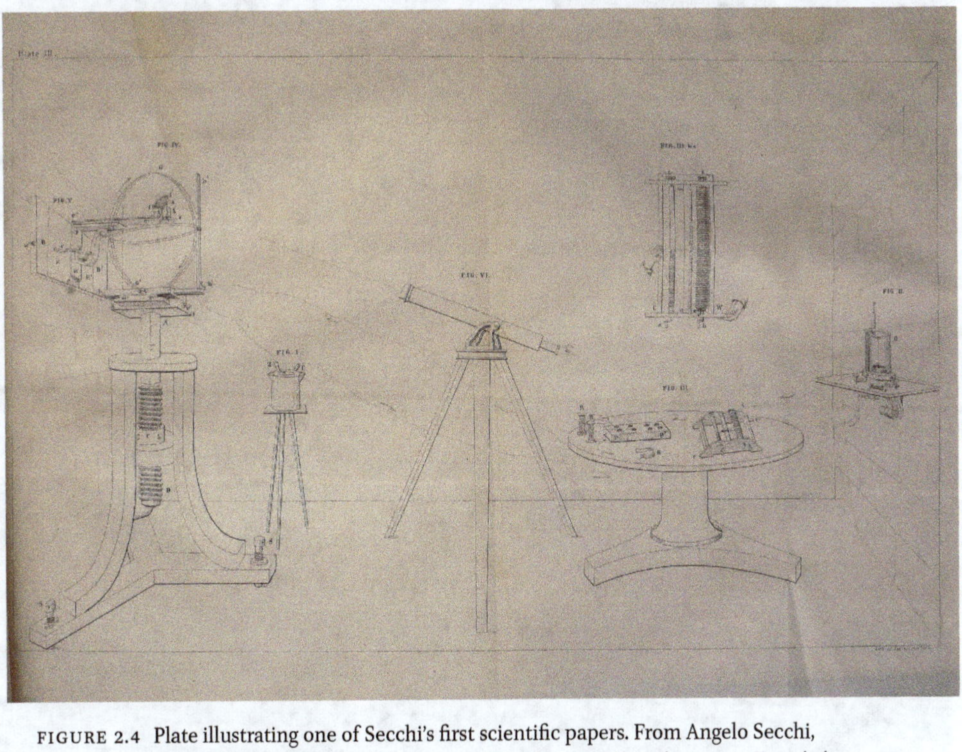

FIGURE 2.4 Plate illustrating one of Secchi's first scientific papers. From Angelo Secchi, "Researches on Electrical Rheometry," *Smithsonian Contributions to Knowledge* 3, no. 2 (1852): 1–59, pl. 3.

34 John Gilmary Shea, *Memorial of the First Centenary of Georgetown College, D.C.* (New York: Collier, 1891), 159.

FIGURE 2.5 Description of the Georgetown College Observatory in an archival document. Courtesy of INAF-OAR.

The article appeared in the third volume of the *Contributions*[35] because of an unexpected delay: the original manuscript was in fact lost in a fire, and in 1851, when he was again in Rome, Secchi was asked to send another copy. In reporting this accident, Curley insisted that it would have been "a great disappointment" for the editor Henry, not to be able to publish Secchi's work, as it "was the best article in the volume."[36]

From Curley, we have one of the early portraits of Secchi:

> I remember Father Secchi as a man who looked very much like Daniel Webster [(1782–1852), the famous American orator and politician]. He was dark, like an Italian, with a piercing but kindly eye, with projecting eyebrows. He was very energetic and fond of work. I remember one day Father Secchi came to me and asked me if I had not something for him to do. I had just made some observations, and gave them to him to calculate. In a very short time he had the calculations made, and was ready for more. He was only thirty-one when he came here, but looked much older. When he arrived at the College, though unable—on account of want of command of the English language—to take full charge of a class, he was employed in assisting the Professor of Physics. This science was then the specialty of Father Secchi.[37]

As he used to do previously when he taught in Italy, Secchi prepared a public experiment with his class:

> At the close of the scholastic year the class gave a public exhibition. Father Secchi constructed an electrical battery large enough to magnetize a bar one hundred pounds in weight, and on the day of the exhibition this magnet was made to hold 1,600 pounds. The magnet is now at the college.[38]

Curley confirms that "Father Secchi never did any astronomical work of importance at the College"; this is quite to be expected, because it was Sestini who was in charge of teaching astronomy and carrying out astronomical observations, having been appointed assistant in the observatory. Secchi wrote afterward to

35 Angelo Secchi, "Researches on Electrical Rheometry," *Smithsonian Contributions to Knowledge* 3, no. 2 (1852): 1–59.

36 Curley to Secchi, Georgetown College, May 1, 1851; APUG, FS 22.III.C.

37 Shea, *Memorial of the First Centenary*, 159.

38 Shea, *Memorial of the First Centenary*, 159. Further studies on electricity would be published by Secchi in 1878 (see Chapter 10).

Curley that "he regretted not having made himself more familiar with [the] college instruments, as he entered upon his new duties with little knowledge of the practical workings of a large telescope."[39] Attentive to the themes debated by the American scientific community and seeking to make his own contribution, at the beginning of April 1849, Secchi saw in the newspapers that Henry was "engaged in collecting observations on the Aurora Borealis, to fix, if it be possible, the laws of this interesting phenomenon so little understood".[40] This report led him to write an interesting letter, addressed to Henry and dated Georgetown, April 12, on the aurora he had observed in Stonyhurst around six months earlier, describing the phenomenon and analyzing some hypotheses on its electrical nature, as "it seems now out of question that this phenomenon is electrical; but how this electricity is produced is yet a mistery".[41] The letter was chosen to be read at the opening of the meeting of the American Association for the Advancement of Science (Henry being the President) on August 14, 1849. Henry was in fact struck by its content: "The paper of Professor Secchi seems to me to be one of considerable interest." – he commented – "It contains a number of ingenious suggestions, which may lead to new results" especially regarding "the connection of the Aurora with electricity".[42] Secchi would almost certainly have forged a brilliant career in the United States; however, a new, crucial, change of route in his life was now approaching.

3 Back to Rome

In July 1849, the Roman Republic came to an end with the French army's occupation of Rome. The Jesuits in exile could now return to Rome, but the situation at the Collegio Romano—and especially at the observatory—had changed. De Vico had died suddenly in November 1848 in England while preparing the transfer of his companions to Georgetown College. Secchi said this about him:

> I certainly did not think that in Liverpool, at the time of my boarding to America, I had to see him for the last time: but the dickey state of his health made me strongly fear for him, and I highly recommended him to take care of his health. He promised me to go immediately to London and once there, free from any effort, he wanted to take time to rest; but

39 Shea, *Memorial of the First Centenary*, 160.

40 Secchi, "Aurora Borealis," 2.

41 Secchi, "Aurora Borealis," 7.

42 Joseph Henry, "On the Aurora Borealis," *Proceedings of the American Association for the Advancement of Science* 11 (1849), 11–12, here 11.

> once arrived there, and taken rest, his health deteriorated dramatically, and despite the cares that were lavished upon him by his and our friend, the famous Dr. Kiernan [Francis Kiernan (1800–74)], it was not possible to have him back.[43]

During his first trip, De Vico had brought his most precious object to Georgetown College: the six-inch objective lens of the Cauchoix telescope that was in use at the Collegio Romano Observatory. Later, Sestini, with the assistance of Curley, mounted this lens into a wooden tube manufactured at Georgetown to continue his research program on colored stars, and upon the restoration of the Jesuits in Rome, he pleaded with the superior general[44] for permission to remain at Georgetown College Observatory as assistant to Curley. He ultimately remained at Georgetown for twenty years, first continuing his research on the colors of stars and studying sunspots, then publishing textbooks on mathematics.

Secchi was thus called to direct the Collegio Romano Observatory, a major turning point in his life. Though some sources suggest that Secchi was designated as successor by De Vico, the circumstances of his appointment remain unclear. In fact, it is more likely that the name of Secchi would have been suggested by Sestini, who intended to remain in the United States.

Secchi left Georgetown College on September 22, 1849 with fond memories of his American confrères; he still remembered them with affection more than twenty years later.[45]

Secchi asked to bring the six-inch objective lens of the Cauchoix telescope back to Rome, as it was one of the main instruments of the Collegio Romano Observatory and as it also had symbolic value, having been brought to the

---

43 *Io non mi credeva certamente, che in Liverpool all'epoca di imbarcarmi per l'America dovessi vederlo per l'ultima volta: ma lo stato mal ridotto di sua salute mi fece fortemente temere di lui, e caldamente gli raccomandai la cura della sua sanità. Mi promise egli di andare immediatamente a Londra e d'ivi franco d'ogni fatica volere attendere a ristorarsi; ma appena giunto colà, e messosi in riposo, sentì più violenta svilupparsi la malignità del suo morbo, e ad onta delle cure prodigategli dal suo e nostro amico il celebre Dottor Kiernan non fu possibile riaverlo.* Angelo Secchi, "Ragguaglio intorno alla vita e ai lavori del p. Francesco de Vico," *Memorie dell'Osservatorio del Collegio Romano: Anno 1850* (1851): 131–49, here 148.

44 See Curran, *Bicentennial History of Georgetown University*, 1:143.

45 See Angelo Secchi to Francesco Marchetti, Augusta, December 23, 1870; APUG, FS 9.II. A photograph of Georgetown College, which was probably donated to Secchi by Curley, is preserved in the archives of Rome Observatory (see fig. 2.6). The photograph bears the caption: "Georgetown College Taken from the Observatory." It is dated 1857 and signed by Titian Ramsay Peale (1799–1885), a famous American photographer who took a series of photographs of Georgetown College; see Julie Link Haifley, "Capital Images: The Photography of Titian Ramsay Peale, 1855–1885," *Records of the Columbia Historical Society* 50 (1980): 229–44.

FIGURE 2.6 Photograph of Georgetown College (Titian R. Peale, 1857). Courtesy of INAF-OAR.

United States by his predecessor, De Vico. Before returning to Rome, it is significant that Secchi first visited the observatories at Greenwich and Paris in order to re-establish the relationships between the Collegio Romano Observatory and the most important astronomical institutions of that time. This focus on maintaining international contacts from the very first, which would be a constant theme in his career, shows Secchi's clear intention of re-launching the observatory, giving it visibility and prestige, and introducing himself as a future player in the field of astronomical research.

Once back in Rome, Secchi completed his religious studies and, as provided for by Jesuit rules, upon reaching the age of thirty-three, on February 2, 1852, he made his solemn profession in the Society of Jesus.

CHAPTER 3

Establishing the New Collegio Romano Observatory: An Observatory for "Physical Astronomy"

Despite having a centuries-old astronomical tradition, an observatory was only installed at the Collegio Romano after the suppression of the Society of Jesus in 1773. Construction began in 1787, thanks to the support of several prelates, especially the influential cardinal Francesco Saverio de Zelada (1717–1801), when a tower was built for establishing an observatory, and several astronomical and meteorological instruments were gradually placed on its top.

The director of the observatory was the priest Giuseppe Calandrelli (1749–1827), who oversaw research until 1824, when the Collegio Romano was given back to the Society of Jesus, which had been reconstituted by Pius VII in 1814. Calandrelli was then charged with building another observatory, established in 1827 and located in the Roman seminary, at the Campidoglio palace. The coexistence of the two observatories reinforced the Roman astronomical tradition.[1] In 1824, once the Collegio Romano had been returned to the Jesuits, Etienne Dumouchel, S.J. (1773–1840), De Vico, and Brother Bernardino Gambara (1814–84) were respectively appointed director, assistant, and keeper (acting as second assistant). One year later, Superior General Luigi Fortis (1748–1829, in office 1820–29) donated a Cauchoix telescope, which De Vico would later use to discover the returning Halley's Comet, thereby gaining an international reputation both for the observatory and for himself. De Vico became director in 1839; his studies on planets, satellites, and comets consolidated the scientific renown of the observatory.[2] Now it was up to Secchi to build on the reputation of his predecessors.

---

1 For more information on the astronomical observatories in Rome from the seventeenth to the nineteenth century, see Roberto Buonanno, *Il cielo sopra Roma: I luoghi dell'astronomia* (Milan: Springer-Verlag, 2008).

2 On the early years of scientific activity at the Collegio Romano Observatory, see Sabino Maffeo, *The Vatican Observatory: In the Service of Nine Popes* (Rome: Libreria Editrice Vaticana, 2001), 8–14.

 | DOI:10.1163/9789004387331_005

FIGURE 3.1 The old observatory in the Calandrelli Tower. From Giuseppe Calandrelli and Andrea Conti, *Opuscoli astronomici* (Rome: Stamperia Salomoni, 1808), title page (detail).

1 In Need of a New Observatory

Secchi's first task was to recover the two main instruments belonging to the observatory, which had been moved by De Vico to be preserved during the exile of 1848, namely the above-mentioned Cauchoix telescope, whose objective lens was returned from the United States in November 1850, and another telescope, a meridian circle by Ertel, used for observing the transit of the stars on the meridian and donated in 1842 by Superior General Roothaan. The observatory's equipment also included a Gambey theodolite (an altazimuth instrument, for measuring angles), two Dent astronomical pendulums,[3] and a few

3 The pendulums—series no. 956 and no. 1832—were donated by Gregory XVI and Secchi's assistant, Fr. Rosa (see below).

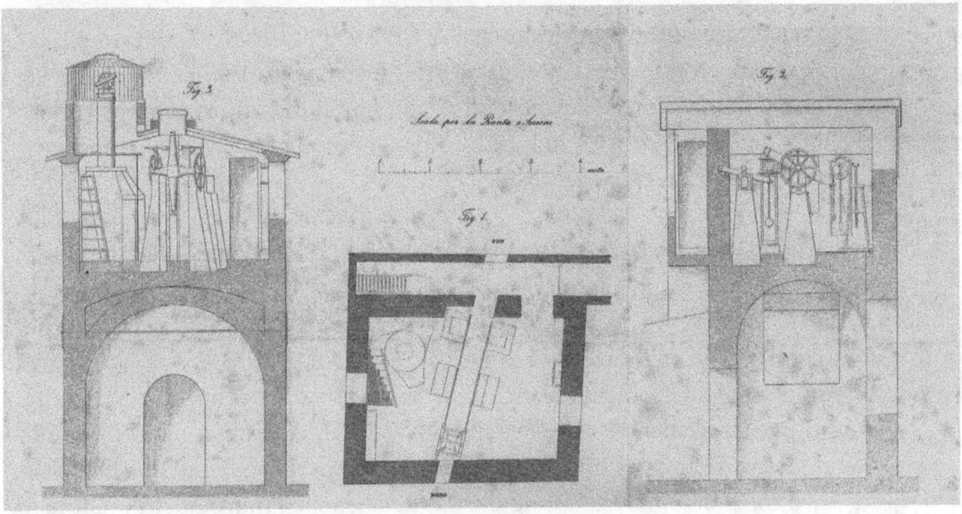

FIGURE 3.2 Plan of the old observatory and instruments in 1850. From [Angelo Secchi], *Memorie dell'Osservatorio del Collegio Romano: Anno 1850* (Rome: Tipografia delle Belle Arti, 1851), pl. [1], figg. 1–3.

eighteenth-century instruments, including a Reichenbach transit instrument, a small Dollond equatorial, and a Bellet repeating circle.[4]

Secchi decided to enlarge the room containing the transit instrument by extending it to a part of the near terrace, and in this room he put the meridian circle, which he checked accurately; the transit instrument, which was removed from the wall and placed on two pillars; and the theodolite, with a new supporting column.

However, when Secchi checked the stability of the tower and tested the effect of thermal expansion on the building's walls, he found that it affected the accuracy of the observations: the error in azimuth varied from 3″ in summer to 11″ in winter:

> These movements of the building entail that the effort of reduction of each observation is almost doubled, and that we must multiply very

4 The transit instrument is a telescope fixed in the meridian; like the meridian circle, it is used to determine the instant of the transit of a star on the meridian. The equatorials are telescopes with a special mounting with an axis (called polar axis) that is parallel to the Earth's axis; in contrast to altazimuth mounting, which allows the telescope to move vertically and horizontally, in the case of the equatorial telescopes, it is unnecessary to apply corrections to the star coordinates, since the telescope follows the same motion of the Earth. The repeating circle was another instrument used for measuring angles and had its main applications in surveying operations.

> much the observations to become independent from these errors [...]. Convinced that veracity must be the first talent of an astronomer, we have not hesitated to expose the flaws found in our instruments and in the building [...]. Certainly, it would be desirable to have an observatory immune from these flaws [...].[5]

The troubles mainly derived from the thermal expansion of the walls but also, to a less extent, from the increasing city traffic, with the astronomer being obliged to adjust the instruments, because of the "trembling produced at the passage of carriages down on the road."[6]

Another major problem was that the observatory did not have an equatorial telescope. The Cauchoix refractor did not have an equatorial mounting, and its use away from the meridian required numerous adjustments. Secchi tried to solve the problem by adapting the old Reichenbach transit instrument to a suitable mounting, made for that purpose:[7]

> The lack of an equatorial instrument in the observatory became, from day to day, more noticeable, but since the current circumstances do not allow us either to have one good enough for our level of science, or to mount the Cauchoix refractor equatorially, we decided to equip a Reichenbach transit instrument, which up to now had been almost useless in the observatory as it had been replaced by the meridian circle, with an equatorial mounting.[8]

5 *Questi movimenti della fabbrica fanno che la fatica di riduzione di ciascuna osservazione sia quasi raddoppiata, e che bisogni assai moltiplicare le osservazioni per rendersi indipendenti da questi errori* [...] *Persuasi che la* sincerità *deve essere la prima dote dell'astronomo, non abbiamo esitato ad esporre i difetti trovati nei nostri strumenti e nella fabbrica* [...] *Certamente sarebbe da desiderarsi all'osservatorio una sede immune da questi difetti.* Angelo Secchi, "Parte astronomica," *Memorie dell'Osservatorio del Collegio Romano: Anno 1850* (1851): 5–45, here 23.

6 *Tremiti pel moto delle carrozze che passano per la sottoposta via.* Secchi, "Parte astronomica", 20. The meridian room, however, was quite stable in comparison with the rest of the building; see Secchi, "Parte astronomica", 21.

7 Secchi had a great deal of experience in building home-made instruments, starting from the time he taught physics in colleges, when he would create his own laboratory equipment (see Chapter 1); in Rome, he could count upon the expert collaboration of the brothers Emilio and Ermanno Brassart, two skillful technicians, whose workshop was an important facility for the Collegio Romano Observatory: many instruments and adjustments, devised by Secchi, were made by Brassart under his supervision.

8 *La mancanza di uno strumento equatoriale nell'osservatorio diveniva ogni di' più sensibile, ma le circostanze attuali non permettendoci di averlo competente allo stato della scienza, né di potere montare equatorialmente il refrattore di Cauchoix, ci risolvemmo di supplirvi con armare equatorialmente uno strumento de' passaggi di Reichenbach che ora restava quasi inutile nell'osservatorio essendovi stato sostituito il circolo meridiano.* Angelo Secchi, "Introduzione:

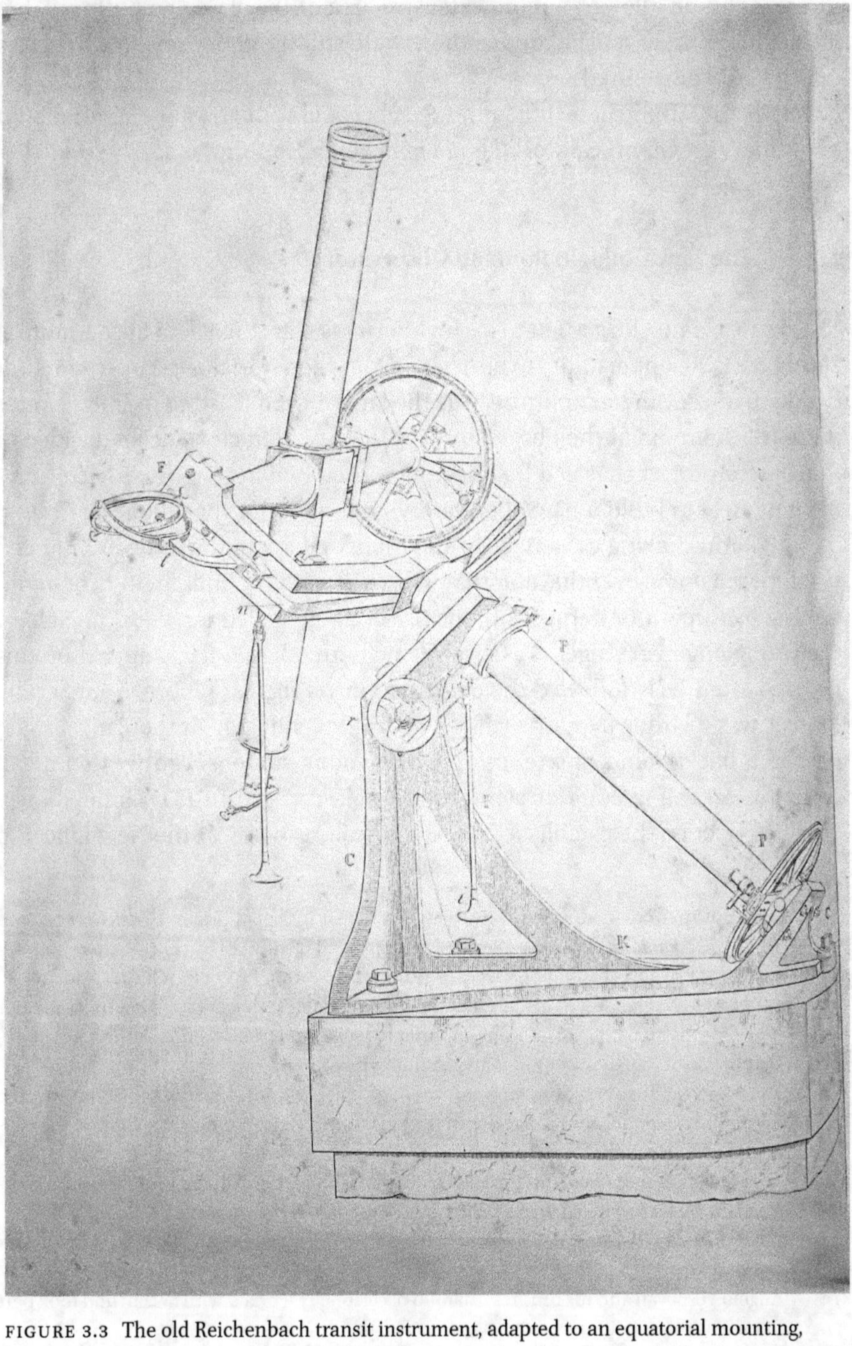

FIGURE 3.3 The old Reichenbach transit instrument, adapted to an equatorial mounting, used for solar observations; it was later replaced by the Cauchoix telescope. From [Angelo Secchi], *Memorie dell'Osservatorio del Collegio Romano: Anno 1851* (Rome: Tipografia delle Belle Arti, 1852), frontispiece.

This solution allowed a more extensive area of the sky to be observed, but it was not entirely satisfactory, as there were still limitations as to where and how it could be pointed.

Notwithstanding these difficulties, Secchi could still carry out a large amount of ordinary astronomical work,[9] but a radical renovation was sorely needed.

2 The New Collegio Romano Observatory

Thanks to a family legacy left to his highborn assistant, Fr. Paolo Rosa Antonisi, Secchi took the opportunity to purchase a Merz equatorial telescope, which, at twenty-five-centimeter aperture, was the largest then existing in Italy.[10] Since there was no room at the observatory for the new refractor, Secchi decided to return to a project that had first been proposed by Ruggiero Boscovich (1711–87),[11] namely to build a new observatory on the roof of the adjoining Church of St. Ignatius, taking advantage of the robust pillars that had been intended to support a dome over the altar that was never actually built. In 1852, after obtaining his superiors' permission, the project was carried out with the help of the young engineer Angelo Vescovali[12] and with the financial support of Pius IX, who attentively followed the construction via regular reports from Secchi. Pius IX was a strong supporter of science and technology in the Papal States; he had a background in science and, as a young student, had written a dissertation on the making of astronomical telescopes.[13] In 1847, immediately after his election, he established the Pontifical Academy of the New Lincei to

---

Sullo stato dell'osservatorio e sui lavori intrapresi nell'anno 1851," *Memorie dell'Osservatorio del Collegio Romano: Anno 1851* (Rome: Tipografia delle Belle Arti, 1852), 1–16, here 7.

9 Secchi was so absorbed in the work at the observatory that he neglected walks and recreation, and, going to sleep very late, he sometimes did not respect the community times for prayers, thus incurring an admonition from his superiors (see Secchi to father provincial, October 20, 1850; Archive of Roman Jesuit province).

10 The Merz Company in Munich produced high-quality optical instruments and was the most renowned maker of refractors in the nineteenth century, and Secchi found the German firm to be a reliable supplier for the equipment he would need. See Ileana Chinnici, "The Maker and the Scientist: The Merz–Secchi Connection," in *Merz Telescopes: A Heritage Worth Preserving*, ed. Ileana Chinnici (New York: Springer, 2017), 39–68.

11 See Renzo Lay, "Il nuovo Osservatorio del Collegio Romano," in Altamore and Maffeo, *Angelo Secchi*, 91–107, here 91.

12 Angelo Vescovali and his brother Anatolio occasionally worked with Secchi and took part in various observations.

13 Pius IX had been a student at a Scolopian College (S. Michele at Volterra) (see Aldo Altamore and Natalia Ptitsyna, "Il primo osservatorio geomagnetico d'Italia," in Altamore and Maffeo, *Angelo Secchi*, 177–96, here 178, 2012). Scolopians had a significant tradition in scientific education at their colleges.

revive the prestigious seventeenth-century academy (of which Galileo Galilei [1564–1642] been a proud member) and, that same year, he donated an Ertel meridian circle to the Campidoglio Observatory. In the following years, thanks to the pope's support, with two active and well-equipped observatories, "Rome [had] nothing more to envy of foreign observatories."[14] The relationship between Secchi and Pius IX was mutually supportive: the pope always supported the Jesuit with benevolence, and Secchi felt encouraged by his protection and goodwill. Secchi wrote to his mother:

> The last time I was with his holiness to submit the work of the observatory, he greeted me together with my assistant in a way that one could not desire to be more courteous, and had the graciousness to lead us by himself to his apartment for us to see certain works and books that had arrived as gifts from America, and ended with giving us a tidy sum of money for the construction of the new observatory, and promised that he would come to see it once finished. The walls are now up over its framework, and I hope that, within the year, everything will be in order.[15]

The new building was ready in about one year. The new Merz telescope and the previously existing instruments were installed in the new premises,[16] while the old Reichenbach transit was taken down and an equatorial mounting was adapted to the Cauchoix telescope.[17]

Because of his background in physics, Secchi's ideas on astronomy differed from those of his predecessors, which led him to adopt a multidisciplinary approach. Thus Secchi conceived the Collegio Romano Observatory as a "multi-tasking" observatory, engaging in astronomical, meteorological, and

---

14 Erasmo Fabri Scarpellini, *La scienza contemporanea nello Stato Pontificio* (Rome: Tipografia della Reverenda Camera Apostolica, 1857), 15.

15 *L'ultima volta che fui da Sua Santità a presentargli i lavori dell'osservatorio, Egli mi accolse insieme col mio compagno in una maniera che non si può desiderare più cortese, ed ebbe la degnazione di condurci Egli stesso nell'interno del suo appartamento per farci vedere certi lavori e libri americani venutigli in dono, e finì col regalarci una bella sommetta di danari per la fabrica del nuovo osservatorio, e promise che sarebbe venuto a vederlo dopo finito. La fabbrica è ormai finita nell'ossatura, e spero che dentro l'anno tutto sarà all'ordine.* Secchi to his mother, Luigia Belgieri, Rome, May 28, 1853; APUG, FS 9. Vbis.

16 To adjust the meridian circle, a sight was placed in a garden on Pincio hill, about 1,236 meters from the observatory; the sight is still preserved and can be identified today by a bust of Secchi that has been placed above it; see Lay, "Il nuovo Osservatorio del Collegio Romano," 103–6.

17 See Angelo Secchi, "Descrizione del nuovo Osservatorio del Collegio Romano," *Memorie dell'Osservatorio del Collegio Romano 1852–55* (Rome: Tipografia delle Belle Arti, 1856), 1–24, here 3.

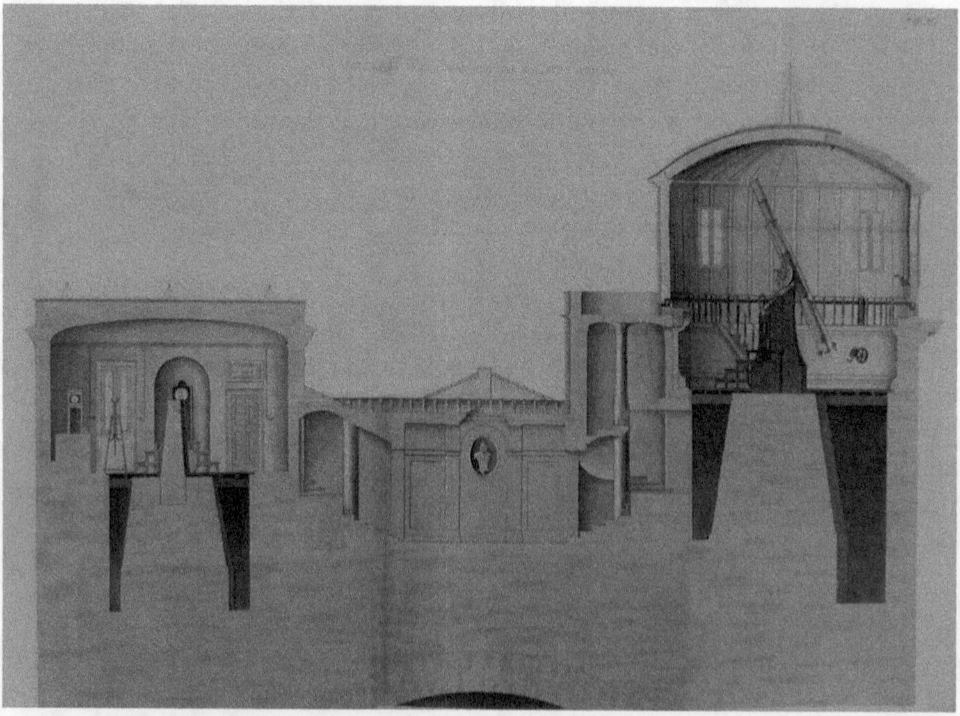

FIGURE 3.4 Section of the new observatory. From Angelo Secchi, "Descrizione del nuovo Osservatorio del Collegio Romano," *Memorie dell'Osservatorio del Collegio Romano 1852–55* (Rome: Tipografia delle Belle Arti, 1856): 1–24, pl. 2.

geomagnetic research.[18] This was a reflection of Secchi's worldview, in which natural phenomena were seen to arise from complex interactions between all physical forces, which were to be analyzed in each of their components.

At that time, regular meteorological observations were usually carried out in astronomical observatories; but Secchi's interest was not limited to their ordinary applications, namely for use in making refraction corrections to the astronomical observations. Instead, his aim in collecting meteorological data was to study their correlation with other phenomena, such as solar activity, earthquake swarms, magnetic disturbances, or polar aurorae. Such observations also had an important application in the service of weather forecasting (see Chapter 4). Secchi decided to maintain the original site of Calandrelli tower for the meteorological section of the observatory. This was an excellent choice because it ensured that the series of meteorological data that had been collected since the observatory had been installed could continue uninterrupted.

18 See Renzo Lay, "Angelo Secchi e il triplice osservatorio del Collegio Romano," *Giornale di fisica* 74, no. 3 (2013): 217–39.

FIGURE 3.5 The Merz equatorial hall. From Angelo Secchi, "L'astronomia in Roma nel pontificato di Pio IX," in *Triplice omaggio alla Santità di Papa Pio IX nel suo giubileo episcopale* (Rome: Tipografia della Pace, 1877), 29–77, pl. 5.

Data about pressure, temperature, and humidity were personally read seven times a day by Secchi and his assistants. As the years passed, this made him increasingly aware of the utility of having recording instruments, and he supplemented the equipment with a "meteorograph," a special device he invented for recording daily meteorological data (see Chapter 6).

In 1853–58, a geomagnetic laboratory was also established as a complement of the meteorological station. It is believed to have been the first geomagnetic observatory in Italy[19] and was described thus by Secchi:

> Complementing the meteorological equipment, the observatory possesses a whole kit of magnetic instruments, which at this point cannot be uncoupled from the meteorological ones [...]. The hall [...] contains

19 See Altamore and Ptitsyna, "Il primo osservatorio geomagnetico d'Italia."

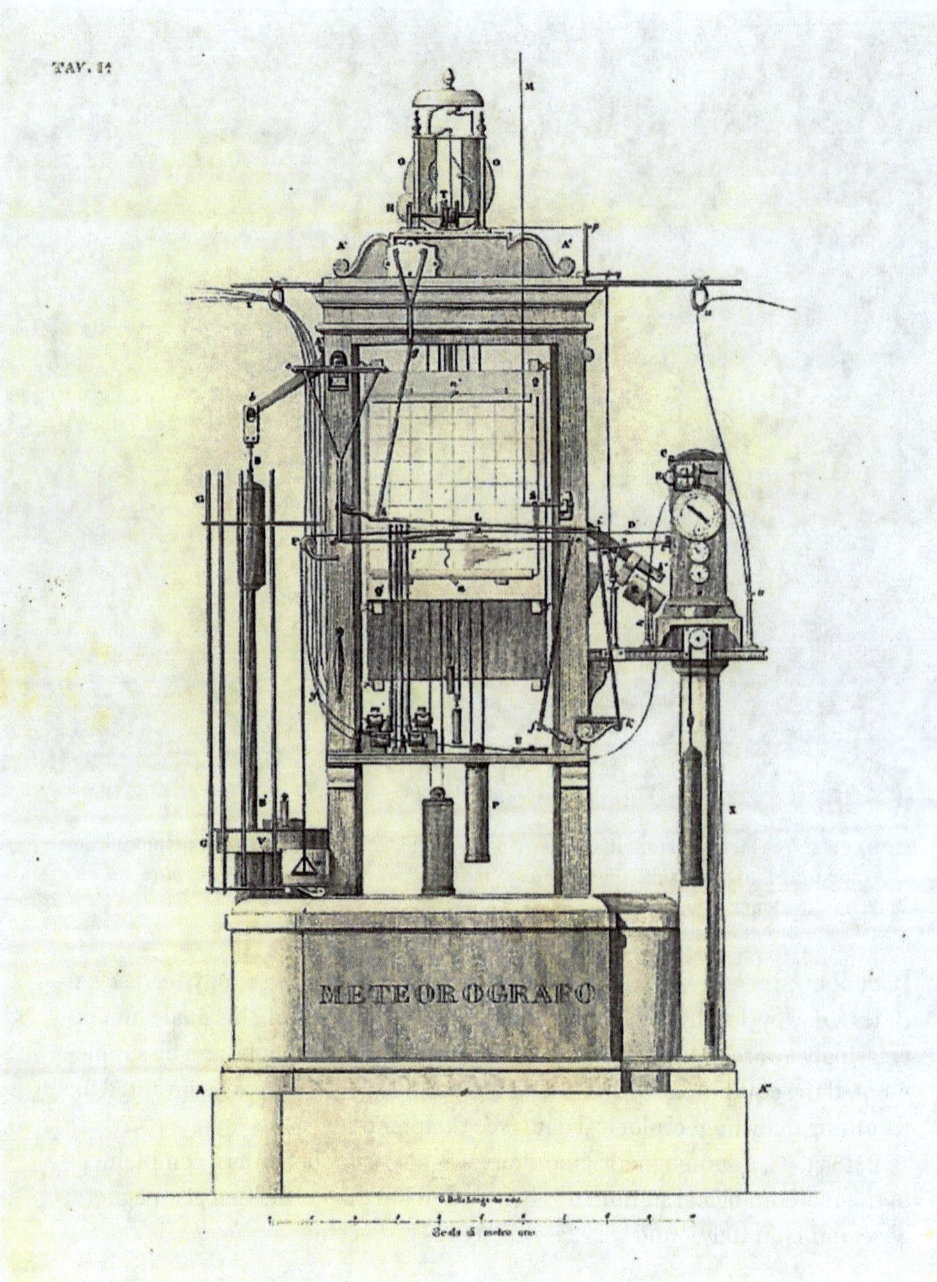

FIGURE 3.6 Secchi's first meteorograph. From Angelo Secchi, "Descrizione di un meteorografo ossia registratore meteorologico universale eretto all'osservatorio," *Memorie dell'Osservatorio del Collegio Romano 1857–1859* (Rome: Tipografia delle Belle Arti, 1859), [1]:1–8, pl. 1.

> six instruments: three fixed and three mobile. The first three are differentials, and give the variations of the Earth's magnetic force in the direction of the three rectangular coordinates: i.e., south–north through the inclinometer, east–west through the magnetometer, the vertical, that is, from the top to bottom, through the balance magnetometer. The three other instruments are designed [to measure absolute values], one to give the absolute declination of the magnetic needle, another the absolute inclination, and the third the horizontal intensity of the magnetic force by the method of Gauss [...]. In the room where these instruments are located, every frame or setting made of iron has been removed and replaced with copper and wood [...].[20]

Secchi was able to use this equipment to record the major magnetic storm[21] that followed the first solar flare ever to have been recorded, observed by Richard C. Carrington (1826–75) in 1859.[22]

There was another laboratory, located close to the magnetic laboratory, with electrical devices to measure atmospheric electricity. This was another innovation, as the Collegio Romano Observatory was the first to carry out regular observations of atmospheric electric currents; Secchi viewed such measurements as an "indispensable complement of magnetic observations."[23]

20 *A complemento della parte meteorologica l'osservatorio possiede un intero corredo di strumenti magnetici, che oramai non possono andar disgiunti dai meteorologici* [...] *La sala* [...] *contiene 6 strumenti: tre fissi e tre portatili. I primi tre sono differenziali, e danno le variazioni della forza magnetica terrestre nella direzione delle tre coordinate rettangolari principali: cioè Sud-Nord mediante il declinometro, Est-Ovest mediante il magnetometro filare, la verticale cioè d'alto in basso mediante il magnetometro a bilancia. I tre altri strumenti sono destinati uno a dare la declinazione assoluta dell'ago magnetico, l'altro l'inclinazione assoluta, il terzo la intensità orizzontale della forza magnetica col metodo di Gauss* [...] *Dalla stanza ove sono questi strumenti fu rimossa ogni armatura o legatura di ferro e supplito con rame e legno* [...] Secchi, quoted in Lay, "Il nuovo Osservatorio del Collegio Romano," 102.

21 See Angelo Secchi, "Sur les perturbations magnétiques observées à Rome le 2 septembre 1859," *Comptes rendus de l'Académie des Sciences* 49 (1859): 458–460.

22 See Meadows, *Early Solar Physics*, 45–47.

23 Secchi used a fixed conductor during storms, a Palmieri electrometer in good weather conditions, and a forty-mile long telegraphic wire connecting the observatory to the port of Anzio, "to explore Earth's electrical currents" (*per esplorare le correnti terrestri*). Angelo Secchi, "L'astronomia in Roma nel pontificato di Pio IX," in *Triplice omaggio alla Santità di Papa Pio IX nel suo giubileo episcopale* (Rome: Tipografia della Pace, 1877), 29–77, here 37. On research into atmospheric electricity in the nineteenth century, see Paolo Brenni, "Prometheus's Tools: Instruments and Apparatus Used in Atmospheric Electricity Research and Experiments," in *Playing with Fire: Histories of the Lightning Rod*, ed. Peter Heering, Oliver Hochadel, and David J. Rhees, Transactions of the American Philosophical Society 99, part 5 (Philadelphia: American Philosophical Society, 2009), 230–255.

The new observatory originally counted a staff including Secchi as director, Fr. Rosa as assistant astronomer, and Brother Bernardino Gambara as assistant for meteorology; the latter was later replaced by Brother Francesco Marchetti (1823–98), who became the best friend and confidant of Secchi.

Over the years, several other assistants came in succession, including Fr. Nazareno Mancini (1822–70),[24] fr. Giovanni Egidi (1835–1897) and Fr. Gaspare Stanislao Ferrari (1834–1903), who served as assistant until Secchi's death.[25] However, Secchi often invited other Jesuits and laypeople to collaborate on some specific projects.[26] The dynamic director clearly trusted his assistants: the first time he left the observatory for a long trip, in 1858, he could verify that he was not indispensable and that they managed to get by in his absence.[27] This experience encouraged Secchi to leave the observatory for several weeks at a time, whenever necessary, without consequences for its scientific activity. However, he probably required from his assistants the same relentless dedication to work that he had. The premature death of Fr. Rosa, for example, was attributed to an excess of work, while Fr. Ferrari sometimes spent much time out of the observatory performing religious duties,[28] probably to escape the exacting routine—and the demanding director, who was known to have a bad temper, even if he was not one to hold a grudge.[29]

---

24 On Fr. Mancini, see Francesco Marchetti, "Cenni necrologici del p. Nazareno Mancini d.C.d.G.," *Bullettino di bibliografia e di storia delle scienze matematiche e fisiche* 3 (1870): 429–38.

25 Ferrari became director after Secchi's death but, one year later, the observatory was confiscated by the Italian government, and Ferrari was forced to leave. In 1882, he then established a small private observatory located on Gianicolo hill for the scientific education of the clergy. In 1894, Ferrari was dismissed from the Society of Jesus for incurring unauthorized debts and left for Paris, where he lived until his death; see Maffeo, *Vatican Observatory*, 24.

26 Among these occasional collaborators were Vittorio della Rovere (1811–?) and father Cappelletti, who respectively assisted Secchi in early photographic works and in making astronomical drawings (see Chapter 6). About Della Rovere, see Piero Becchetti and Maria Francesca Bonetti, "La dagherrotipia a Roma" in Maria Francesca Bonetti and Monica Maffioli, *L'Italia d'argento: 1839-1859. Storia del dagherrotipo in Italia* (Florence: Alinari, 2003), 238-250, here 246. Other collaborators, in the 1870s, were Oratorian priest fr. Giuseppe Lais (1845-1921) and engineer Federico Mannucci (1848–1935), later both included in the staff of the Specola Vaticana (see Conclusions).

27 See Giuseppe Castellani, "Il P. Angelo Secchi S.I. e le sue missioni scientifiche," *La civiltà cattolica* 4 (1933): 606–620, here 619.

28 See Secchi to father provincial, April 21, 1876; ASPEM, Fondo Provincia Romana, "Fascicoli personali," 2604.

29 See Secchi to father provincial, October 20, 1850; ASPEM, Fondo Provincia Romana, "Fascicoli personali," 2604.

FIGURE 3.7 Portrait of Brother Francesco Marchetti. Courtesy of St. Ignatius Jesuit Rectorate; picture provided by Renzo Lay.

3 The Observatory Publications: *Memorie* and *Bullettino*

A second task for Secchi was to relaunch the observatory's original publications on as regular a schedule as was possible. Publications are an important expression of the life of an institution, and Secchi was aware that it was imperative to relaunch them in order to reconnect the observatory to the international community, gain visibility, and circulate the research being carried out by the Collegio Romano's astronomers. A first volume—*Memorie dell'Osservatorio dell'Università Gregoriana in Collegio Romano: Anno 1850* (Memoirs of the Gregorian University at Collegio Romano: Year 1850)—was published by Secchi in 1851. The volume included a detailed description of the work that was being done to renew the observatory, together with astronomical observations, calculations, and studies. The volume also contained an extended meteorological section, including a description of the instruments and the series of thermometric and barometric data from 1843 to 1850. Secchi retrieved the unpublished series of meteorological data collected at the observatory in those years and made haste to circulate them, as the series was almost uninterrupted: in fact, the meteorological data had continued to be read regularly after the 1848 exile, except for a couple of days close to the Jesuits' forced departure. A second volume, covering the year 1851, was published in 1852. It opened with a short dedication to Pius IX from the astronomers of the Collegio Romano[30] and contained the description of some adjustments that had been made to improve the instruments, as well as meteorological data and astronomical observations. The volume also included an appendix containing information about physics experiments, especially on solar heat radiation.

A third volume was published in 1856, after a four-year delay due to the relocation of the observatory. This volume, covering the years 1852–55, had a shortened title, *Memorie dell'Osservatorio del Collegio Romano*; it again opened with a dedication to Pius IX,[31] but this time it was signed solely by Secchi in his position as director, reflecting the fact that he considered this observatory his own "creature," since it was designed and built on the basis of his criteria and instructions. The volume contained a description of the new observatory, as well as many astronomical studies on double stars, clusters, nebulae, zodiacal light, planets, the Sun and moon, colours of the stars, atmospheric effects on astronomical observations (scintillation), and so on. In addition to the meteorological section, the volume also contained a magnetic section with geomagnetic observations, as well as an appendix with studies of comets and planets.

The next volume appeared in 1859, with the same short title, but as a new series—*Nuova serie*—and covered the years 1857–59. The volume contained descriptions of new meteorological and geomagnetic instruments and data reduction, and a collection of astronomical works, mostly under the title "Ricerche fisiche intorno ai corpi celesti" (Physical studies of celestial bodies), including, as usual, observations and discussions about the Sun, moon, comets, and planets.

In 1862, recognizing the importance of timely and up-to-date communication in science, Secchi began to edit the successful *Bullettino meteorologico dell'Osservatorio del Collegio Romano con corrispondenza e bibliografia per*

30 *A Pio IX—Pontefice Massimo—di ogni buona disciplina—favoreggiatore—gli astronomi del Collegio Romano–nel decembre del 1852* (To Pius IX, *Pontifex maximus*, supporter of every good discipline, the astronomers of Collegio Romano, December 1852).

31 *Alla santità di nostro signore—Papa Pio IX—principe ottimo massimo—protettore—e promotore de' buoni studi—per le cui largizioni ed incoraggimenti—un novella osservatorio—ergevasi—in collegio romano—questo primo frutto de' novelli studi—umilmente consacra—A. Secchi d.C.d.G.—dir* (To the holiness of our Lord, Pope Pius IX, *Optimus maximus princeps*, protector and promoter of good studies, thanks to the largesse and encouragement of whom a new observatory is built at the Collegio Romano, this first fruit of the new studies is humbly dedicated by Angelo Secchi from the Society of Jesus).

l'avanzamento della fisica terrestre (Meteorological bulletin of the Roman College Observatory with correspondence and bibliography for the advancement of terrestrial physics); this long title reveals Secchi's intention of taking a comprehensive view of the natural sciences, linking natural phenomena (astronomical, meteorological, seismic, etc.) in a single framework.

This monthly journal was dedicated to the prince Baldassarre Boncompagni (1821–94), who supported many pontifical cultural institutions by printing their publications at the Tipografia delle Scienze Matematiche e Fisiche, which he had created. The bulletin's circulation was so widespread that, in 1867, the famous astronomer Virginio Schiaparelli chose it to publish his discoveries about cosmic dust and the cometary origin of meteors (shooting stars), in the form of five letters to Secchi. He wrote to the Jesuit to thank him: "If I had buried these studies in some academic collection, maybe now the world would not yet know anything about them."[32]

Once the publication of the *Bullettino* had started, the *Memorie* declined in importance. Secchi's most important papers were usually printed in the *Atti della Accademia Pontificia de' Nuovi Lincei* (Acta of the Pontifical Academy of New Lincei) or in other periodicals,[33] which helped to save money on the costs of publishing the *Memorie*. But this was not the only reason. The last volume of the *Memorie*, published in 1863 and covering the years 1860–63, opened with an important warning: Secchi explained that the sequence of the memoirs was irregularly assembled, and that the staff of the observatory was so reduced that it was impossible to explore all astronomical fields and they had thus been forced to select those that were more commensurate with their resources. The 1863 volume was essentially a collection of papers previously published elsewhere, including some works on comets and planets, a paper on the connection between meteorological events and geomagnetic perturbations, and a report on the total solar eclipse of 1860 (see Chapter 5). It is important to mention that the last paper in the volume was Secchi's first paper on star spectral analysis, a subject that would be his principal focus from this point onward.

Through the *Bullettino*, published each month, Secchi informed the readers about the research work that was being carried out at the Collegio Romano Observatory, as well as correspondence with colleagues and discoveries and

32 *Se io avessi seppellito questi studj in qualche collezione accademica, forse a quest'ora il mondo non ne saprebbe ancor niente.* Schiaparelli to Secchi, Milan, January 22, 1867, in Buffoni, Manara, and Tucci, *G.V. Schiaparelli, A. Secchi*, 142.

33 Articles by Secchi can also be found in *Nuovo Cimento, Annali di Barnaba Tortolini, Corrispondenza scientifica di Roma, Atti Accademia Tiberina, Memorie della Società delle Scienze dei* LX, *Giornale Arcadico, Atti P. Accademia dei Nuovi Lincei*, and so on; for a tentative list, see François-Napoléon Moigno, ed., *Le révérend Père Secchi* (Paris: Gauthier-Villars, 1879), 95–120.

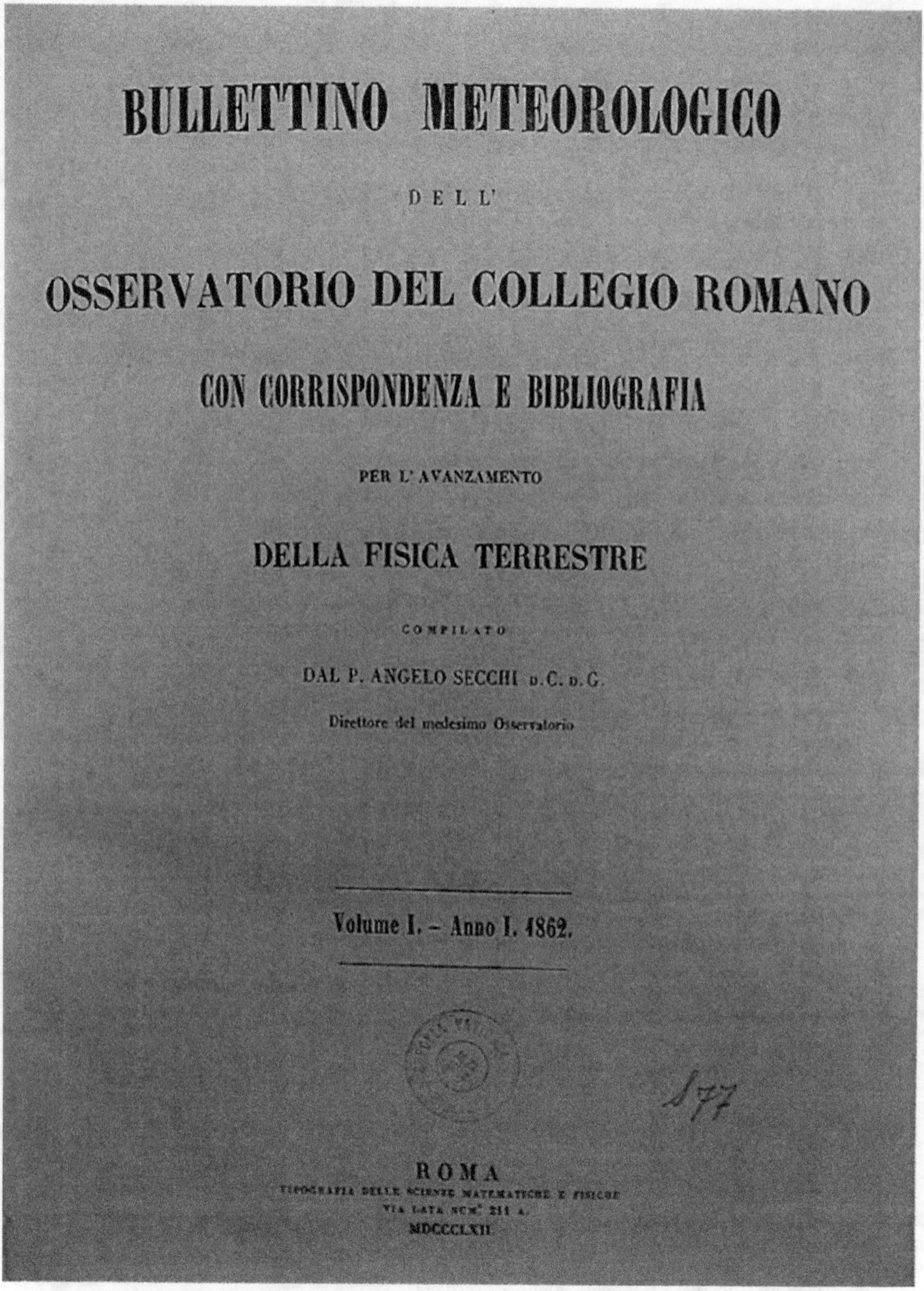

BULLETTINO METEOROLOGICO

DELL'

OSSERVATORIO DEL COLLEGIO ROMANO

CON CORRISPONDENZA E BIBLIOGRAFIA

PER L' AVANZAMENTO

DELLA FISICA TERRESTRE

COMPILATO

DAL P. ANGELO SECCHI D.C.D.G.

Direttore del medesimo Osservatorio

Volume I. – Anno I. 1862.

ROMA

TIPOGRAFIA DELLE SCIENZE MATEMATICHE E FISICHE

VIA LATA NUM.° 211 A.

MDCCCLXII

FIGURE 3.8 Front page of the first *Bullettino meteorologico*, published in 1862. Courtesy of Specola Vaticana.

studies by other scientists, so that it contained an up-to-date review of the main astronomical and meteorological topics of that time.

Secchi regularly corresponded with many international scientific journals, but he especially sent frequent letters, written in French, to the Académie des Sciences in Paris. For this reason, the *Comptes rendus* provide an important mirror of Secchi's scientific production throughout his career.

4 An Observatory for Physical Astronomy: Early Studies

Secchi's early astronomical studies covered a large variety of topics. The Jesuit astronomer appeared to be hungry for knowledge, curious about everything, impatient to use his instruments to examine the universe. Little by little, he became conscious that he would have to narrow his interests and become more selective in his fields of work, given the limits of his human and financial resources. On the other hand, he did not intend to turn back solely to his foundation in physics, and found that combining the two fields of study was the best solution. With the construction of the new observatory, this idea became more and more clear:

> The purpose of the observatory [was] to be dedicated completely to physical astronomy. This choice was suggested to us [...] above all by the quality and the number of people who were available to work there. The fundamental work of the science [of astronomy], such as [making] catalogs of stars and the series of regular observations of the bodies of our system [etc.] requires a large staff dedicated solely to these studies and a series of such well-conducted observations is quite incompatible with the other duties incumbent on the astronomers of the college. On the other hand, this branch of science is more than sufficiently developed in the large national observatories [...] while physical astronomy is less considered there, and left almost completely to the amateurs.[34]

34 *Lo scopo che ci proponevamo nell'erigere l'Osservatorio* [*era*] *di dedicarlo tutto all'astronomia fisica. La scelta di questo scopo ci veniva insinuata* [...] *soprattutto dalla qualità e dal numero delle persone che vi si potevano applicare. I lavori fondamentali della scienza, quali sono i cataloghi di stelle, e le serie di osservazioni regolari de' corpi del nostro sistema* [...] *esigono un numeroso personale a ciò unicamente dedicato, e una serie ben condotta di tali osservazioni riesce per lo più incompatibile cogli altri doveri di cui sono caricati gli Astronomi del Collegio. D'altronde, questo ramo di scienza è più che sufficientemente coltivato nei grandi osservatori nazionali* [...] *mentre l'astronomia fisica in questi è alquanto meno tenuta in conto, e lasciata quasi in retaggio de' dilettanti.* Secchi, "Descrizione del nuovo Osservatorio del Collegio Romano," 5.

Secchi knew that he could not compete in terms of equipment and staff with the larger observatories, but he wanted to stay at the cutting edge of research. Hence he chose to carry out "niche science," applying new techniques and introducing, as he himself said, "the taste of physics into astronomy," a choice that would prove successful and mark Secchi as one of the founders of modern astrophysics.

Secchi's early work at the Collegio Romano Observatory already contained the main scientific interests of his subsequent research. The studies that are mentioned below mostly cover the decade 1850–59. Initially, they were carried out at the old observatory, and then at the new site once the Merz telescope was in operation in October 1852.

4.1 *Early Celestial Photography, Actinometry, Selenography*

Being attracted by new techniques, Secchi was a pioneer in applying photography to astronomical works.[35] In order to investigate the existence of a solar-absorbing atmosphere during the solar eclipse of July 28, 1851, he obtained some daguerreotypes as well as some silver chloride plates reproducing the various phases of the eclipse. He used the Cauchoix telescope in combination with a camera obscura.[36] Although Italy was out of the path of totality, Secchi nevertheless recognized the potential benefits of this approach: "I don't know if the daguerreotype process has already been applied to astronomical observations during eclipses, but it seems to me that it could be highly useful."[37] He was evidently uninformed about the work of Julius Berkowski (1810–92), at Königsberg Observatory in Prussia, who had successfully obtained the first total solar eclipse photograph on the occasion of the same eclipse, which showed the corona and the prominences, typical features of the totality. As a result of the examination of the photographs taken at the Collegio Romano, Secchi confirmed the results obtained at the Paris Observatory in 1845:[38] the photographed Sun

35 See Hughes, *Catchers of the Light*, 766–86, and Laura Gasparini, "Padre Angelo Secchi e l'applicazione della fotografia nelle osservazioni astronomiche," *Fotologia* 12 (1990): 34–47. Secchi, however, was not the first astronomer of the Collegio Romano to make photographic experiments, as he described a daguerreotype (early photograph) of the moon that had been obtained by De Vico in 1843 and was kept at the observatory (see Angelo Secchi, "Fotografie lunari e degli altri corpi celesti," in *Memorie dell'Osservatorio del Collegio Romano* n.s. 1 [1859]: 158–60, here 158). On early daguerrotypes in Rome, see also Becchetti and Bonetti, "La dagherrotipia a Roma," 240.

36 He was assisted by Anatolio Vescovali and Vittorio Della Rovere.

37 *Non è a mia cognizione che siasi finora applicato il dagherrotipo ad osservazioni astronomiche in tempo di Eclisse, ma parmi che potrebbe trarsene notabile vantaggio.* Angelo Secchi, "Sopra alcune osservazioni fatte nell'Osservatorio del Collegio Romano durante l'eclisse del 28 luglio 1851," *Annali di scienze matematiche e fisiche* 2 (1851): 383–393, here 390.

38 See Meadows, *Early Solar Physics*, 7. Secchi did not mention these previous works, which provoked a complaint from François Arago, influential director of the Paris Observatory (see Chapter 9).

showed limb-darkening, which supported the hypothesis of the existence of a solar atmosphere. Secchi would again use photographic techniques on the occasion of the total solar eclipses of 1860 and 1870 (see Chapter 5).

On the occasion of the 1851 eclipse, Secchi also tried to measure solar radiation and to evaluate the Earth's atmosphere absorption.[39] He made several actinometric experiments (measurements of solar radiation) with darkened thermometers[40] and by using a Melloni thermopile,[41] attempting to measure the variation of heat intensity in the various parts of the solar disk. Secchi concluded that the equatorial regions of the Sun were hotter than the polar ones and connected these results with the existence of solar electromagnetic currents.[42] All these attempts eventually led him to devise a thermoheliometer (see Chapter 6).

In 1855, Secchi began making some observations of the moon in order to study what were assumed to be volcanic formations, especially one of the most interesting craters, Copernicus, which was accurately drawn with the help of a professional illustrator.[43] Secchi was not as talented in drawing as he was in observing and often turned to his assistants to produce detailed sketches. This limitation probably encouraged him to make use of photographic techniques to reproduce the results of his observations. In 1856–57, Secchi took some moon photographs, this time using wet collodion plates.[44] The results were encouraging, and in 1857 he produced the first photographic atlas of the moon, with eight photographs of different phases.[45]

---

39 See Angelo Secchi, "II. Considerazioni sulla vera maniera di valutare il raggiamento solare, e ricerche sulla forza assorbente dell'atmosfera terrestre," *Memorie dell'Osservatorio del Collegio Romano: Anno 1851; Appendice alle osservazioni del 1851* (1852): xvi–xxvi.

40 See Angelo Secchi, "Sul modo di valutare la forza del raggiamento solare," *Annali di scienze matematiche e fisiche* 2 (1851): 508–20.

41 See Secchi, "Sopra alcune osservazioni fatte nell'Osservatorio del Collegio Romano durante l'eclisse del 28 luglio 1851," 384–85.

42 See Angelo Secchi, "III. Sull'intensità del calore nelle varie parti del disco solare," *Memorie dell'Osservatorio del Collegio Romano: Anno 1851; Appendice alle osservazioni del 1851* (1852): xxvii–xxxvii; and Angelo Secchi, "V. Nuove ricerche sulla distribuzione del calore alla superficie solare," *Memorie dell'Osservatorio del Collegio Romano: Anno 1851; Appendice alle osservazioni del 1851* (1852): xxxxi–lxvii.

43 See Angelo Secchi, "Selenografia," *Memorie dell'Osservatorio del Collegio Romano 1852–55*, (1856): 134–35, here 134.

44 He was assisted by the Roman pharmacist and amateur photographer Francesco Barelli. For a comparison with other coeval photographs of the moon, see William Crookes, "On the Photography of the Moon," *Proceedings of the Royal Society* 8 (1857): 363–71.

45 See Secchi, "Fotografie lunari e degli altri corpi celesti." Secchi also took photographs of Saturn and Jupiter during the same period (see Secchi, "Fotografie lunari e degli altri corpi celesti," 160).

FIGURE 3.9 Photograph of the moon, taken by Secchi in 1857. Courtesy of INAF-OAR.

Secchi brought several copies of the atlas with him during his journey through France and England in 1858 and displayed the photographs[46] at meetings of the Académie des Sciences and the Royal Astronomical Society, where they were much appreciated (see Chapter 5).

Secchi's photographic atlas of the moon was the final result of his attempts to measure and make more and more detailed drawings of the crater Copernicus,[47] also looking for any seasonal variations. The question of whether changes occurred in the lunar surface continued to be discussed for many years, and Secchi would later return to this subject in 1866–67, when he entered the debate over Linnæus, a small crater that had apparently changed in appearance.[48] Further investigation revealed that the appearance of the crater varied depending on the manner in which it was illuminated by the Sun.[49]

46 See, e.g., Angelo Secchi, "Atlas photographique lunaire," *Comptes rendus de l'Académie des Sciences* 47 (1858): 362–64.

47 A beautiful drawing of Copernicus, made on the basis of Secchi's photograph, was published in 1878 in Newcomb's *Popular Astronomy*; see Simon Newcomb, *Popular Astronomy* (London: MacMillan, 1878), 315.

48 On the debate over Linnæus, see Meadows, *Science and Controversy*, 43–44.

49 See Newcomb, *Popular Astronomy*, 316.

4.2 *Photometric and Astrometric Works*

According to Secchi, "one of the most interesting topics in physical astronomy is getting to know the variations of starlight both in color and in intensity."[50] Sestini had started to study the colors of the stars at the Collegio Romano Observatory before the Jesuit exile of 1848 forced him to interrupt his work. He tried to continue this research at Georgetown, but his duties as a teacher did not allow him to complete the work. In order to complete Sestini's survey, Secchi devised some photometers for stellar colorimetry,[51] namely for determining the colours of the stars, with the aim of recording possible changes in the colors of the stars, a theme that was much debated in those years. Secchi never abandoned the idea of arranging and publishing Sestini's observations as a catalog; this idea came up from time to time in his plans, but he never managed to accomplish it,[52] probably because of his excessive workload, being engaged in the important task of updating earlier work.

Thus, in 1853, once the Merz telescope had been installed, Secchi began the revision of the famous double star catalog published in 1837 by von Struve.[53] Secchi himself described this as "the main work" that was carried out with the large refractor, and it occupied the years immediately following its installment, especially 1856–59. The purpose of this study was "to revise all the immortal work of Struve's *Mensurae micrometricae,* to detect the motions of these systems over the course of the twenty-five years following those measures."[54] Secchi did not in fact update von Struve's entire catalog, as he excluded those systems that could be observable with less powerful instruments and restricted his work to those that were within the reach of his instrument. He included some southern stars visible from Rome, and excluded double stars with very faint companions (magnitude 10–11) because they required a dark sky to be observed; Secchi was not satisfied with the system used to illuminate the crosswire, and in order to get more precise measurements, he preferred to operate

50 *Uno de' più interessanti soggetti della astronomia fisica è il conoscere le variazioni a cui va soggetta la luce delle stelle tanto nel colore che nella intensità.* Angelo Secchi, "Astrometria," *Memorie dell'Osservatorio del Collegio Romano: Anno 1850* (1851): 35–45, here 35.

51 See Angelo Secchi, "Astrometria," also Harold F. Weaver, "The Development of Astronomical Photometry-I," *Popular Astronomy* 54 (1946): 211–230, here 219.

52 Sestini's observations would be published in 1911 by Johannes G. Hagen, S.J. (1847–1930), director of the Vatican Observatory.

53 See Friedrich Wilhelm von Struve, *Stellarum duplicium et multiplicium mensurae micrometricae per magnum Fraunhoferi tubum annis a 1824 ad 1837 in Specula dorpatensi institutae* (Petropoli: Ex Typographia Academica, 1837).

54 *Rivedere tutta l'immortale opera di Struve Mensurae Micrometricae, per riconoscere i moti di questi sistemi nel corso de' 25 anni passati dopo quelle misure.* Angelo Secchi, *Catalogo di 1321 stelle doppie misurate col grande equatoriale di Merz all'Osservatorio del Collegio Romano* (Rome: Tipografia delle Belle Arti, 1860), 1.

with an illuminated field. As soon as the work was sufficiently advanced, on the occasion of his journey to France and England in 1858 (see Chapter 5), Secchi presented a preliminary version of his research to his colleagues at the Académie des Sciences[55] and the Royal Astronomical Society. The work was so greatly appreciated that he felt encouraged to print it apart from the academic proceedings; and so, in 1860, he published a catalog of 1,321 double stars in which he described the method and the criteria used for the revision.

The greatest encouragement came from the astronomer royal, George Biddell Airy (1801–92), who wrote to the secretary of the Royal Astronomical Society, Russell Henry Manners (1800–70): "I have received and looked over Secchi's double star work: it is really a magnificent quantity of splendid work [...]. It is much to be desired that so many extra copies may be printed that the work may be abundantly circulated independently of the academic transactions."[56] Secchi continued this work over the years, and in 1868, he published a second series of micrometric measurements of double stars, along with some nebulae.[57] Some of these double stars were observed together with the astronomer Otto Wilhelm von Struve (1819–1905), son of Wilhelm, during his visit to the Collegio Romano Observatory (see below). Although Secchi intended to complete this updating of Struve's catalog, by the 1860s the study of stellar spectra had become his main occupation, and Secchi preferred to devote himself to compiling another type of catalog, that of spectral types of stars, which would become a milestone in the history of astrophysics. The study of double stars was instead continued by his assistant, Ferrari, who published the results of his work in 1875.[58]

4.3 *Clusters and Nebulae*

In 1852–55, Secchi also carried out a few astrometric studies on star clusters, a topic that was "in its infancy"[59] and largely neglected by other astronomers: Secchi trusted that "it may not be a vain hope to discover someday that these

55 See Angelo Secchi, "Catalogue d'étoiles doubles," *Comptes rendus de l'Académie des Sciences* 47 (1858): 366.

56 Secchi, *Catalogo di 1321 stelle doppie*, 1.

57 See Angelo Secchi, "Serie seconda delle misure micrometriche fatte all'equatoriale di Merz del collegio romano dal 1863 al 1866 inclusive: Stelle doppie e nebulose," *Atti della Accademia Pontificia de' Nuovi Lincei* 21 (1868): 159–73.

58 See Gaspare Stanislao Ferrari, "Terza serie delle misure micrometriche delle stelle doppie fatte all'equatoriale del Collegio Romano dal 22 giugno 1872 a tutto il 1874," *Atti della Pontificia Accademia dei Nuovi Lincei* 28 (1875): 207–28.

59 Angelo Secchi, "Gruppi di stelle," *Memorie dell'Osservatorio del Collegio Romano 1852–55* (1856): 71–79, here 71.

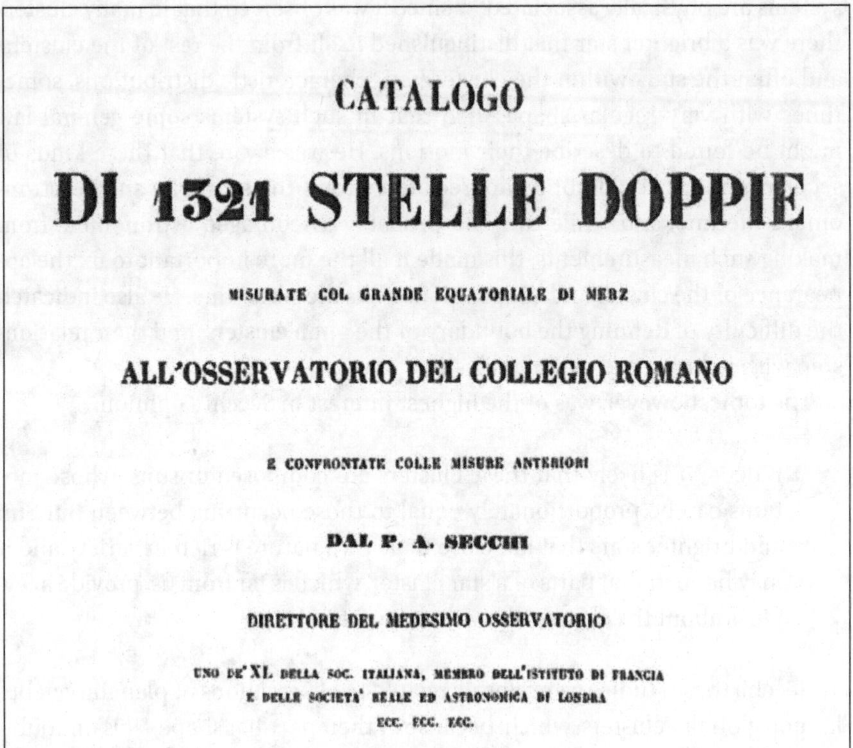

CATALOGO

DI 1321 STELLE DOPPIE

MISURATE COL GRANDE EQUATORIALE DI MERZ

ALL'OSSERVATORIO DEL COLLEGIO ROMANO

E CONFRONTATE COLLE MISURE ANTERIORI

DAL P. A. SECCHI

DIRETTORE DEL MEDESIMO OSSERVATORIO

UNO DE' XL DELLA SOC. ITALIANA, MEMBRO DELL'ISTITUTO DI FRANCIA
E DELLE SOCIETA' REALE ED ASTRONOMICA DI LONDRA
ECC. ECC. ECC.

ROMA
TIPOGRAFIA DELLE BELLE ARTI
1860

FIGURE 3.10 Title page of Secchi's double star catalog. Courtesy of Specola Vaticana.

systems are physically associated"[60] since it was observed that in many clusters there was a brighter star that distinguished itself from the rest of the clusters, and often the stars within the cluster had "characteristic distributions, sometimes with very regular shape,"[61] so that in such systems some general law might be found to describe their motions. He was aware that these kinds of secular motions could not be appreciated within the span of a single astronomer's lifetime, and while this had probably discouraged astronomers from making such measurements, this made it all the more important to fix the appearance of the cluster and the positions of its member stars. He also indicated the difficulty of defining the boundary of the open clusters, and their relationship with nebulae and globular clusters.

The topic, however, was of the highest interest in Secchi's opinion:

> Indeed, it is likely that these clusters are composed of suns, whose motions can be proportionately equal to those occurring between our Sun and brighter stars that are close to us [...]; nature is rich in variety and it may happen that parts of a star cluster, which is far from us, provide a few ideas about the cluster that we are a part of [...].[62]

To Secchi, these studies were closely related to observations of planetary nebulae and globular clusters, which, because of their peculiar shape, "it is undoubtable that they form a family of clusters that is totally different from irregular clusters and planetary nebulae."[63] Secchi was tempted to classify these objects, following the example of Herschel,[64] but he considered his attempts too embryonic for publication.[65] He also noted the existence of diffused luminous matter in some regions of the sky, and dark zones in some others, which he later called "dark nebulae" (see Chapter 6); and he recognized the importance of stellar statistics studies for understanding the structure of the universe.

---

60 *Vi possano essere speranze non vane di poter arrivare un giorno a scoprire in questi sistemi un legame fisico.* Secchi, "Gruppi di stelle," 71.

61 *Singolari distribuzioni in forma talora regolarissima.* Secchi, "Gruppi di stelle," 71.

62 *Invero è probabile che questi gruppi siano composti di Soli, i cui moti possono essere in proporzione eguali a quelli che si manifestano tra il nostro Sole e le stelle di maggior grandezza a noi vicine* [...] *la natura è ricca di varietà e può essere che un sistema stellare parziale a noi estraneo, dia* qualche idea *intorno al sistema di cui noi stessi facciamo parte* [...]. Secchi, "Gruppi di stelle," 72.

63 *Non può dubitarsi che essi formino una famiglia di sistemi totalmente diversa da quella a cui appartengono gli ammassi irregolari e le nebulose planetarie.* Angelo Secchi, "Osservazioni di nebulose," *Memorie dell'Osservatorio del Collegio Romano 1852–55* (1856): 80–96, here 82.

64 See William Herschel, "Catalogue of One Thousand New Nebulae and Clusters of Stars," *Philosophical Transactions of the Royal Society of London* 76 (1786): 457–99.

65 See Secchi, "Osservazioni di nebulose," 82.

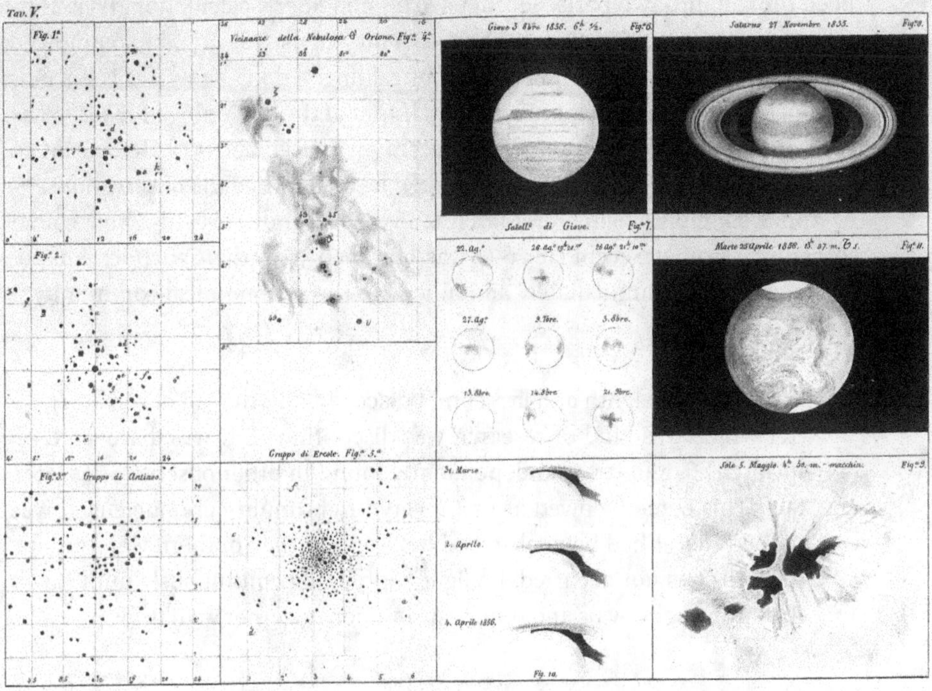

FIGURE 3.11 Observations of clusters, nebulae, and planets. From Angelo Secchi, "Descrizione del nuovo Osservatorio del Collegio Romano," *Memorie dell'Osservatorio del Collegio Romano 1852–55* (Rome: Tipografia delle Belle Arti, 1856), 1–24, pl. 5.

Moreover, while looking for the fragments of Comet Biela (see below), Secchi and Ferrari made a survey of nebulae that had not been recorded by Herschel. Secchi published this survey in 1866 along with his study of the spectrum of the Orion nebula.[66]

4.4 *Cometary Studies*

Secchi initially continued the comet-hunting tradition of the Collegio Romano, where numerous comets had already been discovered by his predecessors.[67] His first achievement in this field was to recover the two parts of Comet Biela (1852 III). In 1846, this periodic comet had appeared double, suggesting it had

66 See Angelo Secchi, "Nébuleuses découvertes avec l'équatorial de Merz au Collège Romain de 11 novembre 1865 au 18 janvier 1866," *Astronomische Nachrichten* 66, no. 1571 (1866): 161. Secchi's findings with regard to the Orion nebula were harshly contested by Huggins (see Chapter 9).

67 For example, in the years 1844–47, De Vico discovered six comets at the Collegio Romano Observatory.

been broken into two parts. Secchi used an ephemeris calculation devised by Giovanni Santini (1787–1877) in 1851; and, on August 26, 1852, he announced the discovery of a small comet in Gemini, adding: "I don't know if this is a new comet, or a portion of that of Biela, which split at the beginning of 1846."[68] The second part was also discovered by Secchi on September 15.[69] He continued to observe them until September 20, when he abandoned the observations.[70]

About six months later, on March 6, 1853, Secchi discovered a new comet in Lepus;[71] Comet Secchi (1853–I) was observed with the Cauchoix equatorial telescope and immediately announced to the astronomical community.[72] Secchi wrote:

> This was the eighth and last comet discovered in the old observatory, after which this kind of research was discontinued. It was discovered on [March] 6, and it was independently found in other observatories later. [...] This comet showed at first a very bright multi-core nucleus. It was hoped that it had been observed before us in the Southern Hemisphere, but this has not occurred. Its light kept fading continuously until it was lost from sight by us and other astronomers together with us.[73]

---

68 *Je ne sais si celle-ci est une nouvelle comète, ou une portion de celle de Biela, qui se divisa au commencement de 1846.* Angelo Secchi, "Entdeckung und Beobachtungen eines kleinen Cometen", *Astronomische Nachrichten* 35 no. 822 (1853): 89–90, here 90.

69 See Angelo Secchi, "Beobachtungen des Biela'schen Cometen und Bemerkungen über densellen und seinen aufgefunde Begleiter", *Astronomische Nachrichten* 35, no. 828 (1853): 191–92.

70 See Angelo Secchi, "Observations de la double Comète de Biela," *Astronomische Nachrichten* 35, no. 832 (1853): 251–52. He had to leave for Sezze, near Latina, where he was called to draft a sundial for the local school of physics; see Angelo Secchi, "Cometa di Biela 1852," *Memorie dell'Osservatorio del Collegio Romano 1852–55 (1856):* 147–49, here 148. For more information on the sundials built by Secchi, see Chapter 4.

71 See Gary Kronk, *Cometography* (Cambridge: Cambridge University Press, 2003), 2:216–218.

72 See Angelo Secchi, "Correspondance," *Comptes rendus de l'Académie des Sciences* 36 (1853): 543 and Angelo Secchi, "Entdeckung und Beobachtungen eines kleinen Cometen", *Astronomische Nachrichten* 35, no. 822 (1853): 89–90.

73 *Fu questa l'ottava ed ultima cometa scoperta nel vecchio osservatorio, dopo la quale questo genere di ricerche fu affatto abbandonato. Questa fu scoperta nel suddetto giorno 6, e indipendentemente da noi venne trovata più tardi in altri osservatorii.* [...] *Questa cometa presentava nelle prime sere un nucleo multiplo assai lucido: si sperava che fosse stata osservata prima di noi nell'Emisfero australe, ma ciò non si è verificato. Essa andò scemando di luce continuamente finchè fu perduta di vista da noi e dagli altri astronomi insieme con noi.* Angelo Secchi, "Cometa del 6 marzo 1853 scoperta all'Osservatorio del Collegio Romano," *Memorie dell'Osservatorio del Collegio Romano 1852–55* (1856): 149–50, here 149.

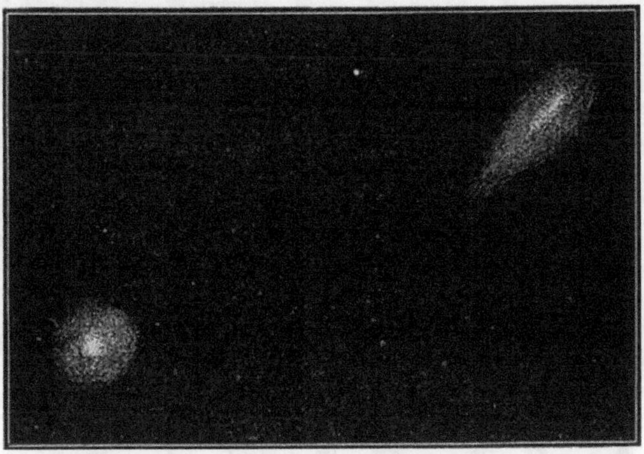

FIGURE 3.12 Comet Biela in 1852, as observed by Secchi. From Amédée Guillemin, *Le ciel: Notions élémentaires d'astronomie physique* (Paris: Hachette, 1877), 554.

Secchi kept his interest for cometary studies alive during these years, as can be seen by some interesting publications from the 1860s.[74] After 1863, spectroscopic observations became much more interesting for him than visual ones, and he became a pioneer in the field of cometary spectroscopy, together with Giovan Battista Donati (1826–1873) in Italy and William Huggins (1824–1910) in England (see Chapter 6).

4.5 *Planet Observations*

Saturn was one of the first planets to attract Secchi's attention. In 1850–51, he engaged in a brief correspondence with William Lassell (1799–1880), an expert observer of Saturn, about some observations of the planet's ring and, in 1854–55, he made some measurements of its size.[75] The mysterious nature of the ring could not but stimulate the scientific curiosity of the Jesuit, who did not miss any opportunity to study it, observing its occultation and its predicted disappearance (because of the changing perspective due to the relative motions of Saturn and the Earth, the ring would have been observed edge-on) during 1861–62.[76] The high-quality optics of the Merz refractor allowed the

74 See, e.g., Angelo Secchi, "Alcune considerazioni su le tre ultime grandi comete," *Memorie dell'Osservatorio del Collegio Romano 1860–62* (1863): 2:18–32; Giuseppe Lais, "Le comete," 1903, in AA. VV., 63.

75 See Angelo Secchi, "Nuove apparenze dell'anello di Saturno," *Annali di scienze matematiche e fisiche* 2 (1851): 39–44.

76 See Angelo Secchi, "Osservazioni di Saturno in occasione della disparizione dell'anello negli anni 1861 e 62," *Memorie dell'Osservatorio del Collegio Romano* n.s. 2 (1863): 73–76.

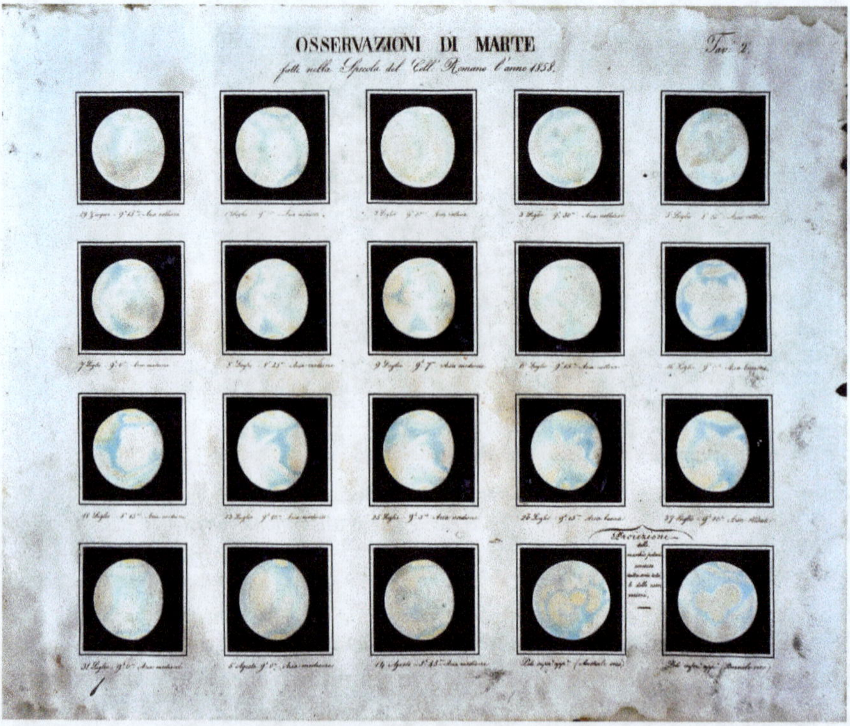

FIGURE 3.13 Color drawings of Mars, dated 1858, made at the Collegio Romano Observatory. Courtesy of INAF-OAR.

Collegio Romano astronomers to improve the quality of their planetary observations. In 1855, with the Merz telescope, Secchi carried out some observations of Jupiter and its satellites, including studies of Jupiter's spots, its diameter, and the flattening of its poles. In 1856, he then began a series of systematic observations of Mars. During the opposition[77] of 1858, he produced a series of forty excellent colored drawings of the red planet[78], probably with the help of his skilled assistant, father Enrico Cappelletti (1831–99). Secchi thus recounted the results of his observations:

---

77 An opposition refers to the period when a superior planet (which orbits outside the Earth's orbit) is on the opposite side to the Sun with respect to the Earth. An opposition is the best moment to observe the planet from the Earth; the opposition may be favorable or unfavorable, depending on the varying distance of closest approach between the planet and the Earth.

78 This series of pastels - the first known color drawings of Mars (see William Sheehan, *The Planet Mars* [Tucson: University of Arizona Press, 1996], 51) - revealed the first evidence of martian dust storms; see Richard J. McKim, "Telescopic martian dust storms: a narrative and catalogue", *Memoirs of the British Astronomical Association* 44 (1999), 19–20.

> The most striking configuration of Mars was to present a large blue spot, almost triangle-shaped, stretched out like a wide channel that extends from one pole to the other. Another narrower channel is located at 120 degrees of longitude from the first, but is not so marked. The rest of the equatorial surface shows a vast reddish and slightly brown-shaded extension, dotted with white and more intense red. The polar spots are white or rather light yellow, and surrounded by blue channels.[79]

Secchi found that the appearance of Mars differed considerably from his own early observations and from the past drawings of other astronomers, especially those of Johann Heinrich von Mädler (1794–1874), leading him to hypothesize that the changes were due to the possible presence of clouds on the planet. Writing on this matter, Secchi suggested that

> Mars is somehow a body of intermediate physical constitution between the moon, the Earth, and Jupiter. The atmosphere of Mars might be thinner in comparison to that of Jupiter. In my drawings of the latter planet, there are spots looking like our hurricanes, and its appearance in this year 1858 was very different from that of 1856. The moon, on the contrary, with no perceptible atmosphere, would be the other extreme, and Mars, with a thin atmosphere, would be intermediate.[80]

In 1862, however, during a very favorable opposition of Mars, Lockyer observed the planet and found a remarkable agreement between von Mädler's drawings and his own observations. This clearly contradicted Secchi's results.[81] The dispute was soon solved in favor of Lockyer's observations, but Lockyer's findings opened a clash between the Jesuit and a large part of the English astronomical

---

79 *La configuration de Mars la plus frappante a été celle de présenter une grande tache bleue de forme presque triangulaire, prolongée comme un vaste canal qui s'étend d'un pôle à l'autre. Un autre canal plus étroit se trouve à 120 degrés environ de longitude du premier, mais n'est pas si marqué. Le reste de la surface équatoriale ne présente qu'une vaste extension de couleur rougeâtre et légèrement ombrée de brun, et parsemée de points blancs et rouges plus vifs. Les tâches polaires sont blanches ou plutôt jaune clair et sont environnées de tous côtés des canaux bleus.* Angelo Secchi, "Études sur la planète Mars," *Comptes rendus de l'Académie des Sciences* 47 (1858): 364–66, here 364.

80 *Mars est à per près un corps de constitution physique intermédiaire entre la Lune, la Terre et Jupiter. L'atmosphère de Mars ne serait que bien légère en comparaison de celle de Jupiter. Les figures que j'ai faite de cette dernière planète montrent des taches qui rappellent nos ouragans, et son aspect dans cette année 1858 a été très différent de celui de 1856. La Lune, au contraire, sans atmosphère sensible, serait l'autre extrême, et mars avec une faible atmosphère serait intermédiaire.* Secchi, "Études sur la planète Mars," 365.

81 See Meadows, *Science and Controversy*, 45.

community, which would eventually have serious consequences for Secchi (see Chapter 9).

Incidentally, the Italian word "canali," first used by Secchi and then by Schiaparelli to describe the details of Mars' surface that he observed in 1877, was mistranslated into English as "canals," instead of "channels,"[82] thus suggesting an artificial origin and reigniting a long-standing controversy about the existence of Martian life.

4.6 *Gravimetric Experiences*

In February 1851, Léon Foucault (1819–68) conducted his famous public experiment that aimed to prove the rotation of the Earth by suspending a weight with a long wire, first at the Paris Observatory and then, most spectacularly, from the dome of the Pantheon. The plane of the pendulum's swing appeared to rotate clockwise, thus confirming the rotation of the Earth. After Foucault had conducted his experiment, a sort of "pendulum mania" spread throughout Europe and the United States, with private repetitions and public demonstrations.[83]

Secchi, who remained a physicist to his core, seized the opportunity to make a significant demonstration in Rome. The symbolic value of this experience in the Roman Catholic context was that of reconciling Copernicanism and Catholicism: as director of the Collegio Romano Observatory, Secchi probably felt it was part of his "mission" to dispel such prejudices between science and church. It was also in keeping with the attention that Secchi consistently paid to the communication and popularization of science. He wrote:

> The experiment of the deviation of the pendulum, made by Mr. Léon Foucault to prove the rotation motion of the Earth, is actually a subject pertaining rather to physicists than to astronomers; nevertheless, we judged it appropriate to deal with it because of its close connection with astronomy, and because it became so popular at the time of its discovery, that we could not resist the urgings of many people who encouraged us to deal with it. All the more willingly we went along with this wish as it offered us a rare chance to set it up in an impressive setting, perhaps inferior only to that of the famous Pantheon in Paris.[84]

---

82 See, for instance, Nathaniel Everett Green, "Mars and Schiaparelli's Canals," *Observatory* 3 (1879): 252, and François Terby, "The Markings on Mars," *Observatory* 3 (1880): 416; see also Nathaniel Everett Green, "On Some Changes in the Markings of Mars since the Opposition of 1877," *Monthly Notices of the Royal Astronomical Society* 40 (1880): 331–332, here 332.

83 See Michael F. Conlin, "Popular and Scientific Reception of the Foucault Pendulum in the United States," *Isis* 90 (1999): 181–204, here 181.

84 *L'esperimento della deviazione del pendolo trovato dal sig. Leone Foucault per provare il moto rotatorio della terra, è in vero soggetto più propriamente di dominio de' fisici che degli*

In May, Secchi publicly repeated Foucault's experiment "in the presence of the most distinguished professors of the city" by suspending a lead ball with an iron wire from the ceiling of St. Ignatius Church; the length of the pendulum was 31.95 meters.[85] It is hardly likely that such an experiment could have been carried out without the pope's permission; indeed, though Secchi did not openly mention him, it is possible that, given his interest in science, Pius IX himself may have suggested carrying out the experiment. Given that the deviation of the plane is influenced by many factors (e.g., the width of the pendulum swing, the quality of the suspension system, air resistance, etc.), the experiment was aimed at studying how physical constraints affect the geometrical law.[86] Secchi thoroughly studied every detail and took every precaution to limit any cause of error. He found that the swings were not in a perfect plane but along a very flattened cone whose horizontal section followed an ovoid spiral; in fact, as the oscillations faded, its major axis tended to rotate in the direction of the rotation of the Earth.[87]

Secchi summarized the experiment in a letter to the Royal Astronomical Society of London:

> I have repeated M. Léon Foucault's experiment with a pendulum thirty-one meters long, and a weight of twenty-eight kilograms; I have measured the angles with the greatest care, and have constantly found the deviation such as it should be according to theory. There is, however, a slight difference in the motion at the beginning and at the end of the series: at the beginning, the angles are a few minutes in excess, and at the end a few minutes in defect. The errors, to be sure, are not larger than the

astronomi, pure giudicammo opportune di occuparcene alquanto per la stretta connessione che ha colla astronomia, e perché ottenne tanta celebrità all'epoca della scoperta, che non potemmo resistere all'istanza di molti che ci stimolavano d'occuparcene. Tanto più volentieri poi secondammo questo desiderio quanto che ci offeriva una rara comodità di eseguirlo in grandiose proporzioni forse inferiori solamente a quelle tanto famose del Panteon a Parigi. Angelo Secchi, "IV. Esperimenti sulla deviazione del pendolo diretti a provare la rotazione della terra," *Memorie dell'Osservatorio del Collegio Romano: Anno 1851; Appendice alle osservazioni del 1851* (1852): xxxviii–xxxx, here xxxviii.

85 See Secchi, "IV. Esperimenti sulla deviazione del pendolo diretti a provare la rotazione della terra," xxxviii. It is possible that Secchi conceived the idea of moving the observatory to the roof of the church during the inspections for fixing the suspension point of the pendulum.

86 Some of these effects were studied theoretically by noted astronomers and mathematicians of that time, such as Jacques Binet (1786–1856), Ottaviano Fabrizio Mossotti (1791–1863), Giovanni Plana (1781–1864), and Domenico Chelini (1802–78).

87 This phenomenon is today known as apsidal deviation; see Giancarlo Scalera, *I moti e la forma della terra* (Rome: Istituto Nazionale di Geofisica, 1999), 135–138.

> probable errors of observation; but as they are constantly the same way, they seem to indicate a physical cause.[88]

Secchi concluded that there was a systematic error: "I believe that physical causes, and particularly the resistance of the wire to flexure, are the main cause of the enlargement of the ellipse."[89]

Secchi did not study this phenomenon again, but in 1930 Fr. Johan Stein (1871–1951) confirmed Secchi's conclusions:[90]

> It seems to us that Father Secchi was fully right. Indeed it may be added that all of the increase [of these angles] can be attributed to physical causes. If these forces did not exist at all, the amplitude of the oscillations and consequently also the minor axis of the ellipse would be constant.[91]

Secchi took advantage of the pendulum experiment to make measurements of the local acceleration of gravity and found a value of g = 9.8042 m/s^2. He considered this result to be an approximation and hoped that it would stimulate other scientists to carry out gravimetric measurements with better instruments: "In fact, in our countries, because of the geological constitution of the soil, it is to be suspected that there are appreciable irregularities in gravity."[92] Secchi's interest in gravimetric measurements

88 Angelo Secchi, "On Foucault's Experiment," *Monthly Notices of the Royal Astronomical Society* 11, no. 8 (1851): 199. In the same communication, Secchi added that he was surprised that "some English journals cast doubt on the correctness of this demonstration," a statement that requires further investigation.

89 *Io credo che le cause fisiche, e specialmente la resistenza del filo alla flessione, siano la causa principale dell'allargamento dell'ellisse.* Angelo Secchi, "Sulle oscillazioni del pendolo avuto riguardo alla rotazione della terra," *Annali di scienze matematiche e fisiche* 2 (1851): 238–242, here 242. In 1917, Hagen also carried out a pendulum experiment in Sydney, where the trajectory of the end of the pendulum was photographed in order to be studied precisely - actually, he was convinced that the elliptical motion of the pendulum was a real phenomenon, not just the effect of systematic errors. See Scalera, *I moti e la forma della terra*, 137-138.

90 In 1917, Hagen also carried out a pendulum experiment in Sydney, where the trajectory of the end of the pendulum was photographed in order to be studied precisely; see Scalera, *I moti e la forma della terra*, 138

91 *A noi sembra che il P. Secchi avesse avuto piena ragione. Anzi si può aggiungere che le cause fisiche perturbatrici sono la causa totale dell'allargamento. Se queste forze non esistessero affatto, l'ampiezza delle oscillazioni e conseguentemente anche l'asse minore dell'ellisse sarebbero costanti.* Stein, quoted in Scalera, *I moti e la forma della terra*, 138.

92 *Infatti, nei nostri paesi, per la geologica costituzione del terreno, vi ha luogo a sospettare d'irregolarità non minime nella forza di gravità.* Angelo Secchi, "Pendolo Foucault," *Annali di scienze matematiche e fisiche* 2 (1851): 325–345, here 345.

would return in 1872, when he was involved in the failed Mont Cenis campaign (see Chapter 5).

5 An Open-to-Visitors Observatory

The Collegio Romano Observatory was far from an ivory tower to Secchi; on the contrary, he would often welcome a whole range of people to the observatory, whether they were scientists, politicians, nobles, diplomats, or common visitors, men or women, boys or girls, from every part of the world. In 1903, the US astronomer William Elkin (1855–1933) still remembered "a day spent with him in Rome when I was a young man."[93]

Among the scientists who visited the observatory, it is worth mentioning the visit by Otto Wilhelm von Struve, director of the renowned Pulkovo Observatory in St. Petersburg. In 1863, he visited the Italian observatories and reported a favorable impression of the Collegio Romano Observatory:

> [The Collegio Romano Observatory] made a very favorable impression on me by the eminently scientific spirit which prevails; the efforts of the director, the Rev. Fr. Secchi, known to the scientific world by many [works] in various branches of astronomy and physics, in whose execution he is assisted by two other members of the Society of Jesus, Mr. Rosa and Marnetti [*sic*, Marchetti]. In the last years, they are occupied preferably with research of physical optics and meteorology, topics for which Mr. Secchi, as a former professor of physics at the college, admits having a predilection. For this purpose, the observatory is provided with a very complete collection of meteorological instruments and other physics devices. Among the astronomical instruments, it is worthwhile to mention a refractor of nine inches, whose excellent optical qualities are proved by many very precise measurements of double stars.
>
> In addition, there is an old Cauchoix refractor of six inches, now used to draw the sunspots, and a meridian circle of two feet in diameter, by Reichenbach and Ertel. The observatory is erected in [*sic*] the top of the Roman College. Despite the extraordinary height of this building, the placement of the instruments is quite strong, as the observation rooms are located within huge thick towers that once were intended to support the huge dome of a church, planned to be built in the seventeenth century. There are four of these towers, only two are now occupied by the observatory. From all sides, they are surrounded, up to the top, by other

93 AA. VV., *Al p. Angelo Secchi nel XXV dalla morte* (Rome: Desclée, Lefebvre et C.ie, 1903), 83.

> buildings of the college, which shelter them against the influence of temperature changes and solar rays.[94]

Twelve years later, in 1875, another report was made by astronomer George Rayet (1839–1905), who visited the Collegio Romano Observatory during a mission supported by the French government to examine the organization of astronomical research in Italy.[95] By then, the prestige of the observatory was well established because of Secchi's renowned spectral studies. Rayet remarked that the observatory, located in the city center, "although being at a great height and in the vicinity of one of the busiest streets in Rome (the Corso), the stability of the instruments is [...] perfectly sufficient for the studies of celestial physics that are so successfully carried out by R[everend] Fr. Secchi."[96]

---

94 *M'a fait une impression très favorable par l'esprit éminemment scientifique qui y domine, grâce aux efforts du Directeur, le Rev. P. Secchi, connu au monde scientifique par des nombreux* [*travaux?*] *en différentes branches d'astronomie et de physique, dans l'exécution desquels il est assisté par deux autres membres de la Société Jésus Mm. Rosa et Marnetti* [sic; Marchetti]. *Dans les dernières années ils se sont occupés par préférence de recherches de physique d'optique et de météorologie, objets pour lesquels M. Secchi, comme ancien professeur de physique au Collège, avoue d'avoir une prédilection. A cet effet l'Observatoire est pourvu d'une collection très complète d'instruments de météorologie et d'autres appareils de physique. Au nombre des instruments astronomiques il fait citer en premier lieu un réfracteur de 9 pouces d'ouverture, dont les excellentes qualités optiques sont prouvées par de nombreuses mesures d'étoiles doubles très resserrées. En outre il y a encore un ancien réfracteur de Cauchoix de 6 pouces d'ouverture, dont on se sert maintenant pour dessiner les tâches solaires et un cercle méridien de Reichenbach et Ertel de deux pieds de diamètre. L'Observatoire est érigé au* [sic] *haut du Collège Romain. Malgré l'extraordinaire hauteur de cet édifice, l'emplacement des instruments est assez solide, comme les salles d'observation se trouvent en dedans des tours d'une épaisseur énorme qui jadis ont été destinés à porter l'immense coupole d'une église, qu'on avait eu l'intention de construire au 17ème siècle. Il y a quatre de ces tours, dont deux seulement sont maintenant occupées par l'observatoire. De tous cotés elles sont entourées jusqu'en haut par d'autres bâtiments du Collège, ce qui les abrite contre l'influence des changements de température et des rayons du Soleil.* Simone Bianchi and Daniele Galli, "Les observatoires astronomiques en Italie: An 1863 Report by Otto Wilhelm Struve," *Nuncius* 30, no. 1 (2015): 195–227, here 222.

95 See Ileana Chinnici, "Le voyage scientifique de Georges Rayet en Italie et la fondation de l'Observatoire de Bordeaux," in *Sur les traces de Cassini: Astronomes et observatoires du Sud de la France; 121ème Congrès national des sociétés historiques et scientifiques, Nice, 1996*, ed. Paul Brouzeng and Suzanne Débarbat (Paris: Editions du C.T.H.S., 2001), 259–269. Rayet had previously carried out spectroscopic research at the Paris Observatory together with Charles Wolf (1827–1918); he had discovered a class of stars in 1867 (then denominated Wolf–Rayet), with strong emission bands in their spectrum.

96 *Quoique situé à une grande hauteur et en voisinage d'une des rues les plus fréquentées de Rome* (*le Corso*)*, la stabilité des instruments y est* [...] *parfaitement suffisante pour les études de physique céleste que poursuit avec tant de succès le R.P. Secchi.* Georges Rayet, "Rapport sur une mission astronomique en Italie," *Archives des missions scientifiques et littéraires: Choix de rapports et instructions*, ser. 3, t. 3–2 [1876]: 529–544, here 534.

Rayet's description of the instruments is more detailed than von Struve's and includes some information about the improvements that had been made to the equipment, like the clock-drive mechanism of the Merz telescope, which had been replaced in 1869:

> The main instrument [...] is one of the masterpieces by Merz. It follows the diurnal [sky] motion using a high conical pendulum system and is enclosed in a cylindrical dome, rolling on spheres. The observatory [...] also has a Cauchoix equatorial [...], which is used daily for the study of the position and the configuration of sunspots, and an Ertel meridian telescope [...], which is used for time [service].[97]

Rayet could not but mention Secchi's spectroscopic equipment for solar observations, which he also had the opportunity to use:

> The R[everend] Fr. Secchi has, himself, used [...] since August 1875, a reflection grating, ruled on mirror metal by Mr. Rutherfurd[98] [...] During my stay in Rome, I spent several mornings to observe the prominences, using this grating, and I must say that these flames appeared with a sharpness much superior to anything I have seen before.[99]

Spectroscopic research had been introduced at the Collegio Romano Observatory in 1862–63, thanks to the visit of another French astronomer, Jules C. Janssen, who spent a few months at the Collegio Romano to carry out studies on the solar spectrum. Thanks to Janssen's spectroscopes, Secchi entered the

---

97 *L'instrument principal* [...] *est un des chef-d'oeuvres de Merz. L'appareil, mobile suivant le mouvement diurne à l'aide d'un système d'horlogerie à pendule conique, est renfermé dans une coupole cylindrique qui roule sur des boulets. L'Observatoire* [...] *possède encore un équatorial de Cauchoix* [...] *qui sert journellement à l'étude de la position et de la configuration des tâches solaire, et une lunette méridienne d'Ertel* [...] *destinée à obtenir l'heure.* Rayet, "Rapport sur une mission astronomique en Italie," 534.

98 Lewis Morris Rutherfurd (1816–1892) was an American amateur astronomer who invented instruments for his studies, including a machine for producing improved ruled diffraction gratings for spectroscopic observations. Secchi possessed two gratings made by Rutherfurd (one in glass and the other in metal), which Rutherfurd donated to him during a visit to the Collegio Romano Observatory in 1874; see Secchi to Tacchini, Rome, February 18, 1874, quoted in Ileana Chinnici and Antonella Gasperini, *Alle origini dell'astrofisica italiana: Il carteggio Secchi–Tacchini* (Florence: Fondazione Giorgio Ronchi, 2013), 332–334.

99 *Le R.P. Secchi a, lui, employé* [...] *depuis août 1875, un réseau à réflexion, trace sur métal des miroirs par M. Rutherfurd* [...] *Pendant mon séjour à Rome, j'ai passé plusieurs matinées à observer les protubérances à l'aide de ce réseau, et je dois dire que ces flammes apparaissent avec une netteté de beaucoup supérieure à tout ce que j'avais vue auparavant.* Rayet, "Rapport sur une mission astronomique en Italie," 535.

field of astronomical spectroscopy, though he soon clashed with his French guest (see Chapter 6).

Among other visits, mention should also be made of Maria Mitchell (1818–89), the first American woman to work as a professional astronomer, director of Vassar College Observatory in the state of New York. She visited the Collegio Romano Observatory in 1858, more than ten years after the discovery of the comet 1847 VI, the discovery of which had initially been attributed to De Vico and then assigned to her.[100]

Among the political figures who visited the observatory, one of the most important was Dom Pedro II, emperor of Brazil (1825–91, r.1831–89), who was a reputed amateur astronomer. He visited the Collegio Romano Observatory in November 1871 and spent many hours engaged in scientific conversations with Secchi.[101] Pedro II awarded Secchi the decoration of Great Dignitary of the Order of the Rose, the highest degree of that decoration, for his scientific achievements.[102] He paid a second visit to Secchi's observatory in February 1877.[103]

Secchi also established a close relationship between the observatory and the local population. Romans were regularly kept informed about the observations carried out at the Collegio Romano Observatory through the *Bullettino meteorologico* and, on special occasions, through articles in the main Roman newspapers. Such communications about astronomical news created high expectations among the public, who would often frequent the observatory. When the spectacular comet Donati (1858 VI) appeared, Secchi was in Paris[104] and wrote to the astronomers of the Collegio Romano: "I am sure that you have crowd of curious people in this circumstance,"[105] and in 1878, almost immediately after Secchi's death, Fr. Ferrari wrote a note about the transit of Mercury, which occurred in June, because "[the] educated public [...] here in Rome is accustomed to wanting to know what is observed and unobserved at the observatory in such cases [...]."[106]

100 See Kronk, *Cometography*, 2:182.

101 See Simone Bianchi, "Un imperatore ad Arcetri," *Giornale di astronomia* 38, no. 2 (2012): 2–12, here 4.

102 It is unclear where and when the decoration was awarded.

103 A manuscript by Secchi on the construction of sundials is kept at the Vatican Observatory Library and was donated by St. Louis College in São Paulo. It is unknown how this document reached Brazil, but it may have been donated to Dom Pedro II as a memento, later given to the college, and, quite recently, donated to the Vatican Observatory. The elegant binding, with the Italian title in golden letters, could reinforce this hypothesis.

104 He had observed the comet just a few days after its discovery, when it was still faint.

105 *Sono persuaso che avrete concorso di curiosi in questa circostanza* [...]. Secchi to Marchetti, Paris, September 24, 1858, quoted in Castellani, "Il P. Angelo Secchi S.I. e le sue missioni scientifiche," 620.

106 *[Il] colto pubblico* [...] *qui a Roma è avvezzo a voler sapere quello che si vede e non si vede alla Specola in simili casi* [...]. Ferrari to Tacchini, Rome, June 3, 1878; ACREA , Fondo Tacchini.

CHAPTER 4

Advisor to the Papal States: Science in the Service of the Public

Secchi was involved with an impressive number of public works during the years he spent at the Collegio Romano Observatory. His involvement in these projects was informed by a strong sense of the common good. In Secchi's view, science should not be hidden away but applied for the benefit of the broader public. He was a committed partisan of "non-profit" science, science in the service of civil society, as is apparent in the opinion he gave in 1859 on the possible introduction of a fee for the calibration of magnetic instruments (e.g., the magnetic compasses used in navigation):

> As far as I know, there is no tax of this kind at other observatories, […] the Collegio Romano Observatory […] will always be ready to adjust [the instruments] for free, and in the event that the government wants to impose some fee, then […] it will not be recognized by the observatory.[1]

A detailed survey of Secchi's contributions to the improvement of public services in the Papal States would require an examination of their previous status at that time, which is beyond the scope of this book. Accordingly, the following simply outlines some of Secchi's most important contributions, namely those that had a significant impact on the advancement of the state, civil society, and/or Secchi's life. As we will see, Secchi's involvement in these tasks complemented his institutional work, with the skills and expertise he gained from his scientific research being applied for the benefit of the Italian people.

1 Improvement of the Time Signal Service

In 1846, Pope Pius IX approved a proposal suggested by the astronomers of the Collegio Romano to use mean time for the regulation of public clocks. The new

1 *Tale tassa per quanto a me consta non esiste negli altri osservatorii* […] *Per ciò poi che spettar potesse l'osservatorio del Collegio Romano, esso sarà sempre pronto a rettificarli gratuitamente, e in caso che pure piacesse al Governo imporre qualche tassa, allora essi sarebbero da noi rettificati* […] *sempre però intendendo che nulla venga percepito dall'osservatorio* […]. Secchi to Minister Camillo Amici, Rome, March 18, 1859; APUG, FS 9.I. Camillo Amici (1802–77) was minister of trading, arts, public works, industry and agriculture between 1859 and 1860.

 DOI:10.1163/9789004387331_006

time signal service facilitated the transition from the Italian to the European time-measuring system in the Papal States. With this system, the time between two successive noons was divided into twenty-four hours of equal duration, and noon was then the reference from which the hours were counted. The change was significant, as previously the measurement of time was marked by the Divine Office, with prayer times at fixed hours, while this new system put the Papal States in line with other European nations.

The time service was entrusted to the astronomers of the Collegio Romano and organized by De Vico. A large black flag with a white star in the middle was hoisted atop the Calandrelli tower a few minutes before noon and lowered at the stroke of twelve, a time that was determined astronomically by observing the transit of the Sun at the meridian. At that signal, a gunner fired a cannon shot from Castel Sant'Angelo, which was audible throughout the city and its surroundings. In 1850, the flag support was destroyed by a strong wind,[2] and Secchi decided to adopt another system using a time ball, on the model of the one installed at Greenwich Observatory. Time balls, which were dropped at a predetermined time, were used mainly in Great Britain and the United States.[3] On the roof of St. Ignatius Church, Secchi placed a pole, almost eleven-meters long, which pierced a black ball made of canvas covering an iron framework

FIGURE 4.1 Photograph of the observatory on the roof of St. Ignatius Church at the beginning of the twentieth century; on the right, the time-ball. Courtesy of INAF-OAR.

2 See Angelo Secchi, *"Ristauri fatti all'Osservatorio," Memorie dell'Osservatorio del Collegio Romano: Anno 1850* (1851): 5–8, here 8.

3 This system was adopted at Greenwich Observatory in 1833 and at the US Naval Observatory in 1845.

about 1.5 meters in diameter, which normally resided at the top of the pole. Unlike a flag, the ball's visibility was unaffected by the strength or direction of the wind and stood out well against the background of the sky. The noon signal was given by a small telegraph connecting the clock with a room where an assistant was charged with dropping the ball with a cleverly devised apparatus that set the ball falling six seconds before exact noon.[4]

The many sundials that Secchi built (mostly in public places—see below) may also have been made with the intention of helping the transition to European time.[5]

2 Measuring of the Geodetic Baseline at Via Appia

The measurement of the geodetic baseline is a fundamental procedure for drawing a map of an area, as the network of triangulation[6] to measure the distances between various geographical points is derived from it. The need for increasingly precise maps was a high priority for political, commercial, and military reasons, and in the second half of the eighteenth century many European states undertook campaigns of triangulations and geodetic measurements on their territories with precision instruments, such as repeating circles and theodolites. In Italy, similar operations were carried out in the territories of Lombardy, Piedmont, and Veneto, which led to the creation of highly detailed maps. In the Papal States,[7] a first attempt was made in 1751 by Boscovich and Christopher Maire (1697–1767), who measured a geodetic baseline on the Via Appia, the ancient Roman road, which is mostly straight and flat. One end of the baseline was near the tomb of Cecilia Metella in Rome, while the other was located at Frattocchie, in the area of Colli Albani; the total baseline length was about twelve kilometers.[8] However, the accuracy of the Via Appia base measurement

4 See Secchi, *"Ristauri fatti all'Osservatorio,"* 8.

5 See Maria Luisa Tuscano, "Angelo Secchi e la gnomonica," in Altamore and Maffeo, *Angelo Secchi*, 255–273.

6 Triangulation is a mathematical construction that ideally links arbitrary points on the terrain to form a network of contiguous triangles, of which all internal angles are measured; the length is calculated from the geodetic baseline, i.e., one side, the length of which is directly measured on the terrain.

7 For a historical *excursus* about the cartography of the Papal States, see Attilio Mori, *Cenni storici sui lavori geodetici e topografici e sulle principali produzioni cartografiche eseguite in Italia dalla metà del secolo XVIII ai nostri giorni* (Florence: Istituto Geografico Militare, 1903), 3–7.

8 It should be mentioned that, in 1815, a shorter baseline—of just over a kilometer—was measured outside Porta Angelica by Calandrelli and Andrea Conti (1777–1840) for the urban cartography of Rome and its surroundings.

was later challenged when a French commission tested the results and found an unacceptably large error of about ten meters,[9] probably because the baseline end at Frattocchie had been lost and some identifications of that end's position had proved to be questionable. Thus a new baseline measurement was required if Boscovich's high scientific reputation was to be vindicated.

A favorable circumstance for identifying a new baseline measurement was given by the archaeological excavations being carried out by Luigi Canina (1795–1856), who was unearthing ancient monuments and landmarks along the Via Appia. Canina got Secchi involved in order to determine the exact extent of the Roman mile; Secchi, who also had a passion for archeology (see below), leapt at the chance and in the process found the right conditions for the requested baseline measurement. In geodetic operations, Secchi proved to have a deep knowledge of the state-of-the-art of geodesy and was able to improve its measurement techniques. The documents bear witness to his organizational skills: choosing instruments, scheduling days and work on the terrain, staff management, publication of the results.[10]

The start of the work, scheduled for the spring of 1854, was postponed several times due to the late arrival and adjustment of the instruments, as well as to muggy summer heat and heavy autumn rains. Operations finally began in November 1854 and continued until April 1855.[11]

The instrument Secchi chose—a standard measuring rod "graduated to tenths of a millimeter, read with powerful magnifying microscopes" (*tesa graduata a decimi di millimetro, letta con microscopi di gran forza*)[12]—was devised by Ignazio Porro (1801–75), who had carried out a preliminary topographic survey for the future Fréjus Tunnel in 1836. Following Porro's method, Secchi used a similar single bimetallic rod made of brass and iron rods housed together in a grooved wooden beam. Thermometers were positioned at the ends of the rods to keep track of their thermal expansion, while a spirit level was placed in the middle of the rod. The ends of the rod were supported by two stands, and micrometric adjustments were made with endless screws so that the end

9 See Angelo Secchi, "Nouvelle mesure de la base de Boscovich," *Comptes rendus de l'Académie des Sciences* 50 (1860): 377–378, here 377.

10 [Secchi] *dimostrò una profonda conoscenza dello stato dell'arte della geodesia e fu in grado di migliorarne le tecniche di misura. I documenti testimoniano anche le sue capacità organizzative: scelta degli strumenti, pianificazione delle giornate, lavoro sul campo, gestione dei collaboratori, pubblicazione dei risultati.* Tullio Aebischer, "Le misure geodetiche," in Altamore and Maffeo, *Angelo Secchi*, 227–240, here 227.

11 See Angelo Secchi, *Misure della base trigonometrica eseguita sulla Via Appia* (Rome: Tipografia della Rev. Camera Apostolica, 1858).

12 Angelo Secchi, "Rapporto della Commissione per la Misura del Meridiano Centrale Europeo," *Atti della R. Accademia dei Lincei* 24 (1871): 232–258, here 235.

FIGURE 4.2 Operations for measuring the geodetic base at Via Appia, near the Tomb of Cecilia Metella. From Angelo Secchi, *Misure della base trigonometrica eseguita sulla Via Appia* (Rome: Tipografia della Rev. Camera Apostolica, 1858), frontispiece.

point of one measurement matched the starting point of the following one. The exact positioning and alignment of the rod was adjusted via sights and small telescopes. For the determination of the control point on the ground, originally done via a plumb-line protected from the wind by a glass tube, Secchi introduced an independent vertical microscope (*meroscopio*), which made it possible to align the control point with the end of the rod. At the end of each day, the last point of that day's measurement was fixed in masonry and buried, to be uncovered the next morning to serve as starting point for the subsequent measurement. Secchi would later write that

> the difficulties posed by this measurement were very considerable, but with the system that we adopted, we could easily overcome them. Since the end points of the baseline, as well as the daily points, are made in masonry and remain on-site, at any time we can use the whole baseline or a portion of it for geodetic purposes.[13]

13 *Les difficultés que présentait cette mesure étaient très considérables, mais à l'aide du système adopté on a pu les surmonter sans peine. Comme non seulement les termes extrêmes, mais*

The work proceeded quickly, notwithstanding the slow initial run-up of the operations and the fact that the standard rod was carried back to Rome by foot every day to avoid possible damage due to the unlevelled road. The baseline was supposed to have been measured twice, but Secchi decided to repeat only the uncertain measurement of the first two kilometers. He estimated the total base length at 12.043,14 meters, with an error of not more than two millimeters.[14] A test on the length of the measuring rod that had been used in 1751[15] revealed that Boscovich's and Secchi's measurements differed by only 0.34 meters: Secchi's identification of the second baseline endpoint was therefore correct.[16]

The Via Appia baseline is one of the longest ever measured. It took six months of work on the ground and three years of calculations. Secchi intended to follow up this work and correct the map of Rome and its surroundings. In 1860, in fact, he announced to the Académie des Sciences that, at an appropriate time, "we will proceed to link the base to Marieni's[17] triangulation, in order to correct the map of the Roman countryside."[18] Yet the project was ultimately abandoned because of the changing political situation.

At that time, the Collegio Romano Observatory had been equipped with a chronograph donated by Pius IX, and Secchi, in cooperation with Lorenzo Respighi (1824–89), was able to measure with precision the difference in latitude between the two Roman observatories (Collegio Romano and Campidoglio). In 1869, in order to coordinate the triangulation networks of the Papal States and southern Italy, Secchi and Emanuele Fergola (1830–1915), astronomer at Capodimonte Observatory, determined the difference in longitude between Rome and Naples by means of telegraphic transmissions of mean time

même ceux de chaque journée sont en maçonnerie et restent sur place, on pourra toujours se servir de toute la base ou d'une portion quelconque pour les usages de la géodésie. Secchi, "Nouvelle mesure de la base de Boscovich," 377.

14 See Secchi, "Nouvelle mesure de la base de Boscovich," 377.

15 The test was carried out by the astronomers of Campidoglio Observatory in the first half of the nineteenth century.

16 The end of the baseline has recently been discovered in the subsoil of the Via Appia, thanks to georadar investigations. See Tullio Aebischer, ed., *La scoperta del Caposaldo B della Base Geodetica di A. Secchi lungo la via Appia Antica* (Rome: Edizioni Nuova Prhomos, 2013).

17 In 1841–43, cartographer Giovanni Marieni was charged with extending the triangulation operations carried out in northern Italy to the Papal States. In 1844, he also connected that triangulation to the geodetic network of the Kingdom of the Two Sicilies.

18 *On procédera à relier la base à la triangulation de Marieni, pour avoir une rectification de la carte de la Campagne romaine.* Secchi, "Nouvelle mesure de la base de Boscovich," 377–378.

measurements in the two observatories.[19] That same year, on the occasion of the meeting of the Italian Commission for the Measurement of the European Central Meridian, Secchi presented a summary of the geodetic works he had carried out over the previous years (see Chapter 5).

3 Meteorological Networks, Weather Forecasting, Earthquakes and Fire Safety Measures

As has been noted above, Pius IX promoted the application of new technologies in the papal territories. In October 1853, a telegraph service was launched, initially for diplomatic purposes, to connect Rome with its neighbor allied states.[20] Two Wheatstone and Morse transmitters were set up to allow the exchange of messages on the service line. The Wheatstone–Morse telegraphic technique was settled on for service between Rome and Terracina (a port on the southern coast of the Papal States, between the promontories of Circeo and Gaeta) and was opened to public use in 1855. Fr. Alessandro Serpieri (1823–85), a member of the Scolopian order, had anticipated the application of the telegraph in the service of meteorology:

> Telegraph transmissions soon [...] will allow us to keep track of the trend of our barometers, how much rain and what winds we are watching [...]. It would be worthwhile to prepare ourselves for this, so that we do not lose this chance to make useful predictions that could be derived from being aware of these phenomena [...]. It would be worthwhile to be ready for the immense new benefits that will profit our agriculture, our business, and public health, promised by such a prodigious rapidity of communication.[21]

19 See Emanuele Fergola and Angelo Secchi, *Sulla differenza di longitudine fra Napoli e Roma determinata per mezzo della trasmissione telegrafica delle osservazioni* (Napoli: Accademia delle Scienze, 1871).

20 Italy preceded the Papal States in installing a telegraph line; in fact, from 1848 on, a telegraph service had been operating in Tuscany, thanks to physicist Carlo Matteucci (1811–68), who had strung an aerial wire along the railway between Florence, Pisa, and Livorno.

21 *Tra poco le trasmissioni telegrafiche di Londra, di Parigi, di Berlino, di Pietroburgo, di Vienna ci sorprenderanno nella tranquilla attività dei nostri studi e delle nostre accademie: ci dimanderanno per quale curva procedono i nostri barometri, di quanta pioggia e di quali venti siamo noi spettatori, a quali condizioni regolari o irregolari di temperatura siamo noi sottoposti. Conviene che ci prepariamo a non porre ostacolo alle utili previsioni che facilmente potranno farsi sulla cognizione dei fenomeni lontani e contemporanei. Conviene che ci prepariamo a profittare per la nostra agricoltura, pel nostro commercio e per la pubblica igiene, degli immensi e nuovi vantaggi che ci vengono promessi dalla prodigiosa rapidità*

Secchi was familiar with telegraphic apparatus as he had made experiments with telegraphs in his early teaching years (see Chapter 1). He considered meteorology one of the most useful sciences for civil society: "Science is useless if it is not beneficial, and meteorology is fortunately one of those sciences from which mankind may receive great and useful service."[22] Meteorological studies were also an important part of the work at the Collegio Romano. In June 1856, Secchi was appointed director of the Meteorological Service of the Papal States and organized a daily telegraphic correspondence of meteorological observations between the stations of the states. These exchanges were mainly aimed at creating a forecast service. The rapid reporting of weather conditions from place to place was of the utmost importance for the prevention of damage due to the arrival of cyclones. Every day, the four initial stations of Rome, Ancona, Bologna, and Ferrara exchanged synchronous observations of pressure, temperature, humidity, wind, and the general condition of the atmosphere with each other in order to obtain useful data for forecasting storms. The data were reduced at the Collegio Romano Observatory, and weather warnings were eventually disseminated via newspapers.[23] Secchi had created the nucleus of the first meteorological network in Italy and took pride in this achievement:

> It is a great question to say <u>who</u> first suggested the telegraphic meteorological service [...] If we are talking about an occasional sporadic service [...], I believe that it had been done in America and England before Europe. [...] In Italy and in Europe, it was I who was first to set it up between Rome, Bologna, Ferrara, and Ancona, but others had already proposed this.[24]

In Europe, the major maritime nations had a strong interest in this field. France soon gained the leadership in the field of meteorological studies and their applications. The energetic director of the Paris Observatory, Urbain

---

delle comunicazioni. Alessandro Serpieri, *Sulla riduzione delle osservazioni meteorologiche e specialmente delle osservazioni triorarie* (Pesaro: Tipografia Nobili, 1854), 8.

22 *La scienza è vana se non è utile, e la meteorologia è fortunatamente di quelle scienze da cui l'umanità può ricevere grandi ed utili servigi*. Angelo Secchi, "L'astronomia in Roma nel pontificato di Pio IX" in *Triplice omaggio alla Santità di Papa Pio IX per il suo giubileo episcopule* (Rome: Tipografia della Pace, 1877), 29–77, here 74.

23 The data were also published in a summary by the *Giornale di Roma*.

24 *La è gran questione quella di dire <u>chi</u> abbia pel primo indicato il servizio meteorologico telegrafico*. [...] *Se parliamo di un servizio momentaneo eccezionale* [...] *credo che in America e in Inghilterra si sia fatto prima che in Europa*. [...] *In Italia e in Europa sono stato io, il primo, ad effettuarlo tra Roma, Bologna, Ferrara ed Ancona, ma altri l'avea già progettato*. Secchi to Denza, Rome, 9 March 1866, quoted in Sabino Maffeo, "Padre Angelo Secchi e la meteorologia," in AA. VV., *Presenze scientifiche illustri al Collegio Romano* (Rome: Ufficio Centrale di Ecologia Agraria, 2001), 21–31, here 26–27.

Leverrier (1811–77), had proposed an extensive meteorological network designed to warn sailors about the arrival of storms. Two catastrophic wrecks of several dozen commercial vessels and warships during the Crimean War (1854–55) accelerated the implementation of this project, which was approved by Napoleon III (1808–73) in 1855 and led to the creation of the *Bulletin international de l'Observatoire de Paris*. Meanwhile, in 1854, Robert FitzRoy (1805–65)[25] headed Britain's first weather office and promoted the creation of a weather bureau, to the great benefit of citizens, farmers, and seamen.

In 1857, the Collegio Romano participated in Leverrier's international network, made up of around sixty meteorological stations: "Each morning, we sent to Paris the telegram of the observations made at 7 a.m. with other requested information, [...] as well as the dissemination of meteorological data in newspapers [...]. The observations received by Leverrier were then lithographed and sent, thus collected, to the observatories."[26]

In 1862, when Secchi started publishing the *Bullettino meteorologico* in order to disseminate meteorological data, he also began to study isobaric and isothermal curves, which he plotted on the map of Europe.[27] From this analysis, "it was possible to detect the well-defined direction that well-delimited cyclonic storms generally have, from N[orth]–W[est] toward S[outh]–E[ast] [...]. These results possibly contributed to make these maps then systematically constructed and published in Paris by Leverrier [...]."[28] In fact, the first weather map of Europe appeared in the *Bulletin international* of 1863.

Secchi later acknowledged that his interest in meteorology was born during his stay in the United States thanks to his contacts with Commodore Maury, whose theories he had helped bring to Italy, and that the initiative to set up a meteorological network stemmed from the application of the principles Maury had suggested:

25 FitzRoy was the famous captain of the *Beagle* during Darwin's second voyage in 1831.

26 *Ogni mattina si spediva a Parigi il telegramma delle osservazioni fatte alle 7 antimeridiane colle altre informazioni richieste, e si continua anche oggidì a spedirlo non solo a Parigi, ma a Firenze e Pietroburgo, oltre alla diffusione dei listini meteorologici nei giornali* [...]. *Le osservazioni raccolte dal Leverrier venivano poscia litografate e rinviate così raccolte agli osservatorii.* Secchi, "L'astronomia in Roma nel pontificato di Pio IX," 71.

27 He was assisted by a young collaborator, Giuseppe Serra-Carpi, who worked with Secchi in making weather charts for the meteorological network and published some works in this field (see Giuseppe Serra-Carpi, *Sulle linee isotermiche dell'Italia de' suoi mari ed isole adiacenti* [Rome: Tipografia delle Belle Arti, 1865]).

28 *Si riuscì a riconoscere la direzione ben definita che hanno in generale le burrasche ben circoscritte di natura ciclonica, da N–W verso S–E* [...] *Questi risultati forse contribuirono a fare che simili carte venissero poscia fatte costruire sistematicamente e pubblicate a Parigi dal Leverrier.* Secchi, "L'astronomia in Roma nel pontificato di Pio IX," 71.

> Meteorology in recent years has entered a new phase; it deals not only with climatology but general atmospheric physics, air currents, and the spread of windstorms. This writer was in America at the time of the great discoveries by Maury. He saw his methods, and from his mouth itself he received his ideas, and it was his care to inform his compatriots upon returning […].[29] Approximately at the same time, Maury went to Europe and in Brussels held a conference where a draft about general studies on the movements of the atmosphere was discussed. The states were not yet officially cooperating, but since then the telegraphic connection for the study of windstorms has been developed. The Collegio Romano Observatory took an active part in this study, and in 1856 the papal government established a daily telegraph communication between the main cities of the state, Rome, Ancona, Bologna, Ferrara that has continued for many years. Meanwhile, Leverrier organized an international telegraphic meteorological correspondence between various states, and we were not slow to accede to it, and from 1857 we started to work on.[30]

The service was a success, and the director of the Collegio Romano could say with satisfaction: "It is true that the scientist can neither impede the raising of storms nor vary the rains' trend, but he can prevent much damage […]; our warnings of distant storms have repeatedly prevented disasters at Civitavecchia and along our coast."[31]

---

29 See Angelo Secchi, "Guida dei naviganti di lungo corso di Vincenzo Gallo," *Annali di scienze matematiche e fisiche* 4 (1853): 245–260. Later, Secchi mentioned his long friendship with Maury in the pages of the *Bullettino*; see Angelo Secchi, "Un omaggio alla memoria del Commodoro Maury," *Bullettino Meteorologico dell'Osservatorio del Collegio Romano* 12 (1873): 41–42.

30 *La meteorologia in questi ultimi anni è entrata in una fase novella; essa non si occupa solo della climatologia, ma della fisica generale dell'atmosfera e delle correnti aeree e del giro delle burrasche. Lo scrivente si trovava in America all'epoca delle grandi scoperte di Maury. Egli vide i suoi metodi, e dalla sua bocca stessa raccolse le sue idee, e fu sua cura informarne al ritorno i suoi compatrioti* […]. *Circa il medesimo tempo il Maury si recava in Europa, e a Bruxelles faceva una conferenza ove discutevasi un progetto di studi generali sui movimenti dell'atmosfera. Gli Stati non concorrevano ancora ufficialmente, ma progettavasi fin d'allora la coalizione telegrafica per lo studio delle burrasche. All'osservatorio del Collegio Romano si prese parte attiva a questo studio, e nel 1856 si otteneva dal Governo Pontificio l'istituzione di una comunicazione telegrafica quotidiana tra le principali città della Stato, Roma, Ancona, Bologna, Ferrara, che continuò parecchi anni. Intanto il Leverrier veniva organizzando tra varii Stati la corrispondenza meteorologica telegrafica internazionale, e noi non fummo lenti ad aderirvi, e si cominciò fin dal 1857 a lavorarvi intorno.* Secchi, "L'astronomia in Roma nel pontificato di Pio IX," 70–71.

31 *È vero che lo scienziato non può impedire la formazione delle burrasche, né variare il regime delle piogge, può però cogli avvisi prevenire molti danni* […] *i nostri avvisi delle lontane burrasche hanno più volte impedito de' disastri a Civitavecchia e al litorale nostro.* Secchi, "L'astronomia in Roma nel pontificato di Pio IX," 74.

Secchi supported structuring the weather service in the same way as it was in France, with meteorological stations differentiated by research field. For example, agricultural meteorology, which studies the influence of meteorological factors on plants in order to improve agricultural production, needed specific facilities compared with the usual meteorological stations; when he was asked to give his advice about the utility of a meteorological observatory for applications to agriculture, Secchi replied with comments that not only reflected technical considerations but also showed that he was attentive to workers' rights, as he recommended a contract adjustment:

> The task of a meteorological observatory does not include agricultural meteorology observations in itself; if this service is to be established, it would require a local station properly surrounded by plants and fields, as it was done in Paris in the Parc Montsouris.[32] Without such a station, to obtain appropriate observations would be unthinkable and without them even the most extraordinarily diligent meteorologist would not be able to get good enough results. I don't think that one can justly ask the observer to do such work, unless you give him an appropriate space, and include this in the employment contract. This does not deny the great merit that such agricultural meteorological observations would have, and does not detract anything [...] about their usefulness and indeed their necessity.[33]

Being convinced that "men of religion too have to contribute to what can be useful to society,"[34] Secchi promoted setting up meteorological observatories in several monasteries in and around the Lazio area (Grottaferrata, Monte Cavo, Frascati), involving their religious communities in the often burdensome data collection service. He also encouraged and supported scientists who intended

32 The famous Parisian park, which was created in 1875, contained a meteorological station.

33 *L'incarico di un osservatorio meteorologico non include di sua natura le osservazioni di meteorologia agraria; se queste istituire si vogliano, si esige un apposito locale fornito di vegetazione e campi opportuni come si è fatto a Parigi nel parco di Montsouris. Mancando di tale accessorio le osservazioni non improponibili e se fosse per istraordinaria diligenza il meteorologista volesse occuparsene, esse non risulterebbero nemmeno abbastanza esatte. Non penso quindi che giustamente possano esigersi dall'osservatore meteorologico tali funzioni, a meno che non gli si dia un conveniente commodo locale, e non gli venga tale ulteriore attribuzione nel contratto d'impiego. Ciò nulla deroga al grande merito che avrebbero tali osservazioni agrarie, e non toglie nulla* [...] *sulla loro utilità e anche spesso necessità di farle.* Secchi to Giacomo di Montereale (mayor of Pordenone), quoted in Stefano Zanut, "L'osservatorio meteorologico di Pordenone e l'ambiente scientifico di fine Ottocento," in AA. VV., "Pordenone, città tra passato, presente e future," *La loggia* 10 (2013): 19–25, here 23.

34 *Anche i religiosi devono occuparsi di ciò che può esser utile alla società*[...]*; Secchi to Denza (copy),* Archivio Specola Vaticana, Fondo Secchi, Ex Moncalieri, 170.

to establish meteorological observatories in that area, as in the case of physicist Ignazio Galli (1841–1920), who erected a meteorological observatory at Velletri (near the Colli Albani).

In addition, mostly in the decade 1850–60, Secchi encouraged François Balley (1827–?), physician to the French army in Rome, to study "sanitary meteorology" (*météorologie hygiènique*): the connection between meteorology and epidemiology, in particular the correlation of trends in temperature and other meteorological variables with that of diseases. Secchi reported the results of Balley's work in the early volumes of the *Bullettino* in order to stimulate discussion on the topic. This work eventually led to a book and an atlas.[35] The Jesuit also remarked that the presence of swamplands near Rome affected the wholesomeness of the climate and influenced the health of the inhabitants of the nearby areas, and he suggested some kind of remediation intervention (drainage, expansion of crops, increase in resident population). These would be carried out many decades later, in the Fascist era.

Weather—especially rain—affected not only agricultural work and the quality of crops in a given climate but also the supply of clean water. Secchi was often involved in advising on and supervising matters concerning the distribution system of drinkable water:

> Rainfall is an element that is [...] strongly connected with the provision of water for public fountains, a primary need for all the small towns surrounding Rome. Concerning this matter, in many cases the advice of a meteorologist would not be without merit. Therefore, the holy father often ordered [me] to inspect some springs that serve the local population and offer suggestions about water mains and service pipes, in order to curb unnecessary expenses.[36]

---

35 See François Balley, *Endemo-epidémie et météorologie de Rome: Etudes sur les maladies dans leur rapports avec les divers agents météorologiques* (Paris: V. Rozier, 1863); and Balley, *Météorologie et météorographie, pathogénie et nosographie, ou: Éléments de recherches sur la connexion entre les divers agents météorologiques et la pathogénie civile et militaire à Rome de 1850 à 1861; Atlas* (Paris: V. Rozier, 1863). The use of diagrams to visualize and correlate the trends was probably suggested by Secchi, who preferred this system for studying the meteorological data (see Chapter 5).

36 *Le piogge sono un elemento che è* [...] *principalmente è legato col regime delle fontane che in paesi come i circostanti di Roma interessano i bisogni di tutti. Su questo il meteorologista può dare consigli non ispregevoli in moltissimi casi. Più volte perciò il Santo Padre* [*mi*] *ordinò di ispezionare alcune sorgenti per servizio delle popolazioni e per dar consiglio sulle condutture e gli allacciamenti, onde non mettersi in ispese che riuscissero poi inutili.* Secchi, "*L'astronomia in Roma nel pontificato di Pio IX*", 74. Patrica, Alatri, Anagni, Ferentino and other small towns, mostly located in south-eastern Lazio, improved their provision of water thanks to Secchi's practical suggestions (see Secchi, "L'Astronomia in Roma nel pontificato di Pio IX," 74–75; also APUG, FS 112.II)

His expertise in such matters, along with his studies on old Roman aqueducts (see below), made Secchi a highly regarded advisor in hydrogeological matters.[37] In 1875, for example, the mayor of the town of Frascati, near Rome, requested his assistance in a complaint about the interruption of a water spring (Tepula) after the channelization of another spring (Algidosa). Because Secchi "[knew] well the work done for the Algidosa water distribution [...], the weight that the advice of Fr. Secchi would have over the judges would be immense and perhaps determining in favor of the municipality."[38] In response to the request, Secchi prepared a long memorandum examining point by point the reports of geologists and engineers, and concluding that there was no causal connection between the two events. Secchi's studies of atmospheric electricity were also aimed at the public good. For example, in 1860, he proposed installing systems of lightning rods to preserve the main monuments in Rome from the dangerous effects of lightning, as had been done for monuments in other countries.[39]

Moreover, Secchi designed advanced fire safety measures for the Roman basilicas. In particular, in the years 1867–69, together with engineer and architect Luigi Poletti (1792–1869), he planned an innovative fire protection system for the Basilica of St. Paul outside the walls, which was being reconstructed after having been largely destroyed by a fire in 1823.[40] He also indicated how to distribute the water network in order to provide the basilicas with water supply points in case of a fire emergency. Secchi also provided early guidelines on seismic prevention. Iin 1859, he was charged with conducting a survey on the effects of a disastrous earthquake at Norcia and its surroundings and, with Poletti, wrote an important report containing crucial suggestions for the reconstruction of the buildings.[41] As regards seismic monitoring, in 1868 an early

37 See APUG, FS 3.IV.

38 *Il rev. Padre conosce bene il lavoro fatto per la conduzione dell'acqua Algidosa, il peso che verso I giudici avrebbe un voto del P. Secchi sarebbe immense, e forse tale da far piegare la bilancia a favore del Comune* (Janari [mayor of Frascati] to Del Grande [a friend of Secchi], Frascati, March 5, 1875; ACF 9813–14).

39 In 1829, Feliciano Scarpellini (1762–1840), astronomer at the Campidoglio Observatory in Rome, applied a lightning rod system at Montecassino Abbey, which became a reference for other Italian countries (see *Istruzione sui parafulmini* [Naples: Reale Tipografia della Guerra, 1842]).

40 The system was controlled by thermometers, set in the vault; when the temperature exceeded a standard, the thermometers activated the opening of water tanks over the roof. This is considered one of the first applications of fire safety engineering to the protection of buildings. For further details, see Monica Calzolari and Stefano Marsella, "La protezione della Basilica di San Paolo fuori le mura dopo la ricostruzione: il primo esempio di ingegneria antincendio?" in *Salvare la storia. Testimonianze di soccorso tecnico e prevenzione incendi nel passato*, edited by Stefano Marsella and Simonetta Monti (Rome: Istituto Superiore Antincendi, 2018), 27–37.

41 See Archivio di Stato di Roma, Ministero dell'Interno, Titolo II [Amministrazione Pubblica], Rubrica 36 (Agricoltura), b.302, Terremoto di Norcia (1869), Sunto della Relazione

network was created by geophysicist Michele Stefano De Rossi (1834–1898), who provided many meteorological stations with seismoscopic instruments, designed by him. Secchi was in disagreement with De Rossi's ideas and methods, but never openly entered into controversy with him, and preferred *not to deal with seismometers* [*ho risoluto di non mi occupare di sismometri*] in order to avoid polemics.[42]

Secchi also studied the climate of Rome and its influence on the flow of the Tiber River. A major flood occurred in December 1870, soon after Rome had been annexed by Italy; a few years later, the Italian government consulted him about protecting Rome from further floods.[43] "I'm very busy and drowned in the rains!,"[44] he wrote to Tacchini, while reporting and examining in detail the data on rainfall, wind, and temperatures in the areas of the Tiber basin.

4 New Lighthouse System

In 1858, a special commission was charged with renovating the system of lighthouses at the ports of the Papal States, as the lamps were obsolete and needed to be replaced with new ones. Secchi was a member of the commission and, according to the pope's will, the principal coordinator of the project.[45] He asked the minister of public works, Cardinal Giuseppe Milesi Pironi Ferretti (1817–73), to go abroad to visit the lighthouses of the main ports to study their systems and to buy modern lamps for the papal ports. Before leaving, Secchi was able to inspect the lighthouse of the port of Civitavecchia personally and to ascertain its state of neglect. He found that docking was dangerous; the entrance was a dump; the building was crumbling, the walls were drooping,

del Sig[nor]e Prof. Comm[endator]e Luigi Poletti, e del R.P. Angelo Secchi della Compagnia di Gesù, intorno alle cause ed agli effetti del terremoto del 22 agosto 1859 nella città di Norcia, co' modi di prevenirne in gran parte le funeste conseguenze, stabilita p[er] roché sia una legge edilizia comunale sulle seguenti prescrizioni di arte, alle quali fanno seguito parecchie disposizioni transitorie, per la specialità delle conseguenze del terremoto sud[dett]o, Roma ottobre 1859; see also Angelo Secchi, "Escursione scientifica fatta a Norcia ad occasione dei terremoti del 22 agosto 1859," *Atti dell'Accademia Pontificia de' Nuovi Lincei* 13 (1860): 63–104.

42 See Secchi to Tacchini, Rome, July 7, 1876, in Chinnici and Gasperini, *Alle origini dell'astrofisica italiana*, 425.

43 See APUG, FS 24.VIII.

44 *Sono occupatissimo e annegato nelle piogge!* Secchi to Tacchini, Rome, May 8, 1874, quoted in Chinnici and Gasperini, *Alle origini dell'astrofisica italiana*, 347.

45 See Alessandro Cialdi, "Cenni storici dei fari antichi più famosi e di alcuni moderni," in *Triplice omaggio alla Santità di Papa Pio IX nel suo giubileo episcopale* (Rome: Tipografia delle Scienze Matematiche e Fisiche, 1877), 4. Secchi also wrote some reports on the ports of Ostia and Terracina, see APUG, FS 5.XI and APUG, FS 110.V.

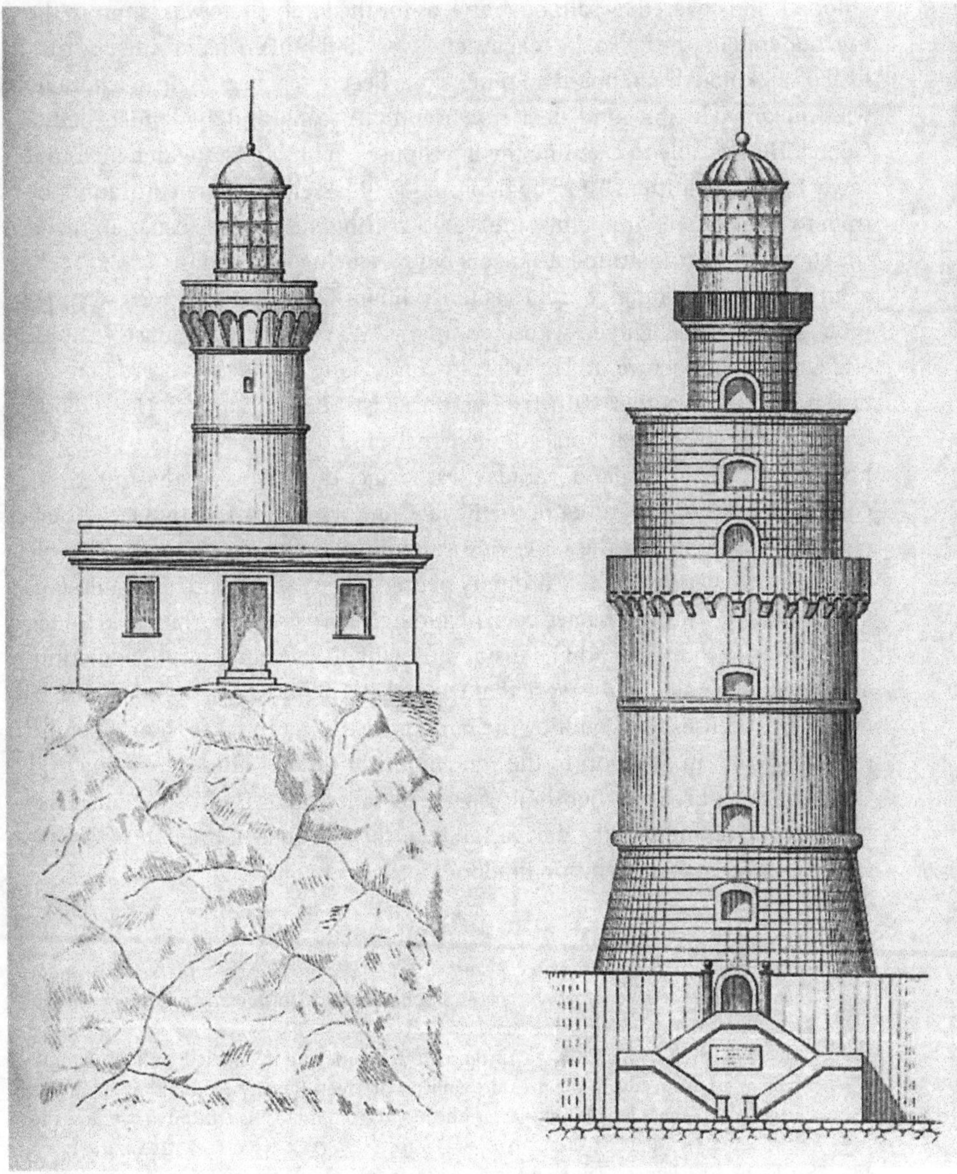

FIGURE 4.3 The lighthouses of the ports of Ancona (on the left) and Civitavecchia (on the right), after their renovation. From Alessandro Cialdi, "Cenni storici dei fari antichi piu' famosi e di alcuni moderni" in *Triplice omaggio alla Santita' di Papa Pio IX nel suo giubileo episcopale* (Rome: Tipografia delle Scienze Matematiche e Fisiche, 1877), pl. 1.

chipped, and covered by saltpeter; and, as for the lamp, there was "smut in the mirrors, smut in the bells; the machinery [was] half broken, half ruined by rust, filth of all kinds."[46] On board a French ship, Secchi reached Livorno where he was welcomed by the "kind, dear, true friendly Tuscans" and then proceeded to Genoa. He was able to examine the lighthouses of both ports, which he found in very good condition. Then he departed for Marseille, before continuing by train to Paris, where he spent some weeks visiting the city (see Chapter 5) and the shops of some instrument-makers before leaving for London.

In the British capital, Secchi visited William Crane Wilkins (1813–84), famous lighthouse engineer, whose company "W. Wilkins Lighthouse Lantern and Lamp Manufacturer to Her Majesty" had a long and highly regarded reputation.[47] Then, together with famous physicist Michael Faraday (1791–1867), he went to inspect a lighthouse under construction at Trinity House, the lighthouse authority in England. Faraday was in fact the scientific advisor for this corporation[48] and a member of the Royal Commission for Lighthouses.[49] Secchi and Faraday shared the same views about the utility of science and how it should be applied to improve ordinary people's lives. At the end of his mission, he concluded: "The lamps have been negotiated at a price of one hundred francs on the spot, without any other consideration for either shipping or installation; and given the quality of the work that will result, I believe it was a good deal."[50]

The lighthouses were built by the famous instrument-maker Henri Lepaute (1809–85) and, in addition to the renovation of the old ones at Ancona and Civitavecchia, three new lighthouses were installed along the Tyrrhenian coast to improve the safety of the navigation near the entrance of the port of Civitavecchia; they started operation in 1866.[51]

46 *Tutto fumo negli specchi, fumo nelle campane; la macchina mezza rotta, mezza rovinata dalla ruggine, dalla sozzura d'ogni specie.* Secchi, quoted in Castellani, "Il P. Angelo Secchi S.I. e le sue missioni scientifiche," 607.

47 In 1837–39, the Wilkins Company invented pneumatic and hydraulic lighthouse lamps, which used, respectively, air pressure and gravity feed (gravity feed is the use of the Earth's gravity to move a liquid without the use of a pump) to raise the oil to overflow the wicks.

48 See Frank A.J. L James, "'The civil engineer's talent': Michael Faraday, Science, Engineering and the English Lighthouse Service, 1836–1865," *Transactions of the Newcomen Society* 70 (1998–99): 153–160.

49 Faraday was well remunerated for this service, and Secchi regretted that the same did not happen in his country.

50 *I fari si sono conclusi per 100 mila franchi in posto, senza nessun altro pensiero nè di trasporto né di installazione; e stante la bontà del lavoro che ne avremo, lo credo un buon mercato.* Secchi to Marchetti, quoted in Castellani, "Il P. Angelo Secchi S.I. e le sue missioni scientifiche," 620.

51 See Alessandro Cialdi, *Cenni storici dei fari antichi più famosi e di alcuni moderni* (Rome: Tipografia delle Scienze Matematiche e Fisiche, 1877).

In the context of Secchi's contributions to the development and improvement of the transport systems, it should also be mentioned that he wrote a report[52] on the introduction of electric railways within the Papal States.[53] The signature on the project was that of Angelo Vescovali, but it was probably Secchi himself who suggested it. The project, however, was too ambitious, required considerable financial resources, and was ultimately abandoned.

5 Religious Consultations

Being both a scientist and a religious, Secchi's advice was particularly sought after in matters affecting both science and religion. Thus some sources indicate that Secchi was also called on to formulate the official answer of the church concerning the rules to be applied in liturgies by priests living near the North Pole, who experienced an unusual day–night rhythm:[54] this is yet to be ascertained, but, because of his competence in astronomy, it cannot be ruled out. Moreover, the knowledge that he acquired in the field of archaeology (see below) meant that his advice was often sought on Christian archaeology as well. According to some sources, Secchi attended the viewing of some findings discovered in 1874 in the underground basilica of St. Petronilla at Tor Marancia,[55] in the outskirts of Rome,[56] and the examination of the remains of the apostles St. Philip and St. James, which were found in 1875 inside an altar in the crypt of the Basilica of the Holy Apostles in Rome.[57] In 1876, he was also invited to attend the inspection of the sarcophagus discovered under the altar of the basilica of Saint Peter in Chains. Secchi edited an accurate report about the site—which was supposed to contain the relics of the seven Maccabees and other saint martyrs—and the opening of the sarcophagus, with a careful examination of all the materials held there.[58]

---

52 See APUG, FS 3.V.

53 Pius IX strongly promoted the development of railways, in contrast to the indifference of previous popes. In 1846, he created a pontifical commission for that purpose, and in 1859 he himself inaugurated the Rome–Civitavecchia line by traveling in a special coach, donated to him; see Maria Grazia Branchetti and Daniela Sinisi, eds., *"La meravigliosa invenzione": Strade ferrate nel Lazio 1846–1930* (Rome: Gangemi Editore, 2003).

54 See APUG, FS 23.III and 23.IV.

55 The basilica was discovered by the Jesuit archaeologist Giuseppe Marchi, S.J. (1795–1860); see Giuseppe Marchi, *Monumenti delle arti cristiane primitive nella metropoli del Cristianesimo* (Rome: Tipografia di C. Puccinelli, 1844), 230–231.

56 The same sources also mention Secchi's involvement in the inspection of the sepulcher of Saints Ambrose, Gervasius, and Protasius in Milan; at present, these circumstances cannot be confirmed but the matter deserves further investigation.

57 See APUG, FS 23.III.

58 See APUG, FS 26.V.

6 Educational and Communication Activities

Secchi was a brilliant and effective communicator. His vast scientific output, which seems "to represent the work of a whole scientific body rather than from an individual,"[59] did not prevent his tireless activity of teaching and popularization.

On the contrary, Secchi felt that part of his duties was to communicate science to the public, and he enjoyed doing so. His public lectures were well attended and received much applause, as is reported by many newspapers of the time. Even his opponents acknowledged his great skill as a communicator of science, noting his ability in public speaking, the fluency of his exposition, and the clarity of his language. Not by chance, on the occasion of the total solar eclipse of 1870 (see Chapter 5), the commission in charge of the preparations for the expedition to Sicily entrusted to Secchi the drafting of

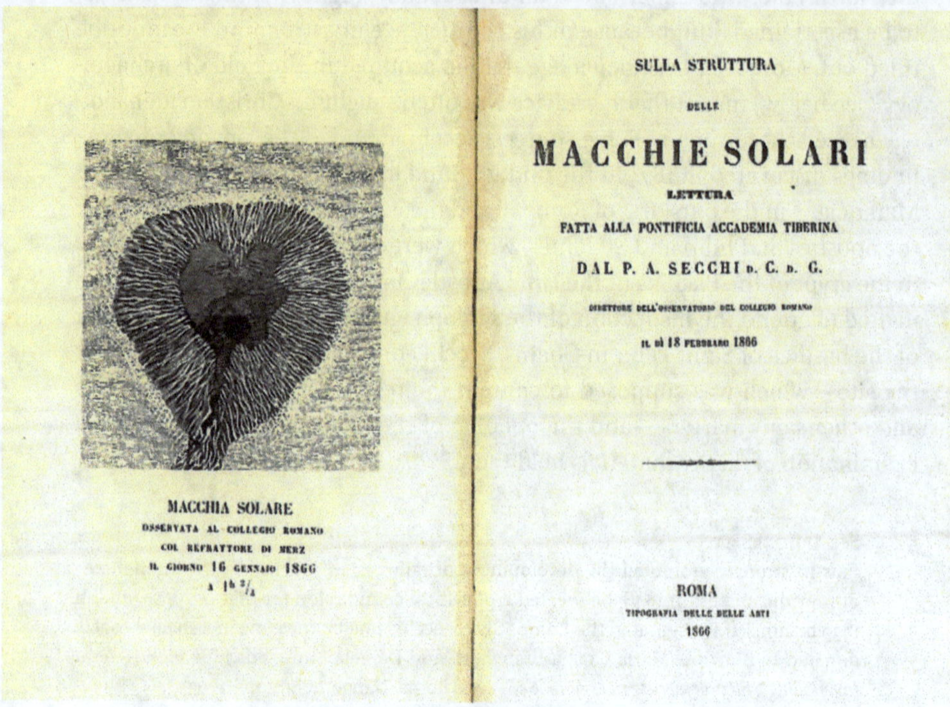

FIGURE 4.4 Title page of a conference given by Secchi to the Accademia Tiberina, with a sunspot drawing in the frontispiece. From Angelo Secchi, *Sulla struttura delle macchie solari* (Rome: Tipografia delle Belle Arti, 1866).

---

59 [*La sua produzione sembra*] *rappresentare piuttosto l'operosità di un corpo scientifico che quella di un solo individuo.* Benedetto Carrara, *Il Padre Angelo Secchi* (Padua: Tipografia del Seminario, 1903), 30.

instructions for the use of amateur astronomers who wished to collaborate in those observations,[60] in explicit recognition of his qualities as a communicator. The reputation he had in academic circles was joined with a great popular esteem, which certainly also resulted from the various public works of which he was in charge as mentioned above. This is not the only reason, however, as Secchi did not hesitate to give public lectures and to disseminate science even among the lower social classes: "I have been busy with some lessons on the Sun given to a workers' association,"[61] he wrote to his friend Tacchini in 1873. In these popularization activities, it is also possible to glimpse an impulse aimed at the human and cultural promotion of social classes that were traditionally excluded from scientific studies. This is evidence of his constant attention to an "inclusive" science, accessible to everyone. In general, his attitude toward the lowest social classes was sympathetic and benevolent, as revealed by some anecdotes reported by Secchi himself.[62] In 1867, for example, having been invited to lunch by a noble family in Paris, twice he was turned away at the door by the concierge; after recognizing the misunderstanding, "she paled and apologized. What buffoons are masters, not informing their servants,"[63] Secchi wrote in his diary, justifying the concierge's behavior. While still in Paris, when Tsar Alexander II (1818–81, r.1855–81) had come to visit the Exposition Universelle, he reported: "He is proud and abrupt and looks *méchant* [bad] as the workers told it,"[64] thereby revealing that he had conversed with the workers and had taken their comments into account.

6.1 *Disseminating Science*

Secchi greatly encouraged amateur astronomers[65] and promoted astronomy education. In many cases, he acted as an intermediary in providing optical instruments to noblemen and churchmen who wanted to learn science, and in supplying many Jesuit colleges of far distant countries with excellent telescopes from the renowned Merz Company in Munich.[66]

60 See Angelo Secchi, *Sull'ecclisse totale del Sole che avrà luogo ai 22 dicembre 1870: Notizie ed istruzioni* (Milan: Vallardi Editore, 1870).

61 *Sono stato occupato in alcune lezioni sul Sole date a benefizio di una società operaia.* Secchi to Tacchini, Rome, March 25, 1873, quoted in Chinnici and Gasperini, *Alle origini dell'astrofisica italiana*, 260.

62 Both anecdotes are reported in one of Secchi's diaries, see APUG, FS 23.II.A.

63 *O che buffoni di padroni che non avvisano i servitori.* APUG, FS 23.II.A.

64 *È fiero e brusco e ha un aspetto méchant come dicono gli operai.* APUG, FS 23.II.A.

65 Even to carry out spectroscopic observations, providing schemes of appropriate optical combinations; see Angelo Secchi, "Schreiben des Herrn Professors Secchi [...] an den Herausgeber," *Astronomische Nachrichten* 67, no. 1608 (1866): 381–382.

66 See Chinnici, "Maker and the Scientist."

Secchi's engagement in the dissemination of science is also apparent from the numerous private courses of physics he taught. He taught nuns who were teachers in some women's colleges, such as the Religious of the Sacred Heart who headed the Colleges of Trinità dei Monti and St. Rufina, the Daughters of St. Joseph, and the Oblate Sisters of the Child Jesus.[67] The high standard of these courses was proved by "the great honor with which, subjected to the rigorous examination demanded by the law, they all obtained their teaching certification so that they could keep teaching in their different institutes, even in many women's colleges of higher education."[68]

Even if times were changing—in the second half of the nineteenth century, women's rights activism was getting started in Britain and the United States—there were still strong prejudices against the education of women. Jesuits encouraged girls' education, to a certain degree, because in their opinion they had to become good wives and mothers; however, "too much intellectual activity not only confused a woman, they asserted, but kept her from completing her household duties."[69] In this context, Secchi's position on women's education, albeit not revolutionary, was nonetheless innovative:

> Catholic girls should be properly educated, in general for the rule of their houses, but also because there are some of them who someday may be placed by providence in a position of managing employees, and it is necessary that this education is suitable to develop self-confidence and to gain some public esteem, so that they will be respected and have authority and they may prove that their faith is enlightened and their pious devotions do not come from ignorance but from the deep conviction of their faith.[70]

---

67 See APUG, FS 8.III.

68 *Quanto poi quelle lezioni furono proficue a quelle buone religiose, si può argomentare dal grande onore con cui, sottoposte tutte a rigoroso esame voluto dalla legge, ne conseguirono tutte la Patente magistrale, e così poterono conservare nei loro diversi istituti l'istruzione, anche superiore, nei numerosi loro femminili conservatorii.* Marchetti, APUG, FS 23.III.19. After the political unification of Italy, the Italian government required teachers belonging to religious orders to obtain teacher certification in order to teach in their colleges.

69 McKevitt, *Brokers of Culture*, 206.

70 *Una conveniente istruzione deve darsi alle giovinette cattoliche per i principii generali di renderle utili alla casa; e dirò anche che certa classe di esse che dalla provvidenza furono poste in istato da avere col tempo delle persone da sé dipendenti, è mestieri, e mestieri che questa istruzione sia tale che possano sapersi regolare ed ottenere nel pubblico una certa stima, onde essere rispettate e avere autorità e fare conoscere che la loro fede è illuminata e che se fanno delle pratiche religiose non è per l'ignoranza delle cose ma per persuasione profonda della loro fede Secchi,* APUG, FS 2. VII.

It is also worth mentioning Secchi's production of physics and astronomy treatises, an activity that was particularly aimed at improving the scientific knowledge of the people and would intensify during the last years of his career (see Chapter 10). Secchi attached great importance to the promotion of "modern" science, and his texts included the latest physical and astronomical theories of the time. The numerous editions and translations into foreign languages of his treatises (see Chapter 6) such as *Le soleil* (The Sun [1870]), one of the most important nineteenth-century works about the Sun,[71] *Le stelle* (The stars [1877]), a treatise on stellar astronomy,[72] and *Lezioni elementari di fisica terrestre* (Elementary Lessons of Earth Physics [1879]), a volume on geophysics, which was published posthumously, like many other translations and editions of his books,[73] prove the extraordinary success of his works. The most popular (and controversial) of his books was his first treatise, *Sull'unità delle forze fisiche* (On the unity of the physical forces), first published in 1864 and followed by many editions and translations.[74] In this book, Secchi intended to disseminate new theories on the ether and the kinetic theory of gases, which were not yet included in physics textbooks (see Chapter 6). Secchi's efforts to promulgate scientific knowledge prove his conviction about "useful science." In his opinion, astronomical studies should not be reserved solely for the intellectual elite:

> The study of astronomy not only is necessary to raise the national culture but it has also other advantages [...]; it is useful for geodesists and engineers, it inspires the habit and the passion for exactness and precision, two qualities, which, once gained, turn out to be very useful for all other studies.[75]

---

71 It was followed by an edition in German (*Die Sonne* [1872]) and by a second, revised and expanded edition, in two volumes (*Le soleil* [1875–77]). Secchi had also prepared an English edition, which was never published; indeed, no book by Secchi was ever published in English (see Chapter 9).

72 It was also translated into German (*Die Sterne* [1878]).

73 The posthumous editions of Secchi's treatises are as follows: *Le soleil* (Spanish 1879; Italian 1884), *Le stelle* (French 1879, 1880, 1884, and 1895), and *Lezioni* (German 1882, 1883, and 1885; Spanish 1886); for a bibliographic review, see Rita Fioravanti, "Per una bibliografia secchiana," in Altamore and Maffeo, *Angelo Secchi*, 277–290.

74 Two other extended Italian editions were published in 1874 and 1885, two French editions in 1869 and 1874, three Russian editions in 1872, 1873, and 1880, and three German editions in 1876, 1884–85, and 1892.

75 *Lo studio dell'astronomia non solo serve a rialzare la cultura nazionale, ma ha pure altri vantaggi* [...] *esso serve ai geodeti ed ingegneri, ed inspira poi l'abitudine ed anche la passione per l'esattezza e la precisione, qualità che una volta acquistate riescono utilissime in tutti gli altri ordini e studi.* Secchi's words in "Processo verbale della terza adunanza della Commissione pel riordinamento degli osservatorii astronomici", quoted in Poppi, Bònoli, and Chinnici, "Il progetto Tacchini e la riforma degli Osservatori italiani," 159.

6.2 *Teaching at the University*

As a teacher, Secchi had regular teaching assignments in astronomy and physics at the Gregorian University.[76] He adopted the waves-in-ether theory of light[77] as a basis of his physics courses, as Pianciani had done in his book *Istituzioni fisico–chimiche* (Physical–chemical principles) about thirty years earlier.[78] Yet Secchi was aware that many textbooks were now obsolete, both in physics and astronomy, and thus in 1852 he published *Quadro fisico del sistema solare* (Summary about the physics of the solar system), a sort of pocket textbook, which had a double purpose, both educational and popularizing.

FIGURE 4.5 Title page of the *Quadro fisico*, with an illustration of the Collegio Romano Merz telescope in the frontispiece. Courtesy of Specola Vaticana.

---

76 See Maffeo, *Vatican Observatory*, 15–16.

77 In the mid-nineteenth century, electromagnetic phenomena were explained by means of mechanical actions transmitted from one body to another by means of a medium (ether) occupying the space between them; the waves theory of light also assumed the existence of the ether as a medium for the waves' propagation.

78 See Giovanni Battista Pianciani, *Istituzioni fisico-chimiche* 1–4 (Rome: Crispino Puccinelli, 1833–35).

In this way, Secchi intended to update astronomy education and disseminate new ideas in the field of celestial science by joining scientific accuracy with clear writing. In the preface to the readers, he wrote that astronomy had made important advances, as recent improvements in the instrumentation had led to many new discoveries:

> But, for the most part, these discoveries remain in the narrow confines of the men of science and, especially among us, are hardly ever communicated in the teaching of young people. Because I have been asked to prepare a summary of such matters, to help in general education, I thought it would be good to collect in a simple summary the more interesting things we now know about the physics of celestial bodies. This format seemed to me more appropriate than others for popular education, and more suitable to promote conversations about the wonders of the universe. No doubt, several similar works do exist nowadays, but this will not be a copy of them; on the contrary, in order to show that science can be popular without losing its dignity and truth, I have tried to bring my descriptions up to the level of the latest discoveries, while still keeping them accessible to students.[79]

Secchi also stated that

> except for some few facts which, by their very nature, it was impossible for me to verify, for the most part I was able to rely on the testimony of my own observations, which I was almost always able to make thanks to the powerful and precise instruments with which the observatory is provided.[80]

This important statement reveals the nature of Secchi's approach to science, which was empirical rather than theoretical. Secchi enjoyed doing

79 *Ma queste scoperte per la maggior parte restano nella angusta cinta degli uomini della scienza, e specialmente tra noi sono pochissimo diffuse nell'insegnamento giovanile. Essendo io pertanto stato richiesto di giovare la commune istruzione con darne una succinta informazione, ho creduto bene raccogliere in un semplice quadro quanto di più interessante si sa ora intorno alla costituzione fisica dei corpi celesti. Tal forma mi è sembrata più opportune delle altre alla popolare istruzione, e più adattata a promuovere la conversazione sulle meraviglie dell'universo. Esistono non vi ha dubbio oggidì diversi di simili lavori, ma questo non ne sarà una copia: anzi perché la scienza con acquistare popolarità non perda la sua dignità e verità, ho cercato di portare questa descrizione al livello delle moderne scoperte, senza farle perdere il vantaggio della facilità dell'insegnamento.* Angelo Secchi, *Quadro fisico del sistema solare secondo le più recenti osservazioni* (Rome: Tipografia delle Belle Arti, 1859), 7.

80 *Tranne alcuni pochi fatti che per la loro natura stessa mi era impossibile di verificare, ho procurato negli altri di avere il testimonio delle mie proprie osservazioni, il che ho potuto fare quasi sempre mercè de' potenti e precisi strumenti di cui è fornito l'Osservatorio.* Secchi, *Quadro fisico*, 8.

experiments, building devices, testing results; he definitely preferred observations over calculations. Yet because of this empirical approach, the results of his observations were sometimes considered too approximate to be reliable, and lacking in mathematical support. However, unlike the tendency of many of the anti-Baconian scientists of his time, Secchi tended to reject theoretical conjectures that were unsupported by observations or experience; on the contrary, he recognized the risk of being influenced by a conceptual framework, which might compromise and distort the objectivity of the observation of natural phenomena.

Secchi then concluded his preface to the readers with a wish: "May this effort benefit young people by stimulating them to follow these studies, and to admire and honor the Creator, whose work I hope they will come to know, better and better, from this book."[81]

The last chapter of the book included a paragraph about stellar astronomy—star clusters, nebulae, the Milky Way, etc.—which ended with some considerations about the distances of the stars[82] and the place of the Sun in the universe: "What place does our Sun have in such a big cluster of bodies? The answer is very difficult [...],"[83] but considering the distribution of the stars and the motion of the Sun toward the Hercules cluster,

> it is not improbable that these stars form a system with our Sun [...]. This system would be the closest to us among all the others and we too would belong to it; and it is likely that many millions of similar systems of clusters together form the Milky Way, which may be nothing more than one of the many nebulae scattered in the universe.[84]

81 *Possa questa fatica riuscire di profitto alla gioventù eccitandola a questi studi, e ad ammirare e onorare il Creatore, la cui opera io spero che imparerà sempre meglio a conoscere da questo libretto.* Secchi, *Quadro fisico*, 8.

82 Secchi wrote that, combining the result of stars' photometry, proper motions, and apparent distribution, it is possible to estimate the various distances of the stars by calculating the time that the light takes to arrive to our eyes, but "for the extreme nebulae, each calculation is useless [...] for many stars components of nebulae outside the Milky Way, light must have traveled millions of years to get to us, so that we are now seeing what were those bodies million years ago!" (*per le nebbie estreme ogni calcolo è vano* [...] *per molte stelle componenti le nebulose fuori della Via Lattea, la luce deve aver viaggiato milioni di anni per arrivare a Noi, onde noi oggi vediamo ciò che furono que' corpi milioni di anni fa!*) Secchi, *Quadro fisico*, 158.

83 *Qual luogo ha il nostro Sole in sì grande ammasso di corpi? La risposta è assai difficile* [...] Secchi, *Quadro fisico*, 158.

84 *Non è improbabile che queste stelle formino un sistema col nostro Sole* [...] *Questo sistema sarebbe adunque il più vicino a noi di tutti quanti e ad esso noi pure apparterremmo, ed è probabile che molti milioni di simili sistemi agglomerati insieme formino la Via Lattea, la quale però non sarebbe essa stessa che una delle tante nebulose sparse nell'Universo.* Secchi, *Quadro fisico*, 159.

Secchi's speculations match our modern understanding of the nature of galaxies, a point that was only resolved in the early twentieth century via the famous Great Debate.[85]

In his concluding remarks, Secchi wrote about matter, forces, and organisms and affirmed the possible existence of extraterrestrial life in the universe:

> We don't know if organic forces are also developed in other bodies of our system, much less if they are to be found at other distant suns, but who would dare to deny it? It would not be a sufficient reason to cite the diversity of climate or other physical conditions. It is true that life does not seem to exist here except under certain conditions of temperature and humidity, but who is to say that these [conditions] are only found on Earth? Given that it has happened in one of many celestial bodies, who can imagine what will happen in the others, where so many other possibilities might occur?[86]

He then made a curious analogy with marine fauna[87] and concluded:

> It is very likely that differences in gravity and other varying conditions modify the nature of the organic forces, and their functions could be

85 The Great Debate was held in 1920; it mainly involved astronomers Harlow Shapley (1885–1972) and Heber Curtis (1872–1942) and concerned the nature of the spiral nebulae and the size of the universe. In Shapley's opinion, distant nebulae were relatively small and lay within the outskirts of the Milky Way (the Earth's home galaxy), while Curtis held that they were independent galaxies, implying that they were large and distant (see *Bulletin of the National Research Council* 2, part 3, no. 11 [May 1921]: 171–217). Further observations confirmed Curtis's conclusions.

86 *Noi non sappiamo se le forze organiche si trovino pure sviluppate negli altri corpi del nostro sistema, e molto meno se lo sieno in quelli degli altri Soli lontani, ma chi oserà negarlo? Non sarebbe ragione sufficiente a ciò né la diversità del clima, ne' delle altre circostanze fisiche. E' vero che la vita non sembra poter sussistere qui da noi che sotto certe condizioni di temperature e umidità, ma chi potrà asserire che queste non si trovino altro che sulla Terra? Da quel che accade in uno de' tanti corpi celesti, chi potrà immaginare quello che accadrà negli altri, ove concorrono tante altre diversità?* Secchi, *Quadro fisico*, 163.

87 "Those who have never seen the ocean or hear about it, how could they believe that in its waters live so many animals? How many people even now ignore the multitude of these beings that inhabit the tropical seas and cover the ocean funds with a real vegetation, almost animal, as herbs in our lawns? Beings that are so small and numerous that millions can be contained in one inch, and that are destroyed just by drawing them from water" (*Chi non avesse mai veduto il Mare o saputo di esso, crederebbe che nelle sue acque viver potessero tanti animali? Quanti anche adesso ignorano la moltitudine di questi esseri che popolano le acque tropicali e che coprono di una vera vegetazione direi quasi animale i fondi dell'oceano, come le erbe i nostri prati? Esseri sì piccoli e in tanto numero che in un pollice ne possono esser contenuti dei milioni, e che per distruggerli basta cavarli dall'acqua*) Secchi, *Quadro fisico*, 163.

> carried out despite circumstances of temperatures that seem incompatible at all with life. Even if, for this reason, life could not actually exist on the flaming stars that occupy the center of these systems, couldn't they find the fruitful and necessary conditions to survive on the other bodies orbiting them? Light, which has no effect on most minerals, seems to be essential to a large number of organic functions and, seeing how generous providence was with the major planets, providing them with satellites [...] we are persuaded that life too exists outside of our sphere.[88]

This belief would be a constant theme in Secchi's thought and can be found elsewhere in his works as well (see Chapter 9).

The book included many illustrations and an appendix containing a table with the orbital parameters of over two hundred comets. Publishing this sort of up-to-date book was so important to Secchi that, at the end of the volume, he even took the time to insert an addendum of a few lines about recent astronomical news (the discovery of the asteroid Pandora and the eclipse of 1858), which had occurred during the printing of the book.

To provide his students with appropriate lecture notes, in 1862 Secchi also wrote *Principii di astronomia compilati per uso delle scuole del Collegio Romano* (Principles of astronomy compiled for use in the Collegio Romano schools), a 168-page text where he collected the contents of his course.[89] The volume can be considered complementary to the *Quadro fisico*—it contains several references to it—and was divided into three sections: (1) Concerning the Apparent Motions of the Celestial Bodies (with an appendix on gnomonics in the subsection on time measurement); (2) Concerning the Motions of the Sun and the Moon; and (3) Concerning the Solar System. The book, not printed with type but by lithography, is well furnished with many instructional figures

---

88 *E' probabilissimo che una gravità e un mezzo diverso modifichino la natura delle forze organiche e le loro funzioni possano eseguirsi malgrado circostanze di temperature che ci sembrano affatto incompatibili colla vita. Che se pure essa per ciò non potesse realmente esistere sugli astri infiammati che occupano il centro dei sistemi, non sarebbero questi abbastanza utili e necessari per conservarla sugli altri corpi che li circondano? La luce che trovasi indifferente alla maggior parte delle operazioni minerali, sembra di prima necessità a un numero grande di funzioni organiche e il vedere come la Provvidenza è stata larga di essa ai corpi principali fornendoli anche di satelliti* [...], *ci sembra persuadere che la vita esiste ancora fuori del nostro globo.* Secchi, *Quadro fisico*, 163–164.

89 See Angelo Secchi, *Principii di astronomia compilati peruso delle scuole del Collegio Romano* [Rome: mimeographed], 1862). The text was then revised in 1864. An interesting compendium of astrophysics lectures, based on Secchi's studies, was later edited by father Ferrari; see Gaspare Stanislao Ferrari, *Compendio di Astronomia Fisica compilato sopra le opere del P. Angelo Secchi* (Rome: Litografia Spellani e Consorti, 1885).

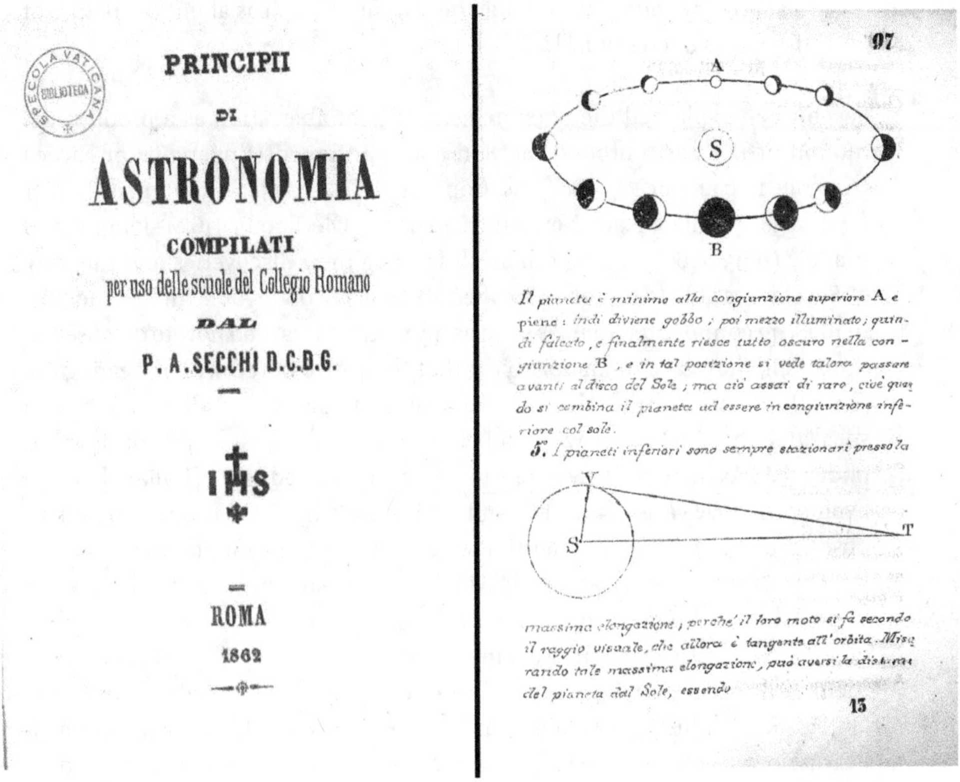

PRINCIPII
DI
ASTRONOMIA
COMPILATI
per uso delle scuole del Collegio Romano
DAL
P. A. SECCHI D.C.D.G.

IHS

ROMA
1862

97

A
S
B

Il pianeta è minimo alla congiunzione superiore A e piene: indi diviene gobbo; poi mezzo illuminato; quindi falcato, e finalmente riesce tutto oscuro nella congiunzione B, e in tal posizione si vede talora passare avanti al disco del Sole; ma ciò assai di raro, cioè quando si combina il pianeta nel essere in congiunzione inferiore col sole.

5.º I pianeti inferiori sono sempre stazionari presso la

V
S
T

massima elongazione; perchè il loro moto si fa secondo il raggio visuale, che allora è tangente all'orbita. Misurando tale massima elongazione, può aversi la distanza del pianeta dal Sole, essendo

15

FIGURE 4.6 Astronomy lecture notes by Secchi. On the left: front page; on the right: page 97 with the illustration of the planets' motion around the Sun. Courtesy of Specola Vaticana.

and contains explanations of theorems, corollaries, and lemmas, as well as the solutions to some problems, as exercises.

In this text, it is interesting to read how Secchi dealt with historic and contemporary proofs of the Earth's rotation.[90] The paragraph dealing with the translational motion of the Earth began with the following words:

> The analogy of other planets makes it almost certain that the Earth must also revolve around the Sun: or else one has to hold that the Sun is the center of all the others and that only the Earth, such a small body, does not move itself around greatest of all the bodies. Then, after Newton encapsulated Kepler's laws in the mere fact of a force acting as the inverse square of the distance, the immobility of the Earth could not be

90 He referred to the experiments of Giovanni Battista Guglielmini (1760–1817), Johann Benzenberg (1777–1846), and Foucault.

> sustained without introducing the most absurd ideas about the nature of the forces of creation.[91]

Secchi then explained the phenomenon of light aberration as a proof of the motion of the Earth around the Sun and mentioned Hipparchus of Nicaea (*c.*190 BCE–*c.*120 BCE), Nicolaus Copernicus (1473–1543), Johannes Kepler (1571–1630), Galileo, Isaac Newton (1643–1727), Ole Römer (1644–1710), James Bradley (1693–1762), among others, describing their discoveries and theories.

Secchi was able to establish an open and sincere dialogue with his students. In 1868, he commented that "as soon as my students learned how to do observations well, they gave them up, saying that they were bored with repeating the same thing incessantly." This comment suggests that the students felt free to speak openly with Secchi, without being afraid of being punished for their impudence. Secchi, with benevolent tolerance, remarked that "[Italians] have a lively spirit, love variety, novelty" and are lacking "the German spirit of patient observation and material repetition of the same thing a million times over."[92] He maintained a close relationship with his former students and liked to involve them in his work. One of them, Colino Kambo (1842–1914), a lawyer, described meeting Secchi and Marchetti during a walk to the Pincio hill in 1862; he joined them, and once they had arrived at the place where Secchi intended to place the landmark for adjusting the meridian circle,[93] the Jesuit told him that, although he was now devoted to the study of law, "he still ought to do a small service to astronomy" and gave him instructions so that, the next morning, he could help a worker properly place the landmark, following the directions he waved to them from the Collegio Romano Observatory, some distance away. After many years, Kambo recalled: "I can't describe my satisfaction at being put in that position: I felt like I was a pars magna in an important astronomical operation!"[94]

---

91 *L'analogia degli altri pianeti rende quasi certo, che la terra deve girare ancor essa attorno al sole: almeno è indispensabile l'ammettere che il sole è il centro commune di tutti gli altri e che sola la terra, corpo sì piccolo, non deve tenere intorno a sé il maggiore di tutti. Dopo poi che Newton epilogò le leggi di Keplero nel solo fatto di una forza che agisce con ragione inversa del quadrato della distanza, non potrebbe sostenersi la quiete della terra senza introdurre le idee più assurde intorno alla natura delle forze della creazione.* Secchi, *Principii di astronomia,* 117.

92 *Manca lo spirito tedesco, di paziente osservazione e materiale ripetizione della stessa cosa milioni di volte. Noi collo spirito vivace che abbiamo, amiamo la varietà, la novità* [...] *appena i miei allievi hanno imparato a fare bene le osservazioni le hanno lasciate andare dicendo che erano seccati di ripeter sempre la stessa cosa.* Secchi to Schiaparelli, Rome, June 30, 1868, in Buffoni, Manara, and Tucci, *G.V. Schiaparelli, A. Secchi*, 177.

93 See Chapter 3, note 16.

94 *Non dico la mia soddisfazione per quell'incarico: mi sembrava d'essere pars magna in una importante operazione astronomica!* Kambo, quoted in Lay, "Il nuovo Osservatorio del Collegio Romano," 103. Reflecting his popularity among the youth, the Catholic Youth

6.3 *Classics and Scientific Hobbies*

Despite his extremely heavy workload, Secchi still found time to engage in his favorite hobbies, one of the most important of which was archaeology, an activity to be practiced outdoors, which he found to be a useful diversion during times when he was stressed by conflicts. In Grottaferrata or in the countryside of Lazio, where he used to rest—often on the advice of doctors—to escape the scorching summer weather of Rome, Secchi greatly enjoyed dabbling in such research, as he wrote to his friend Tacchini:

> If my disease lasts, I will become an antiquarian, and the study of Frontinus[95] begins already to delight me more than that of the Sun [...]. I found the Crabra water aqueduct, the one of the gardens of Lucullus; every day, ancient ways are revealed more and more clearly to me. Thus, the return to Rome would be unpleasant for me, but duty compels me [...].[96]

Secchi's innate curiosity led him to locate the remains of some Roman aqueducts,[97] often while inspecting springs for the water provision of small towns near Latina and Frosinone, and to study archaeological finds related to astronomy. He confirmed that the Egyptian obelisk of Piazza di Montecitorio, which had been found in the archeological site of Campo Marzio, was the gnomon of a large meridian built under Augustus;[98] he also illustrated an ancient sundial, known as Commodus' sundial, which was kept at the Kircherian

---

Society of Palermo chose Secchi as its delegate to pay their respects to Pius IX on the occasion of the pope's election anniversary in 1873. See Circolo di S. Rosalia in Palermo to Father Secchi, June 16, 1873; APUG, FS 23.VI.a.

95 Sextus Iulius Frontinus was a Roman politician of the first century, author of an important treatise on the aqueducts supplying Rome (*De aquaeductu urbis Romae*).

96 *Se dura il mio incommodo, io diventerò antiquario, e già lo studio di Frontino mi comincia a piacere più che il Sole,* [...] *Ho trovato l'acquedotto dell'acqua crabra, quello della villa di Lucullo; le vie antiche mi si scoprono ogni giorno più evidenti. Sicché il ritorno a Roma mi sarebbe poco gradito, ma il dovere me lo impone* [...]. Secchi to Tacchini, Grottaferrata near Rome, September 4, 1873, quoted in Chinnici and Gasperini, *Alle origini dell'astrofisica italiana*, 324.

97 See, e.g., Angelo Secchi, *Intorno ad alcuni avanzi di opere idrauliche antiche rinvenuti nella città di Alatri* (Rome: Tipografia delle Belle Arti, 1865).

98 See Paolo Alberi Auber and Maria Luisa Tuscano, "L'obelisco di Augusto alla luce di due manoscritti di padre Angelo Secchi e dei testi di Giuseppe Vasi e Cornelio Meyer," in *Atti del XIX Seminario nazionale di Gnomonica*, ed. Maria Luisa Toscano (Palermo: Visiva Marketing Tools, 2014), 25–33. In an undated letter to an unknown correspondent—probably the British astronomer Charles Piazzi Smyth (1819–1900), known for his studies of the Great Pyramid of Giza—Secchi wrote (in English): "I have [...] found that the great obelisk of Sesostris now in Rome in the Place of Monte Citorio, has its side inclined exactly so as not cast shadow [= *so as to cast no shadow*] on the day of the solstice, and its pyramid was so cut [= *was cut so*] as to throw a shadow terminated by a strai[gh]t line from the solstice in Cancer to the Equinox so that this [line?] was parallel to the Equator." APUG, FS 25.I.

Museum.[99] Fr. Angelini, who was Angelo's teacher of classical literature, well remembered his pupil's interest:

> In the studies of history and archaeology, he had such extensive and in-depth knowledge that I wondered how he was so advanced in archaeological science, though he usually practiced other [sciences] far from it. [...] How many times it happened that we spoke about the early kings of Rome, the works of Servius Tullius, the ancient aqueducts, and the water channels supplying Rome, and I was astonished by the rich and well-founded erudition with which he dealt with these difficult subjects.[100]

Faithful to his principles of science in the service of society and to the effort of spreading scientific culture among the common people, Secchi often accepted invitations to set up meridians and sundials in public places—gnomonics being another of his favorite hobbies, a recreational activity during his vacation times. Thus far, at least seven wall sundials by Secchi have been identified in various regions.[101] These sundials were made for public use, but it is highly probable that there were others, located in colleges, private palaces, or villas, which have been lost or destroyed.

Another of Secchi's long-standing interests was geology, as demonstrated by the posthumous publication of the above-mentioned volume *Lezioni elementari di fisica terrestre*. Secchi carried out studies on the volcanic activity of the Monti Lepini, a small mountain range located near Rome, which is notable for its karsts[102] and wrote some notes on the findings of prehistoric

---

99 On Secchi's studies on ancient sundials, see Tuscano, "Angelo Secchi e la gnomonica," 269–271.

100 *Negli studi storici e di archeologia avea cognizioni sì vaste e in una sì profonde, che mi destò meraviglia, come con aver tra mano argomenti di tutt'altro genere, fosse così addentro nella scienza archeologica. [...] E quante volte cadde il parlare sopra i primi re di Roma, le opere di Servio Tullio, gli antichi acquedotti, e le vene delle acque condotte in Roma, m'era di stupore la ricca e sicura erudizione, con che svolgeva sì intralciati argomenti.* APUG, FS 23.IV. It is worth mentioning that Angelo's cousin, Giovampietro Secchi (1798–1856), also a Jesuit, was a noted archeologist and was often confused with him. In a letter to Giovampietro, dated 1852, Angelo complained of this frequent misidentification and reported that some French antiques dealers had requested clarifications from him about Egyptian matters; see Flora Parisi, "Lettere di p. Angelo Secchi," in *Angelo Secchi astronomo e fisico: Attualità scientifica e luoghi storici a Roma*, ed. Tullio Aebischer (Città di Castello: Edizioni Nuova Prhomos, 2012), 45–48, here 47.

101 They are located in Lazio (Sezze, 1852; S. Angelo di Bauco, today Boville Ernica, 1865; Alatri, 1875; two at Grottaferrata, 1873 and 1876), Calabria (Cosenza, 1875) and Sicily (Augusta, 1870); see: Tuscano, "Angelo Secchi e la gnomonica," 263–269.

102 See APUG, FS 8.II.r; he also discovered some volcanic craters, see Angelo Secchi, "Comunicazioni", *Atti dell'Accademia Pontificia dei Nuovi Lincei* 30 (1877): 86–87.

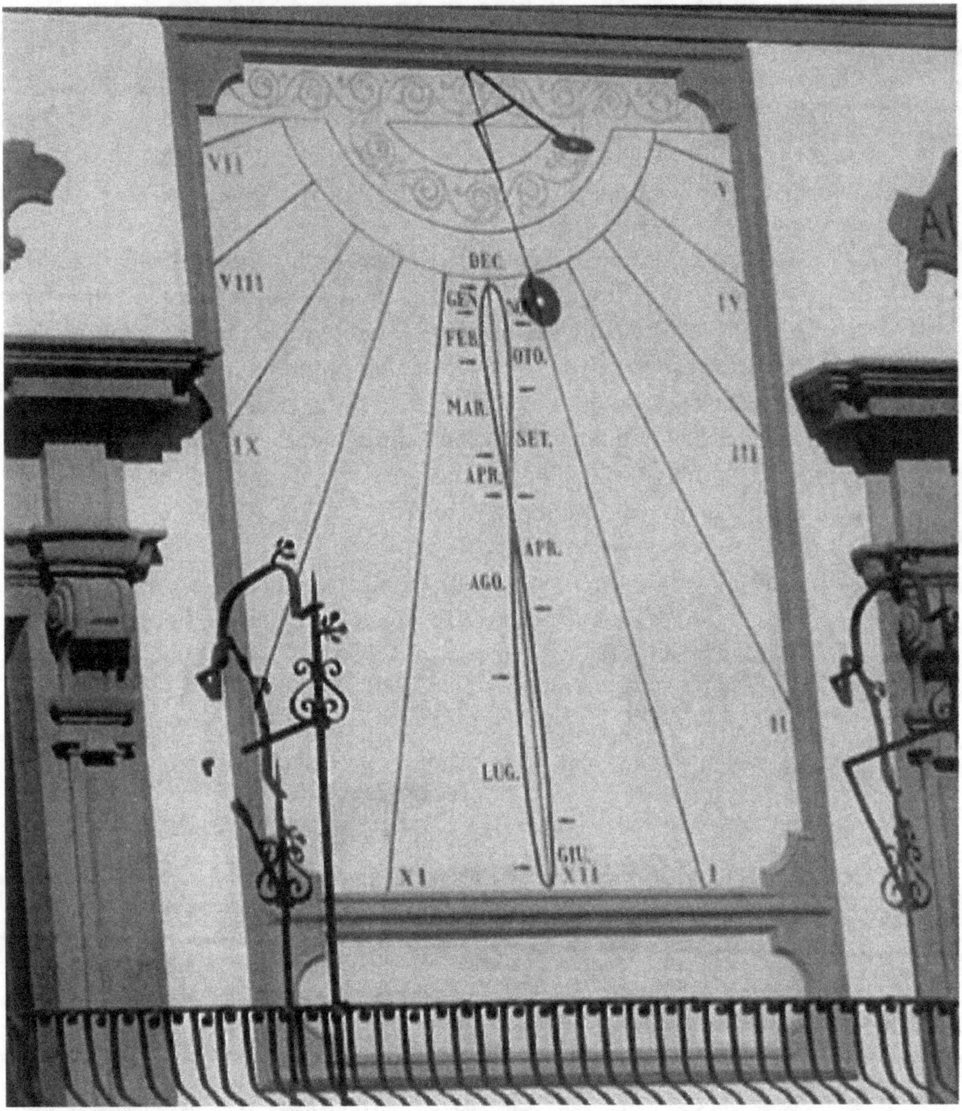

FIGURE 4.7 Sundial built by Secchi and Donati at Augusta. Courtesy of Rotary Club, Syracuse (Italy).

sites,[103] showing that he had wide and accurate knowledge of geologic dating and that he was familiar with the paleontological theories circulating at the time.[104] Because of this interest, he sometimes joined geologists Giuseppe Ponzi

103 See APUG, FS 26B.5.

104 See the debate about the "pliocenic man," APUG, FS 5.VIII.10.

(1805–85) and Augusto Statuti (1829–1911) on trips examining findings in new sites.[105]

Finally, Secchi also ventured into some studies on the history of astronomy, publishing works on Athanasius Kircher[106] (1602–80) and Galileo[107] (see also Chapter 9). According to Molza, he also studied literature and linguistics and, "though not inclined to poetry, he always greatly dabbled in Dante's works which he knew very well."[108]

The intense scientific and public activity that Secchi carried out did not leave him much room for pastoral work. Some confrères recalled that, as a young Jesuit, he had been sent to preach the Spiritual Exercises to the servants of noble families and was so successful that, years later, people still spoke about them. However, once he took on the directorship of the observatory, he had to give up these kinds of commitments to devote himself to the services required as a scientist. He was probably sorry to neglect the care of souls; Brother Marchetti reported that Secchi once confided to him:

> "Oh! You will show up to God with maniples, but I have always been applied to purely scientific things and thought little on the salvation of the neighbor; I won't know what to say to the Lord." Then I comforted him by saying: "Ah! Father, what are you talking about? For was it not but you who held high the torch of science thus closing forever the mouths of malevolent people that continually croak against the ignorance of the clergy?" [...].[109]

---

105 See, e.g., Giuseppe Ponzi, "Sui manufatti in focaja rinvenuti all'Inviolatella nella campagna romana e sull'uomo all'epoca della pietra," *Atti della Pontificia Accademia dei Nuovi Lincei* 20 (1867): 41–52 and Angelo Secchi, "Sulla scoperta delle armi e arnesi in silice, trovati presso Monticelli," *Atti della Accademia Pontificia de' Nuovi Lincei* 20 (1867): 68–70; also Augusto Statuti, "Esame di una calcare ad ippuriti che esiste nei dintorni di Terracina," *Atti della Accademia Pontificia de' Nuovi Lincei* 30 (1877): 106–113, here 108.

106 See Pietro Tacchini, "Un disegno del Sole nell'opera di Kircher Mundus Subterraneus," *Memorie della Società degli Spettroscopisti Italiani* 5 (1876): 135–136.

107 See Angelo Secchi, "La teoria delle macchie solari proposta da Galileo," *Bullettino meteorologico dell'Osservatorio del Collegio Romano* 13 (1874), 45–48; 53–55; also published in *Memorie della Società degli Spettroscopisti Italiani* 3 (1874): 81–89; see also APUG, FS 26.VIII.b. For more information on Secchi's opinion on the trial of Galileo, see Sante Pieralisi, *Sopra la nuova edizione del processo originale di Galileo Galilei* (Rome: Tipografia Poliglotta della Propaganda, 1879), 3–4.

108 *Benché non inclinato alla poesia, si dilettò sempre grandemente di Dante che ottimamente conosceva*. APUG, FS 23.IV.

109 *"Oh! Voi altri vi presenterete a Dio con dei manipoli, ma io che mi sono applicato sempre a cose puramente scientifiche e poco pensai alla salute dei prossimi non so che dire al Signore." Allora lo confortai dicendogli: "Ah! Padre cosa dice? E non fu Ella che tenne alta la face della Scienza chiudendo così per sempre la bocca ai malevoli che continuamente gracitano contro l'ignoranza del Clero"* [...]. APUG, FS 23.III.

CHAPTER 5

A Rising International Career: Travels, Achievements, and Recognition

Secchi's rich and varied relationships with the international scientific community are witnessed by a huge number of letters comprising over 1,600 correspondents, around half of whom were not Italians[1] (a more detailed examination of these documents might substantially add to this number), even without mentioning his extensive and thus far unexplored correspondence with other Jesuits. Among the main correspondents, we find not only the names of famous astronomers of his time such as Leverrier from Paris, Karl von Littrow (1811–77) from Vienna, or pioneers of astrophysics such as Gustav Spörer (1822–95) from Potsdam and Rudolf Wolf (1816–93) from Geneva[2] but even meteorologists such as Christoph Hendrik Buys-Ballot (1817–90) from Utrecht, chemists such as Henri Sainte-Claire Deville (1818–81) from Paris, and famous physicists such as Sir Edward Sabine (1788–1883) from Dublin and Joseph Henry from Washington. The numerous treatises published and translated into several languages, the cordial welcome he almost always received during his travels abroad, and the many prestigious international prizes he was awarded prove the esteem in which Secchi was held across the Alps, perhaps even more so than at home.

Secchi was particularly attentive to the latest scientific debates of his time. In the field of astronomy, he took part in the discussions that developed in 1860–70 on the appearance of the solar photosphere, whose protagonists were William R. Dawes (1799–1868) and James Nasmyth (1808–90); and, along with Tacchini, in the years 1872–73 Secchi entered into a controversy with Hervé Faye (1814–1902) on the cyclonic nature of sunspots (see Chapter 6). Beyond the realm of astronomy, he also became interested in issues concerning the origins of the human species, after the theories by Charles Darwin (1809–82),[3] and other philosophical and theological debates (see Chapter 9).

---

1 See Ileana Chinnici and Wiktor Gramatowski, "Le carte di Angelo Secchi conservate presso l'archivio della P. Università Gregoriana: Un inventario inedito rivisitato," *Nuncius* 16, no. 2 (2001): 571–627.

2 Together with astronomical observations and ephemerides calculations, the correspondence with other astronomers was among the explicit duties of a director, requested by pontifical documents (see Buonanno, *Il cielo sopra Roma*, 124).

3 See APUG, FS 3.VII.

 DOI:10.1163/9789004387331_007

A fellow of the most prestigious international academies, Secchi's name can be found in their official publications as well as in the most prestigious international scientific journals of the time, such as *Les mondes*, *Nature*, *Astronomische Nachrichten*, *American Journal of Science and Arts*, and so forth. Secchi's popularity was the result of the efforts he made to gain visibility and prestige for the pope and the Collegio Romano. During his journeys throughout Europe, he used to visit colleagues, instrument-makers, observatories, museums, and those institutions that stimulated his scientific and cultural curiosity. Secchi's international scientific reputation was a result of his prolific production of articles and texts, dealing not only with his own works but also with new discoveries and theories, which he would comment on in the main academic journals of his time. Despite some opposition (see Chapter 9), Secchi enjoyed a high scientific reputation abroad: he was member of many international committees and an undisputed leader whose advice was considered influential. Being a Jesuit and a personal friend of Pius IX, he was aware of his responsibility to represent the Catholic Church in the scientific community: any personal acclaim was acclaim for the church. Secchi's journeys, as well as his participation in scientific expeditions and international committees, contributed to the success of his career and, for this reason, it is worthwhile examining them in greater detail. Fortunately, his diaries are full of comments, annotations, and details that provide an "unofficial" point of view about the men, institutions, and events he encountered.

1 Travels through France and England, 1858

In 1858, as mentioned in Chapter 4, Secchi was instructed to purchase new lighthouses for the Papal States, and to do so he traveled through France and England. In Marseille, he had the opportunity to visit the old observatory, which appeared to him to have been neglected, as well as to meet the aged director, Benjamin Valz (1787–1867):

> I cannot but pity the poor seventy-year-old astronomer who still works and climbs its telescopes, helped only by the custodian, and who cannot receive even a cent from the government. He is also so deaf that he doesn't understand anybody but the custodian, who basically shouts at him with his nose stuck into his right ear. He says his ears can hear the clock strike, but I can hardly believe that.[4]

4 *Non posso che compatire il povero astronomo di 70 anni, che ancora lavora e si arrampica ai suoi cannocchiali aiutato dal solo custode, e che ancor esse non può avere un baiocco dal governo. Inoltre esso è così sordo che non capisce nessuno, altro che il custode che gli grida proprio col*

Secchi spent a few hours with Valz, showing him his photographs of the moon and his drawings of Mars and giving many explanations to the amazed astronomer.[5] He then traveled to Paris, where he was accommodated at Rue Vaugirard, at the Jesuit College of the Assumption, where he could admire the college's chemical–physical laboratory.[6] In Paris, Secchi met with Charles Hermite (1822–1901), whom Secchi considered "Cauchy's successor in fame and goodness"; Hermite accompanied the Jesuit during his visits to various members of the Académie des Sciences.

During a lunch organized in Secchi's honor by mathematician Michel Chasles (1793–1880), Secchi had the opportunity to meet the main officers of the academy: chemist César Despretz (1791–1863), president; geologist Élie de Beaumont (1798–1874), secretary; mathematicians Joseph Bertrand (1822–1900) and Louis Poinsot (1777–1859), deans; and many other reputed scientists, such as astronomers and mathematicians Paul Laugier (1812–72) and Claude-Louis Mathieu (1783–1875), physicist Heinrich Daniel Ruhmkorff (1803–77), physiologist Claude Bernard (1813–78), and naturalist Alfred Moquin-Tandon (1804–63). Secchi remarked that the atmosphere was "as friendly as it would have been among confrères," and he was edified that the guests did not avoid conversations on religious matters. He also visited the workshops of instrument-makers Maurice Sautter (*c.*1820–*c.*1890) and Lepaute and examined new mirrors made by Foucault. Secchi also made sure to visit the Imperial Observatory, reporting the following impressions:

> [The entrance] is a luxurious room: chandeliers, mirrors, and some old instruments, revamped and gilded to make them serve as special ornaments. A large physics laboratory with large Cauchoix telescopes and other [equipment] is [in] the next room. The meridian instruments are poor and quite obsolete; they are big and impressive to look at, though not very accurate. Under the new dome, which is reached through the outdoor terrace, there is the large eleven-inch equatorial, which is really a nice device, but in some ways not so easy to use; but it is beautiful and

---

naso sull'orecchio destro. Ei dice che sente il cronometro attaccato all'orecchio, ma io poco ci credo. Secchi to Marchetti, quoted in Castellani, "Il P. Angelo Secchi S.I. e le sue missioni scientifiche," 610.

5 He also showed him the drawings of his barometrograph, based on the same principle of the balance barometer and recording the instant values of the atmospheric pressure (see Angelo Secchi, "Sur un nouveau baromètre à balance," *Comptes rendus de l'Académie des Sciences* 44 [1857]: 31–34). During his stay in Marseille, he also encountered a procession of Our Lady of Assumption, which edified him very much; see Castellani, "Il P. Angelo Secchi S.I. e le sue missioni scientifiche," 611–612.

6 See Castellani, "Il P. Angelo Secchi S.I. e le sue missioni scientifiche," 615.

FIGURE 5.1 Paris Observatory Council room in a photograph dating back to the last quarter of the nineteenth century. Courtesy of INAF-OAPa.

very solid and performed with great satisfaction. [...] The lens is stained in two or three spots and magnifies just six hundred times; [the instrument] does not move easily, especially for movement in declination, but should be very stable, more than ours. The clock-driving mechanism is well thought out, but up to now it is not known how well it works, because the position of the instrument has not yet been adjusted. The large dome is empty, and they say that there will soon be a fourteen-inch telescope there. The meteorological instruments are much like ours, but rather neglected. The magnetic section is also half-abandoned; they only operate a magnetic photographic declinometer, and that not well. Aside from that, it is a nice establishment; at least twenty people work there, and they spent a fortune on it. If we had a thousandth [of those resources], we would do as much work![7]

---

7 [*L'ingresso dell'Osservatorio*] *è una sala di lusso: lampadari, specchi e alcuni antichi strumenti, rimodernati e dorati per farne ornati speciali. Un gran gabinetto fisico di grandi cannocchiali di Cauchoix ed altri vengono appresso. Gli strumenti meridiani sono ben poca cosa di moderno; però sono grandi e di aspetto imponente, benchè non di esattezza corrispondente. Sotto la nuova cupola, alla quale si va per la terrazza scoperta, è il grande equatoriale di undici pollici, che è veramente una bella macchina, ma in alcune cose non troppo comoda; però è splendida e ben solida ed eseguita con gran lusso.* [...] *la lente è macchiata in due o tre siti e non porta che 600*

On August 28, Secchi attended a meeting of the Academy of Sciences where he met other scientists such as physicists and chemists Jean-Baptiste Biot (1774–1862), Michel Eugène Chevreul (1786–1889), and Henri-Victor Regnault (1810–78). During the meeting, he presented his work on Mars and double stars, as well as his moon photographs, which were admired by all, especially by the marshal Jean-Baptiste Philibert Vaillant (1790–1872). Following the suggestion of the college's rector, Secchi wore a French priest's cassock, though it was "very ugly-looking" in his opinion.[8]

Secchi also had the opportunity to meet Leverrier, director of the Paris Observatory. Leverrier was known as someone with whom it was difficult to get along, but Secchi was glad of his "cordial welcome, as a colleague, without flattery." They talked about the international telegraphic network for the meteorological service that had been established in those years and the participation of the Collegio Romano in the project. Many other scientists paid visits to Secchi during his stay in Paris, and he considered his time at the *Ville lumière* to have been a great success; his only regret was not having been able to see more of the city because of the rainy weather.

Upon his arrival in England, Secchi first visited the Greenwich Observatory, where he met the director George Biddell Airy and astronomer Edwin Dunkin (1821–98). The new thirty-two-centimeter Merz visual refractor, just acquired, seemed "gigantic" to Secchi, and though it was equipped with all the latest features, it was "deprived of any beauty." Secchi probably still had in mind the elegant setting of the Paris Observatory and the contrast appeared to him quite striking. After some tests, he deduced that the objective lens was poor. Secchi also noticed that thermal expansion affected the enormous metallic masses of the telescope, thereby limiting its precision. The Jesuit appreciated the electrical recording system of the transit instrument. He also examined the time ball and visited the equatorial room, which was still incomplete; he wrote to Marchetti that its flat-topped removable roof appeared "very ugly and tasteless."[9]

---

volte: è duro a muoversi, assai in declinazione, ma del resto deve essere assai stabile e più del nostro. Il moto di orologeria è bene inteso, ma finora non si sa se lavora bene, perchè non è rettificata la posizione dello strumento. La gran cupola maggiore è vuota, e dicesi che presto avrà quello di 14 pollici. Gli strumenti meteorologici sono mezzo negletti, e appena hanno quanto noi. La parte magnetica è pure mezza abbandonata, e solo opera il declinometro a fotografia, e malamente. Del resto, tutto assieme, è un bello stabilimento, e vi sono venti persone almeno a lavorare, e vi si spendono tesori. Se noi avessimo il millesimo, faremmo altrettanto di lavori! Secchi to Marchetti, quoted in Castellani "Il P. Angelo Secchi S.I. e le sue missioni scientifiche," 613–614.

8 See Castellani, "Il P. Angelo Secchi S.I. e le sue missioni scientifiche," 614.

9 *Molto brutto e senza gusto*. Secchi to Marchetti, quoted in Castellani, "Il P. Angelo Secchi S.I. e le sue missioni scientifiche," 616.

The secretary of the Royal Astronomical Society, Manners, invited Secchi to visit the Crystal Palace[10]—in Secchi's words, "a spectacular wonder, wider than Saint Peter's basilica and three times as long"[11]—and to pay a visit to Warren De la Rue (1815–89), a pioneer of astrophotography, the "only one who has cultivated lunar photography with spectacular success."[12] At De la Rue's home, he received an excellent welcome from his entire family. The Jesuit was surprised to find a remarkably large house of cards, made of cards of every kind, created by the same astronomer. "The one whom we know only as an amateur astronomer, is a high engineer, a perfect chemist, and a [...] millionaire in Sterling," he wrote with admiration: "Thus you can do astronomy, without bothering anybody, and take photographs with no hassles. [...] Here, they know that you have to spend money [to do that], and therefore science is cultivated by rich people."[13]

Secchi also visited De la Rue's observatory and was amazed by the splendid photographs of the moon, Jupiter, and Saturn that the Englishman had taken. Secchi admired his "magnificent reflector," which had an equatorial mounting, and its excellent clock-drive: "The work is well done, the engineering is the best, and no expense has been spared."[14] Secchi noted that De la Rue used high-quality collodion plates for his photographs, and in comparison, he was quite satisfied with those obtained at the Collegio Romano using far less resources. In turn, De la Rue appreciated Secchi's drawings, especially those of Mars, finding that they fitted very well with his own drawings of 1854.

Secchi, who was an enthusiast of telegraphs, did not miss the chance to visit the workshops of physicist Charles Wheatstone (1802–75). He was excited to see the many kinds of telegraphs and other instruments that Wheatstone had invented and showed to him:

> But this man is so jealous of his things that he only shows the half of them; indeed, from what I hear, it was exceptional, extraordinary, that I

10 The Crystal Palace was originally built in Hyde Park to house the first Universal Exhibition in 1851; it was moved, expanded, and rebuilt in Sydenham Hill in 1854 and destroyed by a fire in 1936 (see Jan R. Piggott, *Palace of the People: The Crystal Palace at Sydenham 1854–1836* [London: Hurst, 2004]).

11 *Spettacolosa meraviglia, più largo di S. Pietro e lungo tre volte tanto.* Secchi to Marchetti, quoted in Castellani, "Il P. Angelo Secchi S.I. e le sue missioni scientifiche," 616.

12 *L'unico che abbia coltivato con successo spettacoloso veramente la fotografia lunare.* Secchi to Marchetti, quoted in Castellani, "Il P. Angelo Secchi S.I. e le sue missioni scientifiche," 616.

13 *Quello che noi conosciamo solo come amatore di astronomia, è un sommo ingegnere, un chimico perfetto e un* [...] *milionario a sterline* [...] *Così si può fare l'astronomo, senza seccare nessuno, e le fotografie senza fastidi.* [...] *qui si sa che bisogna spendere, e perciò la scienza la coltivano i ricchi.* Secchi to Marchetti, quoted in Castellani, "Il P. Angelo Secchi S.I. e le sue missioni scientifiche," 616–617.

14 *L'opera è ben fatta, l'ingegnere sommo, e non si è badato a spese.* Secchi to Marchetti, quoted in Castellani, "Il P. Angelo Secchi S.I. e le sue missioni scientifiche," 617.

have been allowed to see so many things at all, as he is very distrustful, because he is afraid that someone may steal his inventions.[15]

In London, Secchi also had the opportunity to admire the *Great Eastern*, an iron steamship originally christened with the name *Leviathan*, the largest ship ever built at that time. He called it both a "marvel [*spettacolo*]" and a "colossal flop [*fiasco colossale*]" because of the bankruptcy of its sponsor and the failure of repeated launches.[16]

Secchi also visited the observatory of Stonyhurst College, which had been his old refuge, and the observatories of Dublin[17] and Oxford; he described the observatory in Oxford as "a magnificent establishment, with a superb equatorial and an absolutely unique series of photo-recording meteorological instruments." During his visit to the observatory, he was also able to examine the Robinson anemometer[18] and other devices, pleased by the interest that his barometrograph (see Chapter 6) had engendered.

2 Spain, Total Solar Eclipse of 1860

Secchi's first and most important scientific expedition was the one he made to Spain to observe the total solar eclipse of July 18, 1860. He was eager to photograph the totality of an eclipse to solve some important questions about the nature of solar prominences, and so he obtained a special grant of two hundred ecus from the pope to bring the Cauchoix telescope and other instruments to Spain in order to install a photographic station in the path of the totality. The director of the Madrid Observatory, Antonio Aguilar y Vela (1820–82), who knew about Secchi's intentions, invited him to join the small Spanish expedition. They moved from Madrid to Valencia, where they were welcomed by Antonio

15 *Ma questo benedett'uomo è così geloso delle sue cose, che le mostra solo per metà; anzi, a quel che sento, è stata un'eccezione affatto straordinaria che io abbia potuto vedere tante cose, delle quali è gelosissimo, perché ha paura che gli siano rubate le invenzioni.* Secchi to Marchetti, quoted in Castellani, "Il P. Angelo Secchi S.I. e le sue missioni scientifiche," 618.

16 Three unsuccessful attempts were made to launch the ship in November 1857, which was later launched sideways in January 1858 and had its first sea trial nine months later, in September. On the history of the premature enterprise of the *Great Eastern* and its engineer, Isambard Kingdom Brunel (1806–59), see, e.g., David Hall, *Fred Dibnah's Victorian Heroes* (London: Bantam Press, 2011), 139–172. For a detailed description of the ship, made by the American naval architect William Henry Webb (1816–99), see William Henry Webb, *Descriptive Particulars of the* Great Eastern *Steamship* (London: Marshall & Sons, 1854).

17 See Castellani, "Il P. Angelo Secchi S.I. e le sue missioni scientifiche," 619.

18 In 1846, Thomas R. Robinson (1792–1882) designed an anemometer with four hemispherical cups and mechanical wheels; the same kind of anemometer was later adopted by Secchi for his meteorograph (see Chapter 6).

Rodríguez de Cepeda Garrido (1814–96), "a wealthy and capable lawyer and a very passionate amateur astronomer." As usual, Secchi toured the town, and Cepeda Garrido accompanied him to visit the "rich" botanical garden. On July 3, together with José Monserrat (1814–81), professor of chemistry at the local university, they left Valencia to reach the chosen site, the Carmelite monastery called Desierto de Las Palmas,[19] near Castellón de la Plana. After a stop in this town, where "the inn was awful and it was impossible to sleep,"[20] they reached the convent by mule and visited the place where the astronomical station was originally to have been installed. Secchi immediately recognized the inappropriateness of the site, which had been chosen without having inspected the area ahead of time: "The convent is in a ravine [...] and here it is useless to think to do anything, because the mountains cover the stars."[21]

Near the convent were two small hermitages dedicated to San Juan (St. John the Baptist) and the Virgen del Carmen (Our Lady of Mount Carmel). The Carmel hermitage, though not ideal, was adapted for the transit instrument and the photographic apparatus. They then looked for a more suitable location to make the astronomical observations on a nearby crest, but it took an hour and a half to reach the top, where they found only a very small chapel, dedicated to San Miguel, without any space outside to place the instruments: "The view is superb, but it is really hard to make observations, and there is no shelter for the instruments."[22] Aguilar y Vela was disconcerted by the situation, and Secchi had to dissuade him from returning to Castellón:

> After a bit of deliberation, it was decided that the hermitage station, downward, would be used for photography and for time measurements, while we would be up [the hill] with the [telescopes] for the eclipse observations. Certainly, it was beastly not to have visited the place before, because at least staying downhill we could have been more comfortable, saved so much effort, and had a little open view to the west, while now both the convent and the hermitage are surrounded, especially to the west by high mountains. There certainly is a magnificent view and the approach of the eclipse should be beautiful, but it is very difficult to climb and the site is restricted. These were the considerations that we had

19 At the beginning of the nineteenth century, the same site had been chosen as a station for the triangulation operations regarding to the extension of the French meridian. See Pierre Bayart, "La méridienne de France et sa prolongation jusqu'aux Baléares," *Revue XYZ* 116 (2008): 58–62.

20 *La locanda era pessima e fu impossibile dormire.* AOAR, Fondo Secchi I.D.3, III.36.

21 *Il convento è in un burrone* [...] *e qui è inutile pensare a far nulla, perchè le montagne coprono le stelle.* AOAR, Fondo Secchi I.D.3, III.36.

22 *La vista è superba ma è di caro prezzo l'osservarvi e non vi è sito a dimora.* AOAR, Fondo Secchi I.D.3, III.36.

> made, when the father prior[23] arrived by chance and was very shocked that we wanted to desecrate the Church of St. Michael by placing instruments and staying there. Hence, he told us that we were not allowed to go there. This was a serious affair [...].[24]

Secchi, however, was not lacking in the necessary tenacity and diplomatic ability to solve this unexpected problem. He consulted the father provincial and other priests, and then he decided to try again to obtain the prior's permission:

> The morning after, at breakfast, we [...] launched our attack at father prior, who at first refused [to give permission], but we politely made him understand that he was doing a bad service at the convent with this scruple, that we were not profaning people, and that furthermore if necessary we could get permission from the bishop or the nuncio and even [...] from the government. Without giving us his final answer, he retired to consult, then he concluded that, because of the circumstances of the pope [i.e., his sponsorship of the expedition], etc., we were permitted. Then we began to unpack the boxes and everything was in order. We went to the Carmel hermitage and put the pillar for the equatorial into place and prepared everything so that it was ready to be installed the day after [...].[25]

---

23 From 1852 to 1860, the prior of the Desierto de Las Palmas was Fr. Vicente del Carmelo, O.C.D.; he was replaced from 1860 to 1863 by Fr. Francisco de Jesús, O.C.D. (see Ignacio Husillos Tamarit, "El Desierto de Las Palmas, centro de espiritualidad en la diócesis de Tortosa," in *Teresa y Enrique: Cara a cara; Actas del Congreso internacional teresiano realizado en Tortosa del 16 al 18 de Enero de 2015*, ed. Gemma Bel and Pilar Liso [Ávila: CITeS. Universidad de la Mística, 2015], 35–134, here 63, 66).

24 *Dopo un poco di deliberazione fu deciso che si farebbe la stazione a basso per la fotografia e pel tempo, e che per l'ecclisse si starebbe lassù* [*coi due telescopi*]. *Certo è stata una bestialità di non visitare il luogo prima, perché almeno stando a basso si poteva avere più commodità risparmiare tanta fatica e avere un poco di pianura all'occidente, mentre ora al convento e all'eremo tutto è coperto è proprio a ponente da altissime montagne. Lassù certo è una vista magnifica e l'accostarsi dell'ecclisse essere deve bellissimo, ma è assai scabroso salirvi e ristretto il sito. Questi erano i conti che facevamo, quando giunti a caso il P. Priore si mostrò molto scandalizzato che noi volessimo profanare la chiesa di S. Michele con tenervi strumenti e più collo starvi noi. Quindi ci disse che non si poteva andare laggiù. Questo era un affar serio* [...] AOAR, Fondo Secchi I.D.3, III.36.

25 *La mattina a colazione si diede l'assalto al p. Priore definitivamente il quale dapprima ricusò, ma gli si fece capire con buona grazia il mal servizio che faceva al convento con questo scrupolo, che non eravamo noi persone da profanare, e che se occorreva avremmo ottenuto ciò dal vescovo o dal Nunzio, o anche* [...] *dal governo. Senza che esso ci desse per ciò risposta definitiva si ritirò a consultare, e concluse che attese le circostanze del Papa ecc. si permetteva. Allora si cominciò a disimballare le casse e tutto si trovò bene. Si andò al Romitorio del Carmine e si fondò il pilastro dell'equatoriale e si mise tutto in pronto per montarlo il dì appresso.* AOAR, Fondo Secchi I.D.3, III.36.

FIGURE 5.2 The astronomical expedition at Desierto de Las Palmas in 1860: on the left, the old gatehouse of the convent; in the background, the Carmel hermitage. Courtesy of INAF-OAR.

The expedition members were therefore divided between the photographic station, located at Carmel hermitage, and the astronomical station at San Miguel chapel.[26]

At Carmel, the expedition members installed the Cauchoix telescope (whose hour circle had become broken along the way) with a darkroom and all the necessary materials for photographic operations. The various tasks were handed out: Monserrat and some of his students were to prepare the photographic operations; Fr. Juan Vinader S.J., professor of physics at Salamanca Seminary, was to guide the telescope; Prof. Dionisio Barreda, professor of

26 See Angelo Secchi, "Sull'ecclisse solare totale osservato in Spagna nel 18 luglio 1860," *Memorie dell'Osservatorio del Collegio Romano* n.s. 2 (1863): 33–52, here 36. A photograph of all instruments and members of the expedition was taken on July 15, at Carmel hermitage (see Fig. 5.1).The expedition was then divided into two stations (see Angelo Secchi, "Sull'ecclisse solare totale osservato in Spagna nel 18 luglio 1860," *Memorie dell'Osservatorio del Collegio Romano 1860–62* [1863]: 2:33–52, here 36).

physics at Salamanca University, was placed in charge of spectroscopic and polariscopic observations; and various amateur astronomers were tasked with producing drawings of the corona and records of visible stars.

At the San Miguel station, a Steinheil telescope from the Madrid Observatory and other minor instruments were installed, together with meteorological equipment (thermometers, psychrometer, anemometer), a Melloni thermopile, and a Jones declinometer. The division of work there was as follows:
- Aguilar y Vela: visual observations with the Steinheil telescope;
- Cepeda Garrido: visual observations with a Lerebours telescope, equipped with a Ramsden eyepiece from the Collegio Romano Observatory;
- Secchi: visual observations with a Fraunhofer telescope;
- other assistants: thermopile measurements and declinometer observations.

Secchi decided to use a Morse telegraph as a chronographer, but it did not work reliably and was abandoned.

Secchi made some notes about the uncomfortable accommodation:

> At night, we sleep at San Juan hermitage. I'm really in the wilderness. My hermitage is a small square house consisting of four rooms, hallway, kitchen, bedroom, and chapel, each of three meters. The kitchen is uninhabitable because of the stench, all the other [rooms] have such crumbling walls that from my bed I can feel the wind blow, which fortunately is very hot, from south, so that at 6:00 in the morning [the thermometer] reached 33° and tonight, at midnight, it stands at 30 [degrees].[27]

Secchi decided to test the photographic apparatus during the night, when the temperature was lower. He developed the photographs with a bath of ferrous sulfate, which acted quickly, and then intensified them with mercuric chloride. He then made some attempts during the day, and took instantaneous photographs of the Sun with a dark glass filter.[28] The week before the expected eclipse, all the equipment was prepared and tested in order for it to be ready for use. The prolonged cloudy weather provoked much anxiety for the party, as

27 *La sera si dorme all'eremo di S. Juan. Sicché sono veramente nel deserto. Il mio eremo è una casuccia quadra composta di 4 ambienti, vestibolo, cucina, camera da letto e cappella di tre metri l'una. La cucina è inabitabile pel fetore, le altre sono tutte a mura fracassate in guisa che sento a letto il soffio del vento, che fortunatamente è caldissimo di sud e tale che la mattina a 6h segnava 33° e questa sera a mezzanotte segna 30.* AOAR, Fondo Secchi I.D.3, III.36.

28 In his diary, Secchi also recorded some colorful details, such as the unexpected visit of a madman who rambled in the mountains and threatened them, believing that they were hostile, and how they were able to reassure him and persuade him to go away and not disturb them anymore; see AOAR, Fondo Secchi I.D.3, III.36.

the success of the observations strictly depended on the quality of the sky. The tests that were carried out the day before the eclipse were not encouraging:

> The crest is located at the top of a chain that separates two large valleys, and the wind, coming from the sea and bringing a lot of moisture, condenses over its peak, and forms a permanent cloud that is constantly renewing itself and thus persists despite the blowing wind. Only at midday, when the temperature of the whole mass of air is uniform, did these clouds vanish, and at about one in the afternoon, the weather became beautiful. We then believed that the next day the weather would have been finer, seeing the horizon very clear [...].[29]

To prevent possible failure, Secchi and his companions planned to bring the instruments to the lower plain station the next morning in case the cloud appeared again. On the eclipse day, they had a bad surprise:

> In a quite cruel twist of fate, that morning the wind was light from the northeast, and there was fog on the plain, while the air was clear on top, and it was so up to 9 o'clock in the morning, when the wind began to play the same game as the previous day [...]; now the clouds were a little higher, and therefore their disappearance would be all the later. Our position was all the more critical because, with great regret, we saw the entire plain below becoming completely free from every trace of clouds, with just a little low fog that was fading.
>
> Meanwhile, it was 11:30 and we were in this desolate situation, with a cloud fixed and nailed over our heads, so as to protect us from the solar rays; a service that would have been wonderful in any other circumstance, except the current one. We anxiously watched the barometer and the anemometer, to see if that would rise and the clearing would accelerate, but everything was still, and everything was going on with a very special monotony. I think that, if the barometer could have understood, it would have had compassion on us; and no cloud was ever investigated more carefully [than the one over their heads].[30]

---

29 *Il monte è collocato al vertice di una catena di montagna che separa due ampie vallate, e il vento proveniente dal mare portando umidità in copia, questa si condensa permanentemente nella sua cime, e forma una nube permanente che si rinnova continuamente e persiste così malgrado il soffio del vento. Solo al mezzodì quando la temperatura di tutta la massa d'aria è uniforme, queste nubi svaniscono, e infatti all'una circa pomeridiana il tempo divenne bellissimo. Si credeva quindi che il giorno appresso avremmo avuto tempo migliore vedendo l'orizzonte serenissimo* [...] AOAR, Fondo Secchi I.D.3, III.36.

30 *Ma per fare più crudele la illusione la mattina il vento era N.E. debole, e la nebbia era al piano, e chiaro in alto, e fu così fino alle 9 della mattina quando il vento cominciò a fare lo*

In order to resolve this problem, Secchi decided to go down onto the plain, where he hoped to have enough time to observe at least the totality, but he learned that it would have taken over three hours to get down—too late to make any useful observations. The Jesuit therefore resigned himself to the situation, comforted by his friend Aguilar y Vela, with alternating feelings of hope and desolation:

> Our position was certainly critical, our faces so clearly showed this frustration that people [around them] noticed our anxiety, and someone was beginning to fear that we were so afraid that the eclipse would not occur! Anxiety grew over time and following the thermometer, and it was at its full, half an hour before the start of the eclipse, when we began to see that the large cloud that covered the Sun was thinning, and we hoped for the best; but during the time of the expected contact, while counting seconds with the pendulum, a thin cirrus cloud, the last, passed in front of the telescope!
>
> Such a persecution by clouds should be a lesson for posterity, to never choose mountains for this kind of observation, and to prefer the plains. The mental anxiety is too cruel in those moments and prevents one from preparing themselves calmly for the work and the necessary alignment of the instruments [which, fortunately, had been done the day before]. However, as to compensate us for all that anguish, the sky was beautiful throughout the course of the phenomenon, and almost as a warning of what might have happened, a thunderstorm, previously far away, approached us across the plain so close that during totality, flashes were seen and thunder was heard. We were really close to losing the fruit of so much labor.[31]

---

stesso giuoco del giorno innanzi [...] *le nubi erano un poco più alte, col che esser dovea più tarda la loro disparizione. La posizione nostra era tanto più critica in quanto che con grande rammarico avevamo sott'occhio tutta la pianura libera affatto da ogni vestigio di nubi e con solo un poco di nebbia bassa che si andava dileguando. Erano giunte le 11 ore e 1/2 e si stava in questo stato desolante con una nube fissa e inchiodata sul nostro capo come per proteggerci dai raggi solari; servizio che sarebbe stato ottimo in qualunque altra occasione tranne l'attuale. Guardavamo ansiosamente il barometro e il mulinello del vento per vedere se quello si alzava e questo accelerava, ma tutto era nulla, e tutto andava con una specialissima monotonia Credo che se il barometro avesse potuto capire avrebbe avuto compassione di noi e che non mai alcuna nube fu studiata con maggior attenzione.* AOAR, Fondo Secchi I.D.3, III.36.

31 *Certo la nostra posizione era critica, i nostri visi erano atteggiati a tale corruccio che ben si faceva scorgere la nostra ansietà ai circostanti, e qualcuno cominciava a temere che noi stessimo così per paura che l'ecclisse non avesse più da accadere! L'ansietà cresceva col tempo e col termometro, ed era al suo colmo 1/2 ora avanti il principio, quando si cominciò a scorgere un assottigliamento della gran nube che ci riparava il sole e si concepì speranza di meglio, ma pure durante il momento che si aspettava il contatto, contando i secondi al pendolo, un leggier cirro che fu l'ultimo passò al cannocchiale!*

Secchi also evoked the context of the eclipse observations, including the participation of the villagers, the contrast with the astronomers' feelings, and their reactions:

> Throughout the morning until the end of totality, the mountain had a truly picturesque aspect for everyone except us. All villages around it emptied out and [...] clusters of women, men, kids of all kinds, crammed around the instruments, while others, scattered here and there [...], were having great snacks, carving large slices of roast beef, drinking and singing, so that we who were watching the cloud fixed overhead really envied their good spirits. Breakfast was not the most delicious for us, on that day! The governor had sent police, but they had trouble keeping away the multitude, who were no fewer than about three hundred or four hundred people, and [finally managed] to keep them all at a distance of twenty paces away from us. A few minutes before the beginning [of the eclipse], the crowd was ordered to be quiet and, with an admirable compliance for such a garrulous multitude, all were silent, until our activity showed them that the eclipse had already begun, and [then] the conversations resumed. However, that did not last long because, at the approach of an immersing sunspot, silence was again ordered and obtained with equal compliance. They were now persuaded, by the appearance of our faces, that our anxiety had passed [...].[32]

Sì fatta persecuzione di nubi sia di lezione ai posteri a non sceglier mai le montagne per tali osservazioni, e attenersi alle pianure. E' troppo crudele l'ansietà dell'animo in que' momenti che impedisce di prepararsi con pace agli studi e alle necessarie rettificazioni degli strumenti, le quali se non fossero state fatte il giorno avanti, quel dì non si sarebbero potute fare affatto, con grave danno della precisione nelle osservazioni, e questa fu veramente l'unica ragione che mi fece tardar tanto a risolurmi a partire che non ci fu più tempo. Tuttavia come per indennizarci [sic] *di tanta angustia, il cielo fu bellissimo in tutto il corso del fenomeno, e quasi per farci avvertiti ancora del rischio che avevamo corso un temporale allora lontano si venne tanto approssimando correndo per la pianura che durando la totalità fu visto lampeggiare e sentirsi il tuono. Sì vicini fummo a perdere il frutto di tante fatiche.* AOAR, Fondo Secchi I.D.3, III.36.

32 *Durante tutta la mattinata fino al fine della totalità il monte presentava un aspetto veramente pittoresco a tutti fuorché a noi. Tutti i villaggi di intorno si erano votati, e colà aveano inviati i loro abitanti in abiti di mezza festa almeno; crocchi di donne, di uomini di ragazzi di ogni specie si stipavano attorno agli strumenti mentre altri sparsi qua e colà* [...] *facevano le buone colazioni trinciando buoni pezzi di manzo arrosto, bevendo e cantando che per noi che vedevamo la nube colà inchiodata faceva veramente invidia della loro bonarietà. Per noi la colazione non fu certo la più gustosa quel dì! Il governatore avea mandati i suoi carabinieri, ma questi avean molta pena a tener dietro la moltitudine che non era meno di circa 3, o 400 persone,* [...] *si impegnarono a far ritirare tutti a una distanza di 20 passi circa da noi; formatasi così la corona pochi minuti avanti del principio fu intimato silenzio alle turbe, e con una mirabile docilità in sì garrula moltitudine tutti si tacquero, finché il moto nostro li*

The most important result of the expedition was the photographic record of various phases of the eclipse: fourteen photographs were taken during the partial phases and five during the totality, one being discarded because of its poor quality. The four photographs showing the totality were taken with exposures varying from six to thirty seconds and reproduced the solar corona and the prominences[33] so that it was possible to determine their number and shape, with the details being collected in a drawing. A detailed report about the scientific expedition was sent by Secchi to the Académie des Sciences[34] and later published in the *Memorie* in 1863,[35] after which the comparison of his photographs with those by De la Rue revealed that prominences were features that really belonged to the Sun. This expedition was therefore crucial in settling one of the main debates about the Sun (see Chapter 6). This important result boosted Secchi's international reputation, but soon other major opportunities would give rise to even greater renown.

3 Paris, Exposition Universelle of 1867

One of Secchi's most important international achievements came as a result of his participation in the Exposition Universelle (Universal Exhibition), held in Paris in 1867. These gigantic extravaganzas were a showcase of the triumphs of the industrial revolution that were designed to illustrate the prestige and economic power of the participating nations. Therefore, the decision to display Secchi's spectacular meteorograph, an instrument that recorded daily meteorological data (see Chapter 6), to so vast a public had a political purpose as well as a scientific one: it was "a spectacular and prestigious symbol of the scientific and technological activity of the Papal States whose future, as temporal

avvisò che l'ecclisse era già cominciato e la conversazione ebbe il suo seguito. Essa però non durò molto perché all'accostarsi della immersione di una macchia fu ri intimato il silenzio e ottenuto con pari docilità. Si erano ormai persuasi dall'aspetto delle nostre facce che l'ansietà era scomparsa [...] AOAR, Fondo Secchi I.D.3, III.36.

33 Secchi was not completely happy with these photographs because the corona was only slightly visible on the collodion plates due to an error made by the photographer, who did not follow Secchi's suggestions and used a landscape camera lens instead of a portrait camera lens.

34 See Angelo Secchi, "Observations faites pendant l'éclipse totale du 18 juillet 1860 sur le sommet du mont Saint-Michel au Desierto de Las Palmas en Espagne," *Comptes rendus de l'Académie des Sciences* 51 (1860): 156–162; 276–279; 386–388. The Swiss astronomer Emile Plantamour (1815–82) questioned Secchi's observations of the prominences (see Angelo Secchi, "Eclipse solaire du 18 juillet 1860," *Comptes rendus de l'Académie des Sciences* 51 [1860]: 749–751).

35 See Angelo Secchi, "Sull'ecclisse solare totale osservato in Spagna nel 18 luglio 1860," *Memorie dell'Osservatorio del Collegio Romano 1860–62* (1863): 2:33–52.

entities, appeared uncertain."[36] Pius IX, therefore, supported the construction of a new meteorograph to take part in the competition.

On March 1, 1867—precisely one month before the opening of the exhibition—Secchi arrived in a freezing and snowy Paris to begin the task of assembling the complex instrument. He met with many difficulties and delays, starting with the fact that the cases containing the disassembled instrument sent from Rome had been lost during the journey: fortunately, thanks to his tenacity and perseverance, Secchi was able to retrieve them. At the same time, while working on the installation of the meteorograph, he also participated in the intellectual and academic life of the *Ville lumière*. Secchi attended the sessions of the Institut de France, where he received an "excellent welcome."[37] At the chemistry session, geologist Gabriel-Auguste Daubrée (1814–96), who was the chairman, invited Secchi to sit beside him "much to the wonder of all." "The session is brilliant"—Secchi commented in his diary—"but has little new of interest, and only Henri St. Claire Deville [French chemist] showed beautiful examples of the crystallization of material insoluble in water, like calcium carbonate, obtained by alternately heating and cooling them under pressure."[38] Secchi also attended the formal public session of the institute, where a "distinct" place was assigned to him, along with other members, in the amphitheater: "Audience very numerous and with many ladies,"[39] he noted. He also visited the Bureau des Longitudes; the Gare du Nord,[40] which had been renewed and reopened a few years earlier; and the Conservatoire Impérial des Arts et Métiers, with "really splendid things," especially in the electricity section. There, he could examine the samples of weights and measures and was "welcomed with every good grace and true sympathy" by the director, physicist Arthur Morin (1795–1880), who also gave him useful suggestions on the installation of the meteorograph.

In Paris, Secchi had the opportunity to visit many scientists:[41] instrument-maker Heinrich D. Ruhmkorff, known for having improved and commercialized

36 *Esso avrebbe rappresentato a Parigi un simbolo prestigioso e spettacolare dell'attività scientifica e tecnologica dello Stato pontificio il cui futuro, come entità temporale, appariva incerto.* Paolo Brenni, "Il meteorografo di Padre Angelo Secchi," *Nuncius: Annali di storia della scienza* 1 (1993): 197–247, here 210.

37 See APUG, FS 23.II.A.

38 *La seduta è brillante ma ha poco interesse di novità, e solo Henry St.e Claire Deville mostra dei bei pezzi di cristallizzazione ottenuti sotto pressione per alternative di calore e di raffreddamento, nelle materie insolubili nell'acqua come il carbonato di calce.* APUG, FS 23.II.A.

39 *Uditorio assai numeroso e di non poche signore.* APUG, FS 23.II.A.

40 He was accompanied by engineer Angelo Vescovali, who was involved in the introduction of the electric railways in the Papal States (see Chapter 4).

41 Secchi also reports to have received visits from Elie de Beaumont, Parisian instrument-maker Jean-Baptiste-François Soleil (1798–1878), and other personalities.

the induction coil; Hervé Mangon (1821–88), promoter of a "brilliant" meteorological association; chemist Jean-Baptiste Dumas (1800–84), who presented him with a copy of Antoine-Laurent de Lavoisier's (1743–94) works; Adolphe Ganot (1804–87), "old and lame," author of most popular physics textbooks; Foucault, who showed him the device used for determining the speed of light in 1862;[42] and Hippolyte Fizeau (1819–96), who first proposed the idea of using spectral line shifts for determining the proper motions of stars.[43] Secchi was also able to examine some inventions, such as the *dyalisseum* (an instrument for hydrogen absorption), built by instrument-maker Jules Salleron (1829–97), and the electrostatic generator made by Wilhelm T. Holtz (1836–1913).[44]

Secchi was also invited to some meetings in Catholic circles, but in general he was not much impressed with what he saw.[45] In the house of the nuncio, Monsignor Flavio Chigi (1810–85), he attended some public experiments that were carried out by Abbé François Moigno (1804–84),[46] a famous popularizer and one of the more notable figures on the French scientific scene at that time. Secchi's judgment, however, was quite severe: "Abbé Moigno, trying to be impressive, did some very mediocre spectrometry experiments with electric light."[47]

After much difficulty, the meteorograph was finally installed at the Champ-de-Mars, where the Exhibition Palace had been built. The exhibition was a stunning success, with more than fifty thousand exhibitors, around fifteen million visitors in total, and about one hundred thousand people attending its grand opening. Guns, elevators, and lighthouses were among the most impressive of the technical wonders shown to the public. The meteorograph was classified among the *Instruments de précision et matériel de l'enseignement des sciences* (Precision instruments and materials for scientific education) and placed in the front of the hall hosting the Papal States' art objects. Secchi's recording

42 A description and a sketch of Foucault's device is reported in Secchi's diary; see APUG, FS 23.II.A.

43 See Hyppolyte Fizeau, "Acoustique et optique," *Société Philomathique de Paris, Année 1848* (Paris: Imprimerie de Cosson, 1848), 81–83.

44 The instrument is described in Secchi's diary; see APUG, FS 23.II.A.

45 In his diary, Secchi mentions the soirée of the Holy Land Pilgrims at the house of Baron Joseph Crépin du Havelt (1808–74), the general commissioner of the Papal States at the Universal Exhibition: "A very inconclusive event. I saw Card. [Cardinal] Bonnechose [Henri de Bonnechose (1800–83)] who spoke a lot without saying anything; he made an endless list of holy places and gave the blessing" (*Cosa assai inconcludente. Vidi il card[inale] Bonnechose che parlò molto senza dir nulla; fece una infinita lista di luoghi santi e diede la benedizione*) APUG, FS 23.II.A.

46 Moigno was a mathematician, editor of the journals *Cosmos* and *Les mondes* (later fused together), and a staunch supporter of the idea that there was no conflict between science and religion.

47 *L'Abbé Moigno fece degli esperimenti di luce elettrica assai mediocri sulla spettrometria e voleva fare mirabilia* [...] APUG, FS 23.II.A.

instrument was among the exhibition's main attractions, with the ring of the bell announcing the movement of the mobile arm of the psychrometer arousing the curiosity of the crowd. Sometimes, Secchi would visit the exhibit incognito, mixing with the people to listen to their comments, and he was happy to spend many hours a day answering the visitors' questions. The instrument, which Abbé Moigno described as "the scientific pearl of the Universal Exhibition,"[48] aroused admiration despite it also having some critics (see Chapter 9), and Secchi began to entertain the idea of obtaining the Grand Prix.

The jury members came to visit the instrument on April 13. In his diary, Secchi mentions De la Rue, French physicist–mathematician Jules Antoine Lissajous (1822–80), the Swiss physicist Auguste De la Rive (1801–73), the German engineer Moritz Hermann von Jacobi (1801–74), and Foucault: "All friends, and very well disposed [...]. They came as judges, remained as friends, left as pupils, because they have discovered new things. We'll see; they used extraordinarily sympathetic and gratifying expressions. We'll see."[49]

The meteorograph worked perfectly during the eight months of the exhibition. It was admired by many famous authorities and personalities in Secchi's presence: the duke of Leuchtenberg, Nicolas Maximilianovitch (1843–91); Prince Romanowsky, "intelligent and highly interested," who spent over two hours examining the device; Emperor Napoleon III and his wife; many other scientists, such as the German inventor Werner von Siemens (1816–92), chemist Pierre Hippolyte Boutigny (1798–1884), physicists Alexandre-Edmond Becquerel (1820–91), Charles Wheatstone, and John Tyndall (1820–93). Secchi recorded that Tyndall was "very happy" upon examining the instrument: "He is frenetic, gaunt, a man who is exhausted by work!"[50]

Secchi also found time to visit some of the other exhibits. He was especially impressed by the zincography printing process and a new technique for the conservation of animal tissues.[51] Secchi also visited the main cultural and scientific attractions of Paris; among them, the big balloon manufactured by Nadar, pseudonym of renowned photographer Gaspard-Félix Tournachon (1820–1910), who struck Secchi as being rather strange: a "desperate [guy],

48 François-Napoléon Moigno, "Météorograph du R.P. Secchi (S.J.)," *Les mondes* 14 (1867): 61–66.

49 *Tutti amici e molto ben disposti* [...]. *Sono venuti da Giudici, sono restati da amici, partiti da scolari, perché trovavano cose nuove. Vedremo, han usato espressioni straordinariamente benevole e lusinghiere. Vedremo.* APUG, FS 23.II.A.

50 *È sempre convulso, magro e si vede un uomo che si ammazza per lavorare!* APUG, FS 23.II.A.

51 The technique was developed by Lodovico Brunetti (1813–99), anatomopathologist at Padua University, who obtained the Gold Medal at the Exposition Universelle for his invention.

trying to kill himself honorably. A man without faith, I had some strange conversations with him. He doesn't know when [his balloon] will fly, but there's no escaping that with science."[52]

While waiting for the verdict of the jury, Secchi also visited the famous Cathedral of Notre Dame de Paris ("a beautiful church, but the middle nave is no larger than the Collegio Romano"), the luxurious residence of Versailles (with "superb and wonderful things [...] one understands that people, grown accustomed to such pleasures, neither would nor could attend to business, and the monarchy was doomed to fall");[53] Sainte-Geneviève Library; the Bois de Boulogne Park; and the Cluny Museum, where he could not but notice that

> it is an old convent, full of bits and pieces stolen from churches and monasteries. These relics show that the only protectors of the arts were the monasteries, and this is the gratitude they were shown by the revolution. Most of the stone saints [i.e., statues] have broken heads! [...] Soon the same will happen in Italy![54]

On April 22, Secchi received a note from Beaumont with the expression *Lux facta est*: that was the indication that the jury had awarded the meteorograph the Grand Prix—six thousand francs, and a Gold Medal worth about one thousand francs.[55]

Notwithstanding his personal success, Secchi was not at all pleased by the exhibits representing the Papal States. He cared very much about how the papacy presented itself and did not approve of the scrambled heap of items, selected and displayed with no criteria. Hence it is not surprising that Secchi decided to send a letter to the Ministry of Trade explaining the situation and remarking on the risk that the Papal States were exposed to ridicule and vitriolic criticism. He asked that the number of items be doubled and judged inappropriate the small number of pieces related to the metal industry, clothing, or raw materials:

---

52 *Un disperato, [che] cerca di ammazzarsi onoratamente. Uomo senza fede, mi tenne certi discorsi strani. Non sa quando [il suo pallone] volerà, ma per la scienza non ci si può cavar nulla.* APUG, FS 23.II.A.

53 *Cose superbe* e *mirabili* [...] *Ma si capisce che gente abituata a queste delizie non doveva né poteva badare agli affari e la monarchia doveva cadere.* APUG, FS 23.II.A.

54 *È un vecchio convento pieno di avanzi rubati alle chiese e ai monasteri. Questi cimeli mostrano come i soli asili delle arti siano stati i monasteri e come per gratitudine sono stati trattati dalla rivoluzione. La maggior parte di santi di pietra ha la testa rotta!* [...] *Presto si farà altrettanto in Italia!* APUG, FS 23.II.A.

55 Secchi, as well as other winners, had some difficulties in receiving the money for the prize, because of the financial problems of the exposition; it seems that he obtained it later, but it is unclear when.

FIGURE 5.3 Medal of the Grand Prix of the Exposition Universelle of 1867. Courtesy of INAF-OAR.

> Seven meters of space are occupied by a lion in marmoridea,[56] which should be occupied by agricultural products. That's the result of the overblown idea to send classical things: those who walk by it, laugh at viewing a plaster Egyptian lion in place of agricultural products. [...] These objects were approved by Mr. Schnetz,[57] jury member for the Papal States [...]. I will frankly say that a less appropriate person could not have been chosen. He is deaf, and dotard, and fanatical for his French people [...]. Moreover, it makes it seem as if Rome had no artist of note of its own, because this role is filled by a foreigner. I suppose that this is due to economic reasons, but I wonder how much we have benefited in this case. [...] [About paintings], to tell the truth, [there is] nothing of value, including the last one that was delivered, which, furthermore, was uncommonly indecent [...]. In a papal exhibition, this is an aspect that ought to be considered, and a Madonna should not have to stand beside or in front of a Venus. [...] I have much suffered to see how certain things pass.[58]

56 An artificial form of marble, made of a mixture of alum (or borax), water, and gypsum.

57 Jean-Victor Schnetz (1787–1870) was a French painter, and his academic training may have led him to choose only artistic items.

58 *Sette metri di spazio sono occupati da un leone in marmoridea, mentre doveano essere occupati dai prodotti agricoli. Ecco dove conduce l'idea esagerata di mandare le cose classiche: chi passa ride al vedere un leone egiziano in gesso al luogo delle produzioni di agricoltura, e ho sentito io stesso taluno a dire = Ecco gli italiani che si divertono a fare le figurine invece*

Secchi was also upset by the lack of collaboration from some of the church's personalities:

> It makes me sick to hear a card[inal] apologizing for not sending stuff that would give honor [to] the [Papal] States because he was broke. […] That stuff was his, God knows how, maybe as a gift, but the god of money is what rules him and the rest of these gentlemen: they don't care squat about the interest of the church.[59]

Overall, the exhibits did not paint a positive picture of the Papal States, and even the judgment from the Catholic side was quite severe. In an article[60] illustrating the Papal States' section at the Exposition Universelle, the French writer Prosper Poitevin (1810–84) began with the following words: "Praise Fr. Secchi! Thrice blessed be his meteorograph! Thanks to the learned Jesuit and his work, the honor of the Papal States is saved […]":[61]

> Without Fr. Secchi and his wonderful instrument, what a terrible rout—my God!—would have befallen the army of industry or, rather, the band of irregulars that Rome has dared to put forward at the Champ de Mars! Allowing itself to show its weakness and its helplessness in matters of industry, perhaps the Roman government was showing great Christian

di coltivare i loro campi e i loro territori; ecco come il governo de' preti incoragisca la sola arte che dà da mangiare, e fa la vera ricchezza del paese = […] i signori commissari per lo Stato Pontificio vedendo che non potevano avere sufficiente roba da Roma, si sono rivolti ad italiani residenti in Parigi e sudditi pontifici. Gli oggetti sono stati approvati dal sig. Schnetz giurato per lo stato Pontificio […] dirò francamente che persona meno idonea non poteva scegliersi. Esso è sordo, e rimbambito, e fanatico pei suoi francesi […]. Di più è sembrato che Roma mancasse di un artista di valore da mandare, perché si è servita di uno straniero. Suppongo in questo delle ragioni economiche ma non so quanto ci abbiano giovato in questo caso. […] pei quadri […] bisogna dire il vero siamo senza capo di valore alcuno, non escluso l'ultimo venuto, il quale ha l'aggiunta di una non comune indecenza, e tale che un eguale era già stato escluso. In una esposizione Pontificia era cosa da calcolare, e la Madonna non doveva stare accanto e in faccia a una Venere. […] ho patito assai a vedere come passino certe cose. Secchi to the trade minister, Pier Domenico Costantini Baldini, Paris, May 11, 1867; APUG, FS 9.III.

59 *Fa schifo a sentire il card[inale] che si scusa di inviare roba che faccia onore allo stato perché gli è rivenuta rotta. […] quella era roba guadagnata dio sa come, forse con un regalo, ma il dio pecunia col resto è ciò che governa sti signori: l'interesse della chiesa è un corno per essi.* APUG, FS 23.II.A.

60 See also Heber Marini, *Trente visites à l'Exposition Universelle de 1867* (Paris: A. Lacroix et C.ie Editeurs, 1868), 402–403.

61 *Gloire au P. Secchi! Trois fois béni soit son météorograph! Grâce au savant jésuite et à son œuvre, l'honneur des États pontificaux est sauvé […].* Prosper Poitevin, "V. Les États pontificaux," *L'Exposition Universelle de 1867 illustrée* 44 (1867): 218–219, here 218.

> humility; and for obtaining the respect of mankind, wouldn't it have been a hundred times better to abstain, than to make the world aware of the secret of the deplorable decadence that today the arts and industry have fallen to there? An awful lot of Catholics think so. […] One should show one's best side in public; it is prudent and in good taste to hide and leave the other matters in shade.[62]

Poitevin concluded:

> We have reported the entire list of honors of the Papal States: it is not a rich list; and it can be said that, without Fr. Secchi, this list would not put them in a very good light. Praise Fr. Secchi! Praise the Jesuit who, in the nineteenth century, has made a public reconciliation with science and who came to protest, in the face of Rome, against the odious persecution it suffered once! Glory be to him! Because he avenges Galileo, the great astronomer of Pisa, and the snubs that the Jesuits, his relentless persecutors, poured on him![63]

Along with the other prizewinners, the Légion d'Honneur was bestowed upon Secchi by Napoleon III on July 1. Secchi noted in his diary that he had been "greeted with applause from the audience and with affectionate sympathy from all the winners. A real triumph of science. […] Hurrah for religion and the cassock that was crowned today more than me!"[64] Indeed, the awarding of this

62 *Sans le P. Secchi et son merveilleux instrument, quelle affreuse déroute, bon Dieu! êut subie l'armée industrielle ou plutôt la bande d'irréguliers que Rome n'a pas crainte de mettre en ligne au Champ de Mars! En laissant étaler ainsi sa faiblesse et son impuissance industrielle, le gouvernement romain n'a-t-il pas fait acte d'une trop grande humilité chrétienne; et, par respect humain, n'aurait-il pas mieux fait cent fois de s'abstenir, que de mettre le monde entier dans le sécret de la déplorable décadence où aujourd'hui les arts et l'industrie sont tombés chez lui? Une foule de catholiques le pensent* […] *C'est par ses beaux côtés qu'il faut se produire en public; quant aux autres, il est de la prudence et du bon goût de le dissimuler et de les laisser dans l'ombre.* Poitevin, "V. Les États pontificaux," 218.

63 *Nous avons relevé tout entière la table d'honneur des États pontificaux: elle n'est pas riche en titres; et l'on peut dire que sans le P. Secchi, tous ces titres-là ne l'auraient pas mise en une bien vive lumière. Gloire donc au P. Secchi! Gloire au jésuite qu'au dix-neuvième siècle a fait pacte public avec la science et est venu protester, à la face de Rome, contre les odieuses persécutions qu'elle a jadis subies! Gloire à lui! car il venge Galilée, le grand astronome de Pise, des avanies dont l'ont abreuvé les jésuites, ses implacables persécuteurs!* Poitevin, "V. Les États pontificaux," 219.

64 *Accolto con applausi dal pubblico e con affettuosa simpatia da tutti i premiati. Vero trionfo della scienza.* […] *Viva la religione e l'abito che oggi è stato coronato più che mè stesso!* APUG, FS 23.II.A.

prize to a Jesuit was clearly a jab at the anti-clerical party who were accusing the church of obscurantism and hostility to science.

4 Great Britain, 1867 and 1868

In August 1867, almost a decade after his first invitation, Secchi returned to the country where he had found shelter in 1848. He started by visiting his past refuge, Stonyhurst College, described by Secchi as "one of the most beautiful and well-situated institutions"[65] he had ever seen. The college observatory had just been supplied with a large equatorial telescope[66] and a renovated Kew-type geomagnetic apparatus, with a photographic recording mechanism, thanks to the urging of the observatory's director, Walter Sidgreaves, S.J. (1837–1919), and the support of the past director, Weld, who had become provincial of the English Jesuit province.[67]

After visiting Stonyhurst, Secchi went to London to meet Sir Edward Sabine, a leading expert on the Earth's magnetism: in 1852, Sabine had discovered the connection between geomagnetic variations and the solar activity cycles, a finding that opened a debate in the scientific community. Secchi also went to the workshops of many instrument-makers: optician John Henry Dallmeyer (1830–83), a specialist in the production of photographic objective lenses; the Dent company of clockmakers; the Casella Company, producers of a wide range of scientific instruments; John Browning (1835–1925), one of the leading makers of spectroscopic equipment;[68] Beck & Beck Company, makers of microscopes and other optical instruments; and physicist Wheatstone, to view his new "dynamo-electric machine."

Secchi also visited, as usual, the main sights of the city: Westminster Abbey and Newton's grave; South Kensington Museum,[69] initially known as Museum of Manufacturers, whose collections covered applied arts and science (Secchi wrote that he found "nothing really interesting"); and the British Museum, where the Jesuit was impressed by some Babylonian clay cuneiform astronomical tablets. He also visited the Kew Magnetic Observatory and Greenwich Observatory, in the outskirts of London, where he was "very well" welcomed by

65 *Una delle istituzioni più belle e commode che io abbia veduto.* APUG, FS 23.II.A.
66 See Udías, *Searching the Heavens and the Earth*, 187.
67 Secchi also had a Kew-type geomagnetic apparatus at the Collegio Romano Observatory; see Udías, *Searching the Heavens and the Earth*, 68.
68 Secchi was accompanied by De la Rue when visiting Browning's workshop.
69 It was later divided into two institutions and renamed the Victoria and Albert Museum and Science Museum.

astronomer Edward James Stone (1831–97) and meteorologist James Glaisher (1809–1903), superintendent of the Department of Meteorology and Magnetism. Secchi concluded his two-week stay in Britain with a visit to Cambridge University and its observatory.

Of all these, Secchi's two most important visits were paid to astronomers Huggins and John Herschel on August 8 and 12 respectively. Secchi's reports about these visits are very detailed and deserve a more extended examination. At Huggins's, he received "a very nice welcome." Secchi was impressed by the English astronomer's "nice observatory" at Tulse Hill and his ingenuity in making instruments by himself: "He has two good spectroscopic apparatus, not much to look at on the outside, but excellent inside. [...] He makes the mirrors himself and is very ingenious. He is a very simple man, very friendly and unusually kind, he gave me all the addresses that I asked [from] him."[70]

Herschel lived at Slough, about thirty kilometers west of London, and Secchi "faced three hours of slow trains" to pay him a visit. He wrote:

> I found him at breakfast, and I was received with such enthusiasm and affection that the excitement resulted in a long bout of coughing. This is a falling star, he said with great affability, and immediately offered me a little lunch. Then, we went into the living room, where we compared the Orion nebula [picture] by Lord Rosse[71] with the one [that was drawn] at the Collegio Romano, and the one by him. [...] Afterward, we went to view his garden and, among other things, we discussed the alleged discovery of Pascal and of his letters,[72] to which nobody seems to give credence, no more here than in Paris. By now, Herschel is greatly weakened in energy, but he retains a clear memory and uncommon lucidity of ideas.[73]

70 *Trovo una gentilissima accoglienza.* [...] *Ha un bell'osservatorio* [...]. *Ha due buoni apparati spettroscopici, ma che non sono punto straordinari all'esterno ma ottimi all'interno.* [...] *Lavora da sé i suoi specchi ed è molto ingegnoso. E' molto semplice ed ha molta bonomia e gentilezza veramente rara, mi ha dato tutti gli indirizzi che gli ho domandato.* APUG, FS 23.II.A.

71 William Parsons, earl of Rosse (1800–67), used a giant reflector telescope in Ireland for a systematic survey of nebulae.

72 Some spurious letters supposedly written by Pascal, in which the famous French scientist–philosopher presented the universal gravitation theory before Newton, provoked an international debate in 1867. Secchi mentioned this fraud in his dispute with Volpicelli over Galileo's blindness (see Chapter 9).

73 *L'ho trovato al breakfast e mi ha ricevuto con un tale entusiasmo e affezione che gli ha eccitato un non breve insulto di catarro per l'emozione.* This is a falling star *ha detto con aria amabilissima e immediatamente mi ha dato un piccolo déjeuner. Quindi siamo andati nel salotto, ove abbiamo confrontato la nebulosa di Orione di Lord Rosse con quella del Collegio Romano, e la sua.* [...] *Siamo quindi andati a vedere il suo giardino e abbiamo tra le altre cose*

Secchi was also impressed by Herschel's wide range of interests and his still lively curiosity; indeed, after discussing meteors, the topic of their conversation totally changed:

> Currently, he is totally dedicated to literature. He gave me a Latin elegy, which is a translation from Schiller['s verses]. At the age of seventy-five years, it is curious to study and compose verses: this reveals a man with a strong mind, but brittle. He gave me one of his large photographs and wanted to keep me with him for the evening, but I promised that I would come back if I could, and I returned [to London] after three hours had passed like lightning.[74]

Secchi spent the evening of that day at De la Rue's, talking about photographic techniques in astronomy.

The following year, Secchi again crossed the Channel to take part in the annual conference of the British Association for the Advancement of Science, which was held in Norwich. He was invited to speak about stellar spectroscopy and presented the results of his spectral classification of stars (see Chapter 6).[75] A letter to Secchi from Francis Kerril Amherst (1819–83), bishop of Northampton, refers to the warm welcome Secchi received at the conference:

> I am very happy to know that you have been so well received at Norwich, I can only regret that the general tendency of the scientific meeting is a little in accordance with the principles which are yours, because they are those of the Catholic Church. But your presence cannot fail to do good, it will proclaim to men at least prejudiced that Religion & Science may coexist, & that the Church is not afraid of following science to every logical issue.[76]

---

parlato della pretesa scoperta di Pascal e delle sue lettere, a cui nessuno pare che creda qui, più che a Parigi. Adesso Herschel è molto indebolito di forze, ma conserva la sua memoria limpida e una lucidità di idee non comune. APUG, FS 23.II.A.

74 *Adesso egli è tutto nella letteratura. Mi ha dato una elegia latina che è una traduzione di Schiller. A 75 anni è curioso studiare e far versi: questo mostra un uomo di forte mente ma affralito. Mi regalò una delle sue grandi fotografie e voleva tenermi suo alla sera, ma gli promisi che sarei tornato se avessi potuto, e me ne ritornai dopo 3 ore passate come un lampo.* APUG, FS 23.II.A.

75 See Angelo Secchi, "On Stellar Spectrometry," in *Report of the Thirty-Eighth Meeting of the British Association of Science, Held at Norwich in August 1868* (London: John Murray, 1869), 165–170.

76 Amherst to Secchi, Northampton, August 26, 1868; APUG, FS 8. V. 32–33.

5 Italian Geodetic Commission, 1869

The debate about the shape of the Earth, which had been opened by the Newtonian theory of gravitation, had led to the confirmation in the eighteenth century that the Earth's parallels and meridians were not circular, and probably not even elliptical. In the nineteenth century, the measurement of the meridian arc to try to determine the shape of the geoid became one of the main undertakings of those involved in astronomical research. France, Spain, Britain, and Russia carried out studies and obtained results in this field; but in Central Europe, measurements of sufficiently long arcs of meridian were lacking. Moreover, the few available results were strongly contradictory: "The source of these discrepancies," Secchi would write, "remains uncertain: it can be attributed to inaccuracies in the older work, or to actual variations on the Earth's surface in this part of Europe, crossed by large basins on the one hand, and bristling with towering mountains on the other."[77]

In 1861, a geodetic commission headed by the Prussian general Johann Jacob Baeyer (1794–1885) was established by the German states with the aim of making an accurate measurement of the arc of the meridian passing through Berlin. Other nations crossed by the same meridian were invited to join the commission and, over the years, Russia, Sweden, Norway, Austria, and Switzerland participated. The Papal States were also invited to join, but, because of the uncertain political situation at that time, it was considered imprudent to collaborate, and though there was a general expression of interest from the papal government, an official commitment was delayed, waiting for the turn of events. Italy had also been invited to join the commission, and in August 1869, since Secchi was in Florence for the preparatory meetings of an expedition to Sicily (see below), he was invited to participate in the meetings of the Italian commission for the measurement of the degree of the European central meridian.

Though he received his superiors' permission to participate, Secchi's involvement was in a strictly private capacity, a diplomatic artifice that allowed him to evade the constraints of politics, thereby enabling the Jesuit to contribute to the science. Secchi spoke about the Via Appia geodetic base measurements made in 1854–55 and other initiatives in this field (see Chapter 4). The Italian commission, headed by General Giuseppe Ricci (1811–81), decided to invite the

---

77 *L'origine di queste discrepanze resta incerta; possono esse attribuirsi, o alle inesattezze degli antichi lavori, o alle perturbazioni reali che si hanno sulla superficie terrestre in questa parte d'Europa solcata da vasti bacini da una parte, e irta di altissime montagne dall'altra.* Secchi, "Rapporto della Commissione per la Misura del Meridiano Centrale Europeo," 232.

FIGURE 5.4 Astronomical meeting at Arcetri in 1869, during the conference of the Italian Geodetic Commission; Secchi is seated in the middle. From Simone Bianchi, Daniele Galli, Antonella Gasperini, "Le due inaugurazioni dell'Osservatorio di Arcetri," *Giornale di astronomia* 3 (2013): 19–30, here 21; courtesy of INAF-OAA.

papal government to participate officially in the measurement of the European central meridian, which in fact crossed papal territory.

Secchi handed the invitation letter to Cardinal Giacomo Antonelli (1806–76), secretary of the Papal States, and attached a report in which he explained the advantages of participating, especially that of deriving a second triangulation to extend and complete the map of the Papal States, "which needs it very much, especially at some distance from the capital."[78] In the report, Secchi also proposed an inexpensive method to complete the work (namely entrusting the work to soldiers and astronomers), and estimated the total budget (six thousand lira for the purchase of instruments and ten thousand lira for the execution of the job) and the time it would take (four to five years).

Secchi's proposal was approved, and at the end of 1869, a pontifical commission[79] headed by Secchi was established to contribute to the international geodetic enterprise. The Defense Ministry provided the requisite staff and instruments "so that the work could be done with due care and economy,"[80] and the commission—which held weekly meetings to verify the progress of the work—obtained a detachment of the Corps of Engineers for carrying out the measuring operations.

The official announcement of the start of triangulation works in the Papal States was published at the beginning of 1870. The pontifical commission first oversaw the purchase of instruments (which were to be "exact and light" and were acquired from the Berlin firm of Pistor & Martins) and the selection of locations where geodetic stations were to be installed. Rome was at the center of the triangulation; the city was enclosed by a polygon, which had at its core the geodetic base measured in 1855, to which the polygon was to be linked.

The commission soon had an important decision to make in which pragmatism and science would triumph over ideology and symbolism, thus keeping science and faith on clearly separate planes: "The first question that arose was [to decide] whether the central station ought to remain on the dome of Saint Peter's [basilica]—a station [that was] naturally used by all of our predecessors. But after a thoughtful discussion, that was immediately ruled out."[81] In fact, the International Commission required that for each station the instruments should be placed in a position from where they could be completely

78 *Che ne ha non piccolo bisogno, specialmente a qualche distanza dalla Capitale.* Secchi, "Rapporto della Commissione per la Misura del Meridiano Centrale Europeo," 236.

79 The commission was composed of Lorenzo Respighi, director of Campidoglio Observatory; Colonel Mattia Azzarelli (1811–97), professor of hydraulics at the University of Rome; Major Francis Oberholtzer (1828–91), papal army engineer; and Alessandro Bettocchi, professor of geodesy at the University of Rome.

80 *Onde eseguire i lavori colla debita sollecitudine ed economia.* Secchi, "Rapporto della Commissione per la Misura del Meridiano Centrale Europeo," 237.

81 *La prima questione che sorse fu quella se doveasi conservare la stazione centrale sulla cupola di S. Pietro; stazione usata naturalmente da tutti i nostri predecessori. Ma una ponderata discussione ce la fece senza esitanza eliminare.* Secchi, "Rapporto della Commissione per la Misura del Meridiano Centrale Europeo," 242.

rotated with a full view of the horizon; that would have been impossible from St. Peter's dome, and bringing the instruments there would also have been very complicated. Maintaining such a station, incongruent with the positioning of the instrument, would have required "gimmicks of secondary measurements and calculations, all things that are forbidden in high precision measurements, as they are sources of many inaccuracies. [...] All these and other considerations made us absolutely reject this station."[82]

After an inspection, the central station was chosen to be a trigonometric point on the top of Monte Mario, near the Vatican, within the vineyard of Prince Carlo Felice Barberini (1817–80), duke of Castelvecchio, with whom an agreement was signed after "by no means simple" negotiations. In May 1870, the first stone of the central geodetic station was laid, marking the trigonometric point that is crossed by the meridian of Rome, later the reference meridian of Italy. Secchi wrote:

> The station of Montemario not only replaces the dome of St Peter's [basilica] but it is by far more convenient; and it is not only visible everywhere [...] but is a suitable place to site precisely an astronomical station from which you can directly determine the latitude and azimuth [...] so that it constitutes a real model station.[83]

The new trigonometric point on Monte Mario was then linked via the triangulation network to a section of the geodetic basis of the Via Appia, which met the requirements of the International Commission.[84] Secchi also recommended collecting hydrometric, meteorological, geological, and magnetic data, typical of his multidisciplinary approach to science.[85]

The construction and use of the central station was interrupted by the war and the political events of September 1870. The geodetic work was later

82 *Amminnicoli* [sic] *di misure secondarie e di calcoli, cose tutte proscritte nelle operazioni di alta precisione, perché sorgenti di moltissime inesattezze.* [...] *Tutte queste ed altre considerazioni ci fecero assolutamente rigettare questa stazione.* Secchi, "Rapporto della Commissione per la Misura del Meridiano Centrale Europeo," 242.

83 *La stazione di Montemario è tale che non solo supplisce, ma supera in vantaggio la cupola di San Pietro ed è non solo visibile da per tutto* [...] *ma si presta a fissare con esattezza una stazione astronomica ove si possa determinare direttamente la latitudine e l'azimut* [...] *onde essa costituisce una vera stazione modello.* Secchi, "Rapporto della Commissione per la Misura del Meridiano Centrale Europeo," 244.

84 The first end of the new base, shorter than the original, was moved from the tomb of Cecilia Metella, where the horizon was covered by buildings, to the tower of Capo di Bove.

85 See Secchi, "Rapporto della Commissione per la Misura del Meridiano Centrale Europeo," 255–256.

continued by the Italian government, and Secchi no longer worked directly in this field. However, it should be noted that, after various vicissitudes,[86] the reference point of the meridian of Rome that became the prime meridian of Italy is still positioned today at the point indicated by the astronomers of the pontifical geodetic commission, Secchi and Respighi.

6 Sicily, Total Solar Eclipse of 1870

The Italian government organized its first national scientific expedition on the occasion of the total solar eclipse of December 22, 1870, visible from south-eastern Sicily. A commission was appointed in 1869 to determine the scientific program, and the directors of the main Italian observatories were involved in the project. The idea had originated from an initiative of Tacchini, a dynamic astronomer of Palermo Observatory, who persuaded his director, Gaetano Cacciatore (1814–89), to submit a proposal to the minister of public education. Though he was the most expert astronomer in Italy in this field, given the expertise he had gained on the occasion of the 1860 eclipse expedition, Secchi was initially excluded from the commission due to the tensions between the papacy and the Kingdom of Italy. Many anti-clerical astronomers, such as Donati, director of Florence Observatory, and Cacciatore himself, would had opposed inviting Secchi to be a member of the commission. However, in the interest of science, Tacchini solicited the president of the commission, Santini, his former professor of astronomy in Padua with whom he had retained a close relationship, to formally invite Secchi, as he was "the only one in Italy that can compete with foreigners in spectroscopy."[87]

Tacchini will play a crucial role in Secchi's life and work on numerous occasions: they developed an important scientific partnership and a deep personal relationship (see Chapter 8). He recommended that Secchi accept the invitation and, together with Santini, managed to surmount the financial obstacles that were put forward by Secchi: "You would have nothing to ask from your government," he wrote him

86 The station was demolished in 1878 to build a military fortress as part of the defense of the city of Rome, and the trigonometric point was moved to the terrace of Villa Mellini, now the headquarters of the National Institute of Astrophysics. After the construction of the fort, the trigonometric point was re-established on the ground in 1882 and, in the same year, the station was reconstructed in the same place. See Aebischer, "Le misure geodetiche," 235–236.

87 *L'unico in Italia che possa gareggiare cogli stranieri in fatto di spettroscopia* [...]. Tacchini to Secchi, Palermo, May 29, 1869, quoted in Chinnici and Gasperini, *Alle origini dell'astrofisica italiana*, 99.

> except to be allowed to come to Sicily and bring the observatory's instruments, among which it would be interesting to have the telescope, namely the equatorial that you used in Spain; it is well understood that transportation, packing, and travel costs would be the responsibility of our government and not the pontifical.[88]

Secchi obtained permission from his superiors to join the commission and to participate in the expedition, provided that all expenses were charged to the Italian government. In the preparatory meeting held in Florence in September 1869, the commission decided on the distribution of the work; Secchi insisted on the importance of taking photographs of the phenomenon and persuaded

FIGURE 5.5 Minutes of the preparatory meeting for the 1870 expedition to Sicily; the page on the right reports Secchi's proposal for photographic operations. Courtesy of INAF-OAN.

88 *Ella non avrebbe nulla a chiedere al suo governo, se non il permesso di venire in Sicilia e di portare con sé istrumenti dell'osservatorio, fra i quali sarebbe interessante l'avere il cannocchiale, cioè l'equatoriale di cui ella si servì in Spagna; ben inteso che trasporto, spese di imballaggio e viaggio sarebbero a carico del governo nostro e non pontificio* [...]. Tacchini to Secchi, Palermo, September 13, 1869, quoted in Chinnici and Gasperini, *Alle origini dell'astrofisica italiana*, 106.

the commission members (who questioned the use of photographic techniques on the grounds that they were too expensive) to assign to him the task of the photographic operations.

This resistance to using photographic techniques reveals a general shortsightedness on the part of the Italian astronomical community and, perhaps, a dismissive attitude toward the proposals of a Jesuit.

The eclipse was in winter, and given the likelihood of it taking place in poor weather conditions, the commission, at Tacchini's suggestion, chose two different stations, located in the western and southern coasts of Sicily (in the towns of Augusta and Terranova, now Gela, respectively). Secchi was assigned to Augusta,[89] where the best-known Italian astronomers were expected. Tacchini was charged with providing the staff and the instruments for both stations and traveled some months to visit the observatories that could contribute to the expedition and to negotiate their participation.[90]

The outbreak of the Franco-Prussian War (1870–71), a crucial political event that had serious consequences for Secchi's life and career, changed all that. The war led Napoleon III to recall the French garrisons that were defending Rome, as a result of which the Italian troops of the Bersaglieri[91] could easily enter the city through Porta Pia. On September 20, Rome was occupied. Secchi annotated in the meteorological register of the Collegio Romano Observatory: "Fine [weather]. Gunfire in the morning, rogueries until evening [...]."[92]

Rome was annexed by the Kingdom of Italy and became its capital, while Pius IX was allowed to remain in the Vatican, in a sort of captivity. Public opinion was divided: those Italians who were loyal to the pope considered this a humiliation, while the anti-clerical wing applauded the restriction of the pope's authority and power. The conflict between the papacy and the Kingdom of Italy became more and more heated in the following years, before developing into a full diplomatic war, with reciprocal counteractions finally leading to the confiscation of the church's properties (1873) and the pope's decree *Non expedit* (1874), urging all Catholics to take no part in the new regime and its politics.

The situation had serious consequences for Secchi, who was caught up in the middle of this diplomatic war. At a stop in Naples while he was traveling to Sicily, a report about a possible expulsion of the Jesuits from the Collegio Romano led him to send his accompanying assistant, Brother Marchetti, back to Rome to

89 The astronomical station was installed in the old castle of the town.

90 See Ileana Chinnici, *L'eclisse totale di sole del 1870 in Sicilia* (Palermo: INAF-Osservatorio di Palermo, 2008).

91 A specialized, high-mobility light infantry unit in the Italian army.

92 *Bello. Cannonate al mattino, furfanterie fino a sera* [...]. Quoted in Franca Mangianti, "L'Ufficio Centrale di Ecologia Agraria e la sua sede nel Palazzo del Collegio Romano," *Agricoltura: Speciale "120° Anniversario dell'UCEA"* 277 (1996): 15–26, here 23.

save the instruments if need be. Secchi arrived in Palermo on November 10, together with the Barnabite priest Francesco Denza (1834–94),[93] who was charged with carrying out spectroscopic observations of the eclipse. This was an important task, which Secchi would have carried out himself, given his experience in astronomical spectroscopy (see Chapter 6) and the interesting findings made by his foreign colleagues during the two previous total solar eclipses (see introduction). However, in the absence of someone attending to the photographic operations, he was forced to leave the spectroscopic observations to others.

Secchi visited the Palermo Observatory and was put up in the apartment of the director, Cacciatore, a fervent anti-clerical but also a true Sicilian who felt the duty of hospitality. The Jesuit, whose popularity was as widespread in Sicily as everywhere else, was welcomed in every regard:

> I have been in Palermo for a few days, where I have received many honors and compliments. Mr. Marquis Spedalotto[94] is very courteous to me and shows me every kindness, he sends a carriage to go about in his company. This morning, I was in the cathedral, where the canons gave me an amazing welcome; they showed me the urn of Saint Rosalia, which is on display only twice a year and in the presence of the royal family! They told me that there had been no small difficulty to be overcome for this, since the keys are held by the municipality and the chapter. I knew nothing about this trouble, nor had I asked for the favor. Then they wanted me to see the church and all the curiosities of the sacristy, which are really superb, there are works of art of all ages, most notably a diadem of a queen of the eleventh century[95] [...]. There are the tombs of the kings, certificates, and other things of great importance, golden goblets, etc. The chasuble that I used was decorated with corals and the golden chalice [was] studded with gems.[96]

93 Denza was astronomer and meteorologist at Moncalieri, near Turin; he later founded the Specola Vaticana (see Maffeo, *Vatican Observatory*).

94 Achille Paternò di Spedalotto (1815–88) was a nobleman and activist of the clerical Bourbon party in Palermo; see Paolo Alatri, *Lotte politiche in Sicilia sotto il governo della Destra: 1866–74* (Rome: Einaudi Editore, 1954), 268. It seems that, during his sejourn in Palermo, Secchi spent some days at Villa Spedalotto, in Bagheria, near Palermo: a marble buste of Secchi was placed in the entrance courtyard of the villa in 1871 (see Spedalotto to Secchi, Palermo, January 1871; APUG FS 20) and is still preserved in its original site.

95 The queen was Costanza d'Altavilla (1154–98), mother of Emperor Frederick II (1194–1250, r.1220–50).

96 *Da alcuni giorni sono a Palermo, dove ricevo molti onori e complimenti. Il sig. marchese Spedalotto è molto cortese verso di me, e mi fa ogni gentilezza, inviandomi la carrozza pel passeggio in sua compagnia.*

Sono stato questa mattina nella cattedrale dove i canonici mi hanno fatto un'accoglienza straordinaria, giacché hanno fatto scoprire l'urna di S. Rosalia, cosa che non si fa che due

In Palermo, Secchi carried out some work at the observatory:

> We started to mount the meteorograph [see Chapter 6], and time passes in this work, and in adapting the equatorial to the spectrometer. The [spectra of the] main stars were shown to the professors [of Palermo University], who were stunned; other than that, nothing else could be done, as the equatorial[97] was so hard to put in operation that even a machine with a horse could not move it. We did our best to operate it by hand.[98]

In the meantime, the marine steamship *Plebiscito*, carrying the instruments for the expedition, arrived in Palermo, and the party left on the evening of November 27 for their final destination of Augusta.[99] They stopped in Messina en route, where Secchi admired "the magnificent lighthouse strait, where the whirlpools are so beautiful and visible that they are a marvel." He was also received by the bishop and visited the cathedral:

> Once my arrival was known, a large crowd of people ran in, and representatives of the university, from whom I repeatedly received compliments. I visited the university, which was certainly not a big event, but I received many attestations of esteem and respect. On the morning of the twenty-ninth, we departed for Augusta and, from about eleven miles offshore, we

volte l'anno, e poi in presenza di casa reale! Vi sono state difficoltà mi dicevano non piccole ma superate, essendo le chiavi tenute dal Municipio e dal capitolo. Io non sapeva nulla di queste mene, né avea domandato il favore. Poi vollero farmi vedere la chiesa e tutte le curiosità di sagrestia, che sono veramente superbe, essendovi oggetti d'arte di tutti i secoli, e principalmente è notabile un diadema di una regina del 11° secolo [...] *Vi sono i sepolcri de' Re, diplomi, e altre cose antiche di molto valore, calici d'oro, ecc. La pianeta che usai era di corallo e il calice d'oro gemmato.* APUG, FS 23.II.C.

97 The Merz equatorial of Palermo Observatory was almost identical in size to the one at the Collegio Romano (see Chapter 8); see Ileana Chinnici and Paolo Brenni, "The Palermo Merz Equatorial Telescope: An Instrument, a Manuscript, Some Drawings," *Nuncius* 30 (2015): 228–279.

98 *Si è cominciato a montare il meteorografo, e il tempo passa in questo lavoro, e in accomodare l'equatoriale per lo spettrometro. Si fecero vedere ai professori le stelle principali, che ne rimasero storditi, del resto niente altro fu possibile di fare essendo l'equatoriale così duro che né anche una macchina d'un cavallo lo potrebbe muovere. Si fece alla meglio movendolo a mano.* APUG, FS 23.II.C.

99 The journey was troubled by some errors made by the captain, who "failed the route, because he sails without a corrected compass and does not apply any correction! In front of [the island of] Vulcano he became aware of the error, and after many consultations it was remedied" ("Il capitano sbagliò di nuovo il rombo, poiché naviga senza bussola corretta, né vi applica correzione alcuna! In faccia a Vulcano si accorse dell'errore, e dopo molte consulte fu rimediato"). APUG, FS 23.II.C.

> admired the beautiful view of Etna, which is enchanting for its beauty. [...]; small towns and villages are scattered on the slopes. From half up, there are woods; and at the top there is snow! Unique wonder in the universe. The arrival was in the evening [...], we safely arrived and in the same evening the cases were landed.[100]

The instruments were then installed, adjusted, and protected from rain and wind, "which arrived furiously for some days." Secchi made some attempts to photograph the Sun and the moon; however, the weather conditions were unstable and, consequently, a source of concern for the astronomers hoping to observe the eclipse. Secchi evaluated the possibility of moving the station to an abandoned "magnificent" convent of Minims,[101] but he then gave up on that idea, realizing that, for the many anti-clerical members of the party, "breathing that air hurts them." The conduct of some of his astronomer colleagues was sometimes quite embarrassing for Secchi: "Tonight, there is a great bunch of women, among whom Cacciatore is quite frisky. I realize what bounders these liberals are."[102]

During his stay in Augusta, Secchi's mind was full of dark thoughts about the future of his observatory. He was especially anguishing over his decision to renounce the chair of astrophysics at La Sapienza University, which had been offered to him by the Italian government immediately after the occupation of Rome (see Chapter 7). In Palermo, he had reached the decision to renounce the chair and sent the letter announcing this decision, and at Augusta he received a letter from the superior general declaring that the renunciation had been officially communicated. "Thus gossip will stop," Secchi commented. However, fearing retaliation from the Italian government, he wrote to his confrères in Rome asking them to remove his instruments and books from the observatory:

100 *Ai 28 fummo a Messina, e con tempo discreto vidi il magnifico stretto del faro ove i vortici sono così belli e manifesti che sono una meraviglia. [...] sparsasi la voce del mio arrivo, corse una turba grande di popolo, e i rappresentanti dell'Università da cui fui a più riprese complimentato. Visitai l'Università dove certamente non è gran cosa, ma ebbi molto attestati di stima e rispetto. La mattina del 29 si partì per Augusta ed ammirammo il bel prospetto dell'Etna a largo di mare di circa 11 miglia, che è un incantesimo di bellezza. [...] sono tutte borgate e paesi sparsi alle falde. La metà in su è boschi; e la cima è neve! Spettacolo unico nell'Universo. Alla sera [...] arrivammo in sicuro e la sera stessa furono messe a terra le casse.* APUG, FS 23.II.C.

101 During festive periods, Secchi and Denza would officiate in the main church of the town, where they were cordially welcomed, as usual.

102 *Un sito magnifico per la commissione sarebbe stato il convento dei paolotti abbandonato e disabitato, ma a costoro fa male quell'aria. Questa sera vi è grande combriccola di femine in cui Cacciatore è tutto vita. Mi accorgo che cosa sia questa canaglia di liberali. Non sono vivi che alle femine.* APUG, FS 23.II.C.

> Even without an official expulsion, it may become incumbent for me even to withdraw from the observatory, if there is no means to assure an absolute independence from the government. [...] My renunciation of the chair puts me in a position that if these [people] are still in Rome, I will not be tolerated there.[103]

An American expedition was based close to the Italian astronomical station in Augusta, and cordial relations were established with the expedition's members.

Given the uncertain fate of the Collegio Romano, Secchi thought about returning to the United States:

FIGURE 5.6 Staff of the astronomical station installed at Augusta, in Sicily, for the 1870 eclipse; in the middle, Secchi; on his left, Cacciatore; between them, standing, physicist Pietro Blaserna (see below) last on the left, seated, Donati; second on the right, Denza. Courtesy of INAF-OAPa.

103 *Anche senza pensare ad una espulsione ufficiale, ora per me è divenuta una cosa di convenienza anche il ritiro dall'osservatorio, qualora non si verifichi un mezzo di una assoluta indipendenza dal Governo.* [...] *la mia rinunzia alla cattedra, mi mette in una posizione tale che se questi restano ancora a Roma io non vi sarò tollerato*. Secchi to Egidi, Augusta, December 18, 1870. APUG, FS 9.I.

> These good gentlemen[104] [...] live at our Georgetown College at Washington. [...] They told me so many things about Fathers Curley, Stonestreet,[105] and Maguire,[106] all my contemporaries there. I spent an evening in Georgetown! [...] What good people! Perhaps this is providential, because if I cannot stay in Rome, I might return to my old asylum.[107]

From that moment onward, such concerns regularly crossed Secchi's mind and troubled his last few years (see Chapter 10).[108]

On the day of the eclipse, the observations at both Italian stations were compromised by cloudy sky conditions. Secchi did not succeed in capturing the totality with his photographic apparatus,[109] and the event could only be partially recorded.[110]

7 Failed Participation: Expeditions to Mont Cenis (1871) and Bengal (1874)

In 1871, Secchi was also invited to prepare a program of geomagnetic and gravimetric measurements to be carried out at the Alpine crossing of Mont Cenis. In fact, once the construction of the Fréjus Tunnel was completed, the tunnels of Mont Cenis provisional railway (used during the construction of those works) were well suited to a campaign of gravity measurements. The project

104 Secchi is referring to Asaph Hall (1829–1907), William Harkness (see below), John R. Eastman (1836–1913) and his wife, members of the American expedition, settled in Syracuse.

105 Charles H. Stonestreet, S.J. (1813–85) was the president of Georgetown University from 1851 to 1852.

106 Bernard A. Maguire, S.J. (1818–86) was the president of Georgetown University from 1853 to 1858.

107 *Questi bravi signori* [Hall, Harkness, and Eastman with his wife] [...] *abitano proprio presso al nostro collegio di Georgetown presso Washington.* [...] *Mi hanno detto tante cose del P. Curley, del P. Stonestreet, e del P. Macguire tutti miei contemporanei colà. Ho passato una serata in Georgetown!* [...] *Che buona gente! Questo forse è provvidenziale perché se non potrò restare a Roma tornerò al mio vecchio asilo*. Secchi to Marchetti, Augusta, December 23, 1870. APUG, FS 9.I.

108 Later, he was also accused by the pope's political supporters of having taken part in a toast to the Italian government during an official banquet that was offered by the municipality of Augusta to the members of the expedition.

109 See Angelo Secchi, "Rapporto del prof. P. Angelo Secchi," in *Rapporti sulle osservazioni dell'ecclisse totale di sole del 22 dicembre 1870 eseguite in Sicilia dalla Commissione italiana*, ed. Gaetano Cacciatore (Palermo: Stabilimento Tipografico Lao, 1872), 15–30.

110 In memory of the event of the eclipse, a sundial was drawn by Secchi and Donati on the façade of the town hall of Augusta (see Chapter 4).

was promoted by Demetrio Emilio Diamilla Müller (1826–1908), a highly controversial figure later implicated in some political scandals; Denza proposed that Secchi serve as the director of this measurement campaign. In preparation for the campaign, Secchi wrote an interesting report with a description of the Fréjus works, which he was able visit personally.[111] He did some preliminary experiments with Müller and Denza to make sure that neither the passage of trains nor the iron materials in the tunnel influenced the observations, and they made studies of the temperature of the rocks. Since gravimetric research at Fréjus had been proposed by French astronomer Hervé A. Faye in November 1871, Secchi presented the provisional research program to the Académie des Sciences,[112] inviting it to send him suggestions and proposals. Unexpectedly, however, the Italian government was dissuaded from providing the necessary financial support, and the project was abandoned. "As for the works of Cenis, I am not too much displeased, because it would have been a truly hard corvée and I had only agreed to do it as a favor for friends,"[113] he wrote to Tacchini. Secchi suspected that Respighi had derailed the project (see Chapter 9),[114] but it was probably the dubious reputation of Müller that had influenced the decision of the government, which was suspicious of his scientific credentials. Moreover, in those years the relationship between Secchi and the Italian government was increasingly strained because of his renunciation of the chair that had been offered to him (see Chapter 7).

The Jesuit astronomer would also have liked to have traveled to India to observe the transit of Venus across the Sun, which occurred on December 9, 1874. This rare astronomical event was particularly important for determining the solar parallax, a parameter from which the measurement of the distance between the Sun and the Earth could be obtained; however, it is necessary to observe the phenomenon from multiple stations, located at a great distance from each other, in order to carry out parallax measurements. An Italian expedition was proposed by Tacchini, but the proposal met with the unfavorable counsel of the most influential members of the national astronomical community (namely Schiaparelli and Donati). Consequently, the government allocated such limited resources that Tacchini had to reformulate

111 Angelo Secchi, *Il traforo delle Alpi nella Catena del Moncenisio al Colle di Fréjus* (Rome: Tipografia e Libreria Alessandro Befani, 1872).

112 See Angelo Secchi, "Sur les expériences du pendule qui vont être entreprises dans le tunnel des Alpes occidentales," *Comptes rendus de l'Académie des Sciences* 73 (1871): 1192–93.

113 *Per i lavori del Cenisio non mi preme gran fatto, perchè era una corvée ben dura che avea accettato solo per fare servizio agli amici.* Secchi to Tacchini, Rome, April 13, 1872, quoted in Chinnici and Gasperini, *Alle origini dell'astrofisica italiana*, 192.

114 See Secchi to Schiaparelli, Rome, April 13, 1872, quoted in Buffoni, Manara, and Tucci, *G.V. Schiaparelli, A. Secchi*, 196.

the expedition's objective: rather than measuring the solar parallax, the expedition would now compare the ordinary visual and the spectroscopic methods, and determine which was the more appropriate for the observation of the phenomenon. However, in March 1874, disputes arose with Giuseppe Lorenzoni (1843–1914), an astronomer in Padua, on the best technique to be used for transit spectroscopic observations; this was a good pretext for Secchi to abandon his participation: "I would not feel healthy enough to undertake so controversial an enterprise,"[115] he wrote to Tacchini. Secchi's withdrawal goes along with the skepticism of the Italian astronomical community, the controversies with Lorenzoni, and his precarious health situation, but it probably also reflects his desire to keep himself at a distance from the Italian government, as well as the difficulties of getting permission from his superiors in the troubled years following the annexation of Rome.

8 International Meter Commission: Paris, 1870–72

The first definition of the standard meter dates from the late eighteenth century, when it was determined that the length of a meter was equal to a ten-millionth part of the length of the terrestrial meridian, as measured by French astronomers Jean-Baptiste Delambre (1749–1822) and Pierre Méchain (1744–1804).[116] On this basis, a standard meter called *Mètre des Archives* was selected and kept at the National Archives, Paris. However, with the increasing international adoption of the meter, copies of the standard meter acquired by the countries that adopted this measurement did not assure the necessary uniformity for ever more precise measurements. In fact, the ends of *Mètre des Archives*, which consisted of a platinum bar of the required length, were prone to wear with use.

In 1867, the International Geodetic Conference consequently proposed the creation of a new international prototype meter and the arrangement of a system where national standards could be compared with it. The international prototype would also be a "line standard"; that is, the meter was defined as the distance between two lines marked on the bar, thus avoiding the problems of wear. The French government, with the urging of the Ministry of Trade and Agriculture, supported the creation of an International Meter Commission. The commission met in Paris at the Conservatoire National des Arts et Métiers from August 8 to 13, 1870. Secchi participated in the meeting as a

115 *Io non mi sentirei in gambe a intraprendere una impresa tanto controversata.* Secchi to Tacchini, Rome, March 5, 1874, quoted in Chinnici and Gasperini, *Alle origini dell'astrofisica italiana*, 339.

116 On the measurement of the terrestrial meridian, see Ken Alder, *The Measure of All Things* (London: Simon & Schuster, 2002).

member of the commission. The commission established a Committee for Preparatory Research (Comité des Recherches Préparatoires), which met in April 1872 and prepared the next meeting of the commission, from September 24 to October 12, 1872, with the participation of about thirty countries.

Between the two meetings, however, the political situation changed both in France and Italy. The Franco-Prussian War ended with the capitulation of Napoleon III at Sedan in 1870 and the fall of the second empire; and indirectly, as mentioned above, it also led to the annexation of Rome by the Kingdom of Italy. Nevertheless, the committee of 1870 continued its work and, in 1872, "expressed the wish that, despite the changed conditions of the Holy See, the International Commission would not be deprived of the personal contribution and studies of Fr. Secchi, which were of great benefit to the work of the committee in the previous session."[117] Secchi was thus invited by Thiers, president of the Third French Republic, and by the commission's president, mathematician Mathieu, who was one of the main supporters of reinstating the metric system. Secchi's participation now had political significance and, this time, the Jesuit astronomer had no problem in obtaining authorization—and subsidy—from Cardinal Antonelli in order to participate as a delegate of the Holy See—a state whose existence was in dispute in those years. This invitation to Secchi, in fact, "was regarded by the Vatican as an act in which the French government recognized the right of the pope to send a delegate to the commission."[118] As a result, the money for travel expenses was immediately allocated by Antonelli to Secchi, who "asked for six hundred and received one thousand francs."

In Paris, Secchi served as president of Sub-Commission I ("Etude des bouts du Mètre des Archives" [Study of the ends of the *Mètre des Archives*]), which was charged with examining the ends of the *Mètre des Archives*. This sub-commission included Gilberto Govi (1826–89), professor of physics at Turin University; General Carlos Ibáñez de Ibero (1825–91), director general of the Geographical and Statistical Institute of Spain; and the famous French engineer Henri Tresca (1814–85), professor of mechanics at the Conservatoire des Arts et Métiers. Secchi was also member of Sub-Commissions V ("Température moyenne du mètre et poids du kilogramme dans le vide ou dans l'air" [Mean temperature of the meter and weight of the kilogram in the vacuum or in the

117 *Espresse il desiderio che, malgrado le cambiate condizioni della S. Sede, la Commissione internazionale non fosse privata del concorso personale e degli studi del Padre Secchi ch'erano stati di grande utilità ai lavori della Commissione nella sessione precedente.* Giuseppe Castellani, "Storia di una spedizione del P. Angelo Secchi (agosto–ottobre 1872)," *La civiltà cattolica* 11 (1944): 159–162, here 159.

118 *Fu riguardato dal Vaticano come un atto in cui il governo francese riconosceva il diritto del S. Padre di inviare un delegato alla Commissione.* Castellani, "Storia di una spedizione del P. Angelo Secchi," 159.

air]) and VI ("Comparateurs"), whose presidents were General Giuseppe Ricci and William H. Miller (1801–80), professor of mineralogy at Cambridge University, respectively. The related reports were published in the proceedings of the meeting,[119] whose final issues were to deal with the manufacture of new metric prototypes; the meter convention, signed in 1875 by the nations participant to the commission; and the creation of the International Bureau of Weights and Measures (Bureau International des Poids et Mesures).

Secchi's participation, however, greatly distressed the two above-mentioned Italian delegates, General Ricci and Gilberto Govi. In that meeting, in fact, a diplomatic incident occurred, which finally resulted in a brilliant personal triumph for Secchi, whose popularity was at its peak in those years.

At the opening of the session, the list of the delegates per nation was announced, and Secchi's name appeared on the list as a delegate of the Holy See.[120] The reaction of the Italian government was immediate. On September 28, the Italian minister plenipotentiary in Paris, Costantino Nigra (1828–1907), a leading Mason, sent a letter as well as a telegram to the foreign minister, Emilio Visconti Venosta (1829–1914), to inform him that Secchi's name had appeared in the *Journal officiel de la République Française* as a delegate of the Holy See:

> Fr. Secchi was sent here to take part in the work of the commission of the meter. He was received as a representative of the Holy See. Please, make known to me if there is place to ask for explanations from the French government, to express our reservations. However, I do know that Father Secchi, speaking with Mr. de Rémusat, spoke in very suitable terms about the government of the king.[121]

Charles de Rémusat (1797–1875) was foreign minister of the French Republic and a close friend of Thiers; naturally, he did his best to prevent this diplomatic incident escalating into a crisis. It is true that Secchi's attitude toward the Italian government was quite prudent and that he was often accused by the clericals of being too compliant.[122] He probably wanted to avoid compromising the

---

119 See Commission Internationale du Mètre, "Rapport de la première commission," *Réunions générales de 1872: Procès-verbaux* (Paris: Imprimerie nationale, 1872), 122–131.

120 Commission Internationale du Mètre, "Rapport de la première commission," 8.

121 *Le père Secchi a été envoyé ici pour prendre part aux travaux de la Commission du Mètre. Il y a été reçu comme représentant du St. Siège. Je Vous prie de me faire connaître s'il y a lieu de demander explications au Gouvernement français, faire réserve, ou réclamer. Je sais du reste que le père Secchi, en parlant a M. de Rémusat, s'est exprimé en termes très convenables au sujet du Gouvernement du Roi. Ministero degli Affari Esteri, Documenti diplomatici italiani, seconda serie: 1870–1896* (Rome: Istituto Poligrafico dello Stato, 1976), 4:143–144.

122 See note 107.

possibility of cooperation and receiving support from the Italian government, in the interest of science over and above the political situation.

While waiting for the minister's reply, Nigra consulted the two Italian delegates and telegraphed:

> I asked the Italian delegates about the presence of Father Secchi within the commission. They said that his presence can be explained because the present commission is the continuation of that of 1870. They added that any comments thereon would produce a bad impression on the commission and the learned world.[123]

The wise warning from the Italian delegates to avoid mixing science and politics was ignored. The minister replied by telegram that the incident appeared to him "quite serious [*assez grave*]"[124] because of the publication in the *Journal officiel* and wrote to Nigra:

> I know that this distinguished scholar of physical and mathematical sciences had already taken part in the earlier work of the commission itself, and certainly no one could make remarks about it if Father Secchi represented at that meeting the science of which he is such a distinguished professor in his homeland, Italy. But the government of the king cannot let pass in silence the designation, made in the French Official Gazette, of a state that no longer exists in European public law.[125]

Visconti Venosta acknowledged to "the holy father, as head of the Catholic Church, all privileges of sovereign inviolability," but

> an imperious reason of dignity demands that I cannot confine myself to express a simple reservation. The government of the king cannot be open

123 *J'ai interrogé les Commissaires italiens sur la présence du Père Secchi au sein de la Commission. Ils m'ont dit que cette présence s'explique parce que la Commission actuelle est la continuation de celle de 1870. Ils ont ajouté que toute observation à ce sujet ferait mauvaise impression sur la Commission et sur le monde savant. Ministero degli Affari Esteri, Documenti diplomatici italiani,* 144.

124 *Ministero degli Affari Esteri, Documenti diplomatici italiani,* 145.

125 *Io so che questo illustre cultore delle scienze fisiche e matematiche avea già preso parte ai lavori precedenti della stessa commissione, e certo niuno potrebbe fare osservazioni a questo proposito se il Padre Secchi rappresentasse in quel consesso la scienza di cui è sì distinto professore d'Italia sua Patria. Ma il Governo del Re non può lasciar passare in silenzio la designazione fatta nel foglio ufficiale francese di uno Stato che più non ha esistenza nel diritto pubblico europeo. Ministero degli Affari Esteri, Documenti diplomatici italiani,* 145.

> to sign agreements with one of its own subjects, who would act on behalf of an authority that has, in our view as in those of the powers that want to be our friends, neither existence in law nor in fact, in non-ecclesiastical matters.[126]

The minister's decision was drastic: "Therefore, I found myself in need to give orders to our delegates, by telegraph, to refrain from signing any act in which Father Secchi intervenes, assuming a title that, referring to Italian territory, could not be conferred to him by anyone other than the government of the king."[127]

In other words, Secchi should have been invited as a delegate of Italy, not as a delegate of the Holy See. The French minister, verbally informed by Nigra about this decision, assured him that there was no political intent: the incident had simply resulted from an error—Secchi's participation had been requested by the committee but, in the invitation, there was no indication that the Jesuit was to have been a delegate of the Holy See. Unfortunately, the publication of the list, being an unofficial document, had been prepared by the Ministry of Trade and Agriculture without any communication with the Foreign Ministry. However, now it was necessary to amend it and remove all doubts; Nigra clarified:

> This is neither a matter [...] of making a demonstration in favor of the sovereign pontiff or against him; nor [in favor or against] the person of Father Secchi; that is absolutely out of the question. This is solely to correct the indication of a situation that the French official journal mistakenly gave to provide for this same situation to be well determined within the International Commission.[128]

---

126 *Una ragione imperiosa di dignità m'impone di non limitarmi a fare una semplice riserva. Il Governo del Re non può esporsi a dover firmare accordi con un proprio suddito, il quale li stipulerebbe a nome di un'autorità, che non ha ai nostri occhi, né a quelli delle Potenze che vogliono esserci amiche, alcuna esistenza di diritto, né di fatto, in materie non ecclesiastiche.* Ministero degli Affari Esteri, *Documenti diplomatici italiani*, 145.

127 *Egli è perciò ch'io mi sono trovato nella necessità di dare per telegrafo ai nostri delegati l'ordine di astenersi dal firmare alcun atto al quale il Padre Secchi intervenga, assumendo una qualità ufficiale che, riferendosi al territorio italiano non avrebbe potuto essergli conferita da altri che dal Governo del Re.* Ministero degli Affari Esteri, *Documenti diplomatici italiani*, 146–147.

128 *Non può trattarsi qui di fare, a proposito di questioni di rettifica ed applicazione del sistema metrico, una dimostrazione in favore del Sovrano Pontefice o contro di esso; non può trattarsi nemmeno della persona del Padre Secchi che è assolutamente fuori di questione. Si tratta unicamente di rettificare l'indicazione di una situazione che il giornale ufficiale francese ha dato erroneamente e di provvedere perché questa stessa situazione sia bene determinata in seno alla Commissione internazionale.* Ministero degli Affari Esteri, *Documenti diplomatici italiani*, 149.

Rémusat promised to prepare an official response to the Italian government; an announcement of the error in the *Journal officiel* appeared to him "difficult and dangerous," because an official amendment would have given rise to an inconvenient discussion in the press, which had been largely silent on the subject until then.

In the meantime, the Italian delegates were placed in a very difficult position by the minister's orders: they protested, unofficially, with the French members of the commission and tried to determine Secchi's intentions. In fact, since the commission had established that voting would be expressed by nation, the Holy See was *de facto* recognized as a state having the right to vote, through its delegate, Secchi. The question now was whether he intended to vote or abstain.

The Jesuit had received precise orders from Cardinal Antonelli: he should not withdraw for any reason, and if the question was put to a vote, he should retire with a note of protest. Moreover, Secchi could count upon the support of many of the commission's members, including most of all the influential director of the Paris Observatory, Leverrier.

With the situation becoming increasingly complicated, Nigra wrote to the minister:

> In my diplomatic career, I cannot remember [...] a more unfortunate incident than the one to which Father Secchi has now joined his name. The unpleasant character of this incident is not exceeded but by its inappropriateness. There was undoubtedly no political intention in it from the French government [...]. But there was negligence and a lack of care and concern from the French Ministry of Foreign Affairs. There was ignorance or more culpable negligence in the French Ministry of Trade [...]. Mr. de Rémusat recognized immediately [...] that the publication of the French official journal contained a misnomer, [...] [and] agreed that appropriate negotiations with the Ministry of Trade were to be taken so that, in the proceedings of the commission, the name of Fr. Secchi would not appear as delegate of the Holy See.[129]

129 *Nella mia carriera diplomatica non mi ricordo* [...] *d'un incidente più spiacevole di quello a cui il Padre Secchi ha oramai unito il suo nome. Il carattere sgradevole di quest'incidente non è superato che dalla sua inopportunità. Non vi fu senza dubbio nessun'intenzione politica in esso per parte del Governo francese* [...] *Ma vi fu negligenza e mancanza d'attenzione e di riguardi per parte del Ministero francese degli Affari Esteri. Vi fu ignoranza o più colpevole negligenza nel Ministero francese del Commercio che inserì il rendiconto dell'apertura delle conferenze della Commissione del Metro nel giornale ufficiale, senza riferirne al Ministero*

Rémusat's official response was considered unsatisfying, and Nigra also met Thiers, to be assured about their desire to resolve the incident. The solution, however, did not come from the diplomats, but from the scientists. The two Italian delegates, in fact, had to make a difficult and crucial decision, well aware of its consequences.

At the end of the sessions, on the penultimate day of the meeting, when it was time to vote on the resolutions of the commission, the two men read an official note of protest and asked that the text be reported in the proceedings.[130] Besides the official reports, however, it is interesting to tell the story from Secchi's point of view, quoting the summary of that event from his diary:

> Then Gen[eral] Ricci stood up, and said he had an announcement to make before the vote, and read the protest [...] in which he said that, having seen Fr. Secchi on the list of members as the representative of the Holy See, he was instructed by his government to protest against that qualification. The view of the Italian government was that the Holy See, being a purely spiritual power, could not have a place at that meeting, and so long as such a state of affairs would last, the Italian delegates would refrain from taking part in any decision of the commission. This reading was greeted with icy silence and, seeing that no one moved, the president said "I'm sorry that this statement has been made, but since it has been made, I cannot but consider it as a *fait accompli*." After another short silence, Gen[eral] Morin [one of the leading figures in the work of the commission] said "Gentlemen, it is very unpleasant to see that issues unrelated to science come to trouble the good intelligence that reigns in the commission and to divert helpful members. It is to be hoped that, in the future, our colleagues will come with more conciliatory statements

---

degli Affari Esteri. Finalmente i nostri Commissarii avrebbero, credo, dovuto informare, appena giunti, la Legazione o il Governo della difficoltà a cui la presenza del Padre Secchi in seno alla Commissione poteva dar luogo. Comunque sia, l'incidente ebbe luogo e bisognava risolverlo. Il signor de Rémusat riconobbe subito nella prima conversazione che ebbi con lui, che la pubblicazione del giornale ufficiale francese conteneva una designazione erronea, che il Governo francese s'era limitato a domandare ai superiori ecclesiastici del Padre Secchi il permesso perché questi potesse assistere alle Conferenze in conformità del desiderio espresso dal Comitato preparatorio, il quale contava sul concorso di questo scienziato, e finalmente che conveniva che fossero prese le opportune intelligenze col Ministero del Commercio perché il Padre Secchi non figurasse negli atti della Commissione come Delegato della S. Sede. Ministero degli Affari Esteri, Documenti diplomatici italiani, 153–154.

130 See Commission Internationale du Mètre, "Rapport de la première commission," 140–141.

> [...]." Then everybody fell silent again, and I stayed fixed in the chair like a statue.[131]

The commission then proceeded with the vote, from which the Italian delegation abstained in accordance with the note of protest that had been presented. Secchi's description could almost be a script for a satirical comedy:

> The president said: "Well, gentlemen, [we have to] do nothing but put to a vote the proposals from the secretary: those who approve, raise your hand." I raised my arm as high as I could, though without intending any mockery, and cast an eye on the Italians. They held their hands under the table. A little later, we went to another vote, and I told Ibáñez, who was near me, to watch what the Italians were doing: I saw, he said, they didn't vote.[132]

During a break, Ricci apologized to Secchi, telling him that the protest was nothing personal but was simply his duty. Secchi's reply was severe:

> I replied that I could not act as a diplomat, and I considered that what was done to me was an insult done to the pope, and especially from him, whom I considered a good Catholic [...]; he insisted on saying that it was his duty, and I responded: "Commands are to be examined and, if they

131 *Allora si alzò il Gen. Ricci; e disse che avea una comunicazione da fare avanti al voto, e lesse la protesta che sta nel processo verbale, in cui diceva che avendo veduto il P. Secchi nella lista de' deputati come rappresentante la Santa Sede egli era incaricato dal suo governo di protestare contro tale qualifica. La Santa Sede esser un potere puramente spirituale, e non poter aver luogo in tale riunione, così vedeva la cosa il governo italiano, e fino a tanto che tale stato di cose sarebbe durato, i delegati italiani si sarebbero astenuti dal prender parte a qualunque decisione della Commissione. Questa lettura fu accolta con glaciale silenzio, e dopo un buon 40 secondi di silenzio, vedendo che nessuno si moveva, il Presidente disse = Mi dispiace che questa dichiarazione sia stata fatta, ma dacché è stata fatta non posso riguardarla che come un fatto compiuto = Dopo un altro poco di silenzio il Gen. Morin disse = Signori, è molto dispiacente di vedere che delle questioni estranee alla scienza vengono a sturbare la buona intelligenza che regna nella commissione e a distogliere i membri utili. E' da sperare che nelle prossime conversazioni i nostri colleghi verranno con delle istruzioni più concilianti (il testo a stampa dice un poco diversamente, ma è certo che disse così). Allora tutti zitti di nuovo, e io fermo nella sedia come una statua.* APUG, FS 23.II.D.

132 *Il presidente disse, ebbene signori non resta dunque altro che metter a voti le proposizioni lette dal segretario: quelli che le approvano levino la mano = Io alzai il braccio più che potei, senza caricatura però e tenevo d'occhio gli italiani. Essi tennero le mani sotto la tavola. Poco dopo si passò ad un'altra votazione, e dissi a Ibanez che mi era vicino, guardate quello che fanno gli italiani: Ho visto, disse: essi non votano.* APUG, FS 23.II.D.

are wrong, to be declined: you say to have understood that this thing was wrong and to have done everything possible to avert it—and then you did it: too bad. [...] The pretext of duty and command, in this case, is the excuse of Pilate."[133]

Secchi could write with satisfaction: "Once the session was over, several [colleagues] came to shake my hand [...]. The thing, hence, went smoothly and with disgrace [of the Italian delegation]. They thought to arouse a debate and nothing came of it."[134]

The Italians' debacle was completed the day after, when, because of their abstention, the Italian delegation was excluded from the standing committee. The serious consequences of the Italian government's position provoked embarrassment in the Italian delegation and criticism from inside the commission.

In the meantime, the diplomatic exchanges continued. On the very same day of the establishment of the standing committee, Visconti Venosta sent a telegram about the Secchi incident to the Italian ministers plenipotentiaries in Berlin, London, St. Petersburg, Vienna, The Hague, and Madrid and invited them to solicit the foreign ministers of those countries to send their advice to the French government for facilitating a solution to the incident. The government of Vienna and Berlin supported the Italian request, while that of The Hague abstained.[135] It is unclear if the other governments were actually involved in the affair. For the Italian government to have taken so drastic a position certainly created a diplomatic bottleneck: however, once informed about the initiative of the Italian delegates, the incident was considered solved. The publication of the note of protest in the proceedings would have made the position of the Italian government officially known.[136] As expected, the Italian

133 *Io risposi che io non poteva fare come i diplomatici e che faceva mio l'insulto fatto al Papa, e soprattutto da lui che io teneva per buon cattolico, [...] e insistendo esso che era suo dovere, gli replicai = i comandi si esaminano e quando sono cattivi si declinano, ella dice di capire che non andava bene questa cosa, che ha fatto il possibile pur di stornarla e poi l'ha fatta: peggio per lei. [...] la scusa del dovere e del comando in questo caso è la scusa di Pilato.* APUG, FS 23.II.D.

134 *Finita la seduta, parecchi vennero a stringermi la mano [...] La cosa passò quindi liscia e con disonore [degli Italiani]. Credevano di suscitare una discussione e non vi fu nulla.* APUG, FS 23.II.D.

135 See *Ministero degli Affari Esteri, Documenti diplomatici italiani,* 4:163–166.

136 Nevertheless, in November 1872, a question about the incident was raised at the Chamber of Deputies, and the minister had to defend his position; see "Discussioni: Tornata del 25 novembre 1872," *Atti Parlamentari, Camera dei Deputati, 1872–73* (Rome: Eredi Botta, 1874), 3311–32, here 3322.

FIGURE 5.7 Secchi and other "*coryphaei* of astronomy": clockwise: Schiaparelli, Secchi, Huggins, and his collaborator, William Allen Miller (1817–70). From *Nuovo giornale illustrato universale* 23 (1869): 181. Portraits of Secchi often appeared in popular magazines of his time.

protest gave rise to strong disapproval in the scientific community, and this resonated in the newspapers.[137] The event was not considered solely a personal success for the Jesuit astronomer but also an indirect political achievement for the Vatican in that international context. At last, a convenient diplomatic solution was reached: the commission would now assume the role solely of a scientific institution, having no political function. Therefore, at the final convention, signed in 1875, only political authorities attended, without delegates from the Holy See.

137 The president also invited Secchi to lunch.

Once back in Rome, Secchi was received in an audience by Pius IX:

> The pope welcomed me with a hand raised, saying "I vote for Fr. Secchi"; this pretty way and kindness in pronouncing these words, as soon as I went in and genuflected, showed his pleasure. He also assured me that he had never had such consolation in those days as that which I had procured. I replied that the best reward for the few annoyances that I had endured was the satisfaction of his holiness.[138]

138 *Il Papa mi ricevette a mano alzata dicendo = Io voto pel P. Secchi = colla quale graziosa maniera e col garbo con cui disse queste parole appena entrai e feci la genuflessione, mostrò il suo gradimento. Esso pure mi assicurò che non avea mai avuto consolazione più sincera in que' giorni che quella che io gli avea procurata. Io risposi che la miglior mia mercede per li non pochi fastidi sostenuti era la soddisfazione di sua Santità.* APUG, FS 23.II.D.

CHAPTER 6

Contributions to Astrophysics and other Sciences

The popular books by Agnes M. Clerke have already provided a comprehensive exposition of Secchi's contributions to astronomy and astrophysics, placing Secchi in the general scientific context of his time and noting the value of his contributions and discoveries. Accordingly, this chapter does not aim to retrace those contributions in comprehensive detail but focuses on the main ones, connecting them with Secchi's life and scientific interests.

1 Early Studies on Stars' Spectra

Secchi's first introduction to spectroscopic instruments dates back to November 1862, when he received a visit from French astrophysicist Jules C. Janssen at the Collegio Romano Observatory.

Janssen's mission, supported by the French government, was aimed at studying the "telluric lines" of the solar spectrum, namely the absorption lines due to the terrestrial atmosphere. In order to do so, he needed to observe the solar spectrum in a very favorable climate, like the Italian one.[1] Secchi was happy to open his observatory to his French colleague, as he was intrigued by some recent applications of spectroscopy to astronomy. In fact, his scientific curiosity was stimulated by the results that were obtained by Donati in Florence, who, in 1860, for the first time, analyzed starlight with an early spectroscopic apparatus and found that stars of the same color showed similar spectral patterns.[2] Starting from Donati's results, Secchi would come up with the idea of a spectral classification of stars.

1 See David Aubin, "Orchestrating Observatory, Laboratory, and Field: Jules Janssen, the Spectroscope, and Travel," *Nuncius* 17, no. 2 (2002): 615–633; see also Françoise Launay, *Un globe-trotter de la physique céleste: L'astronome Jules Janssen* (Paris: L'Observatoire de Paris-Vuibert, 2008), 35–41.

2 See Giovan Battista Donati, "Intorno alle strie degli spettri stellari," *Nuovo cimento* 15 (1862): 292–304; 366–376. According to instrument-maker Georg Merz (1793–1867), however, after Fraunhofer, the first person to spectroscopically analyze starlight was astronomer Johann von Lamont (1805–79) at Munich Observatory in 1838; see Georg Merz, "Objectiv-Spektralapparat," *Repertorium für Experimental-Physik* 6 (1870): 164–165, here 164; and Johann von Lamont, "Ueber die Anordnung und physische Beschaffenheit der Planeten," *Jahrbuch der Königliche Sternwarte bei München für 1838* (1838): 175–202, here 190–191.

 | DOI:10.1163/9789004387331_008

Lacking a spectrometer himself, Secchi asked Janssen to adapt his Hofmann pocket spectroscope[3] to the twenty-five-centimeter Merz refractor of the Collegio Romano Observatory.[4] The observations were successful: the spectra of numerous stars were clearly detected, and Janssen began to publish the results of these studies in Secchi's *Bullettino*.[5] However, a few months after giving him a warm welcome, Secchi barred Janssen from access to his observatory—a decision that was interpreted by the French astronomer as being rooted in Secchi's envy of the scientific results he was obtaining at the Collegio Romano.[6] Later, Secchi privately disclosed that he had been deceived by Janssen's uncooperative attitude;[7] probably, once in possession of his own Hofmann spectroscope, he dismissed Janssen and decided to continue the stars' spectra investigations on his own, having tested the best configuration of the instrument. Actually, Hofmann's direct-vision spectroscope was mainly constructed for solar observations, while Secchi intended to use it for analyzing starlight: therefore, it needed to be modified, which he did successfully.[8]

In 1863, at the Académie des Sciences of Paris, Secchi presented the first results of his spectroscopic observations of planets and, especially, of thirty-five stars, noting the connection between the color and the spectrum of the stars.[9]

3 Janssen described his five-prism pocket spectroscope in a short note to the Pontifical Academy of Nuovi Lincei in Rome (see Jules C. Janssen, "Note sur de nouveaux spectroscopes," *Atti della Pontificia Accademia dei Nuovi Lincei* 16 [1862–63]: 73–75, here 74).

4 Secchi stated that he had commissioned a Hofmann spectroscope before Janssen's visit—later, he pointed out that the Hofmann direct-vision spectroscope had been ordered from instrument-maker Marc-François-Louis Secretan (1804–67) in Paris in May 1862, once he had learned about Donati's results (see Angelo Secchi, "Sugli spettri prismatici delle stelle fisse," *Memorie della Società Italiana delle Scienze*, serie 3, t. I-I [1867]: 67–104, here 69), but the instrument was only delivered in December 1862 (see Angelo Secchi, "Sugli spettri prismatici della luce de' corpi celesti," *Bullettino meteorologico dell'Osservatorio del Collegio Romano* 2 [1863]: 108–110; 114–116; 124–126, here 108). In fact, he also mentioned a Duboscq spectrometer attached to his Cauchoix telescope and used for solar research (see Secchi's note in Jules C. Janssen, "Note sur la présence du sodium dans l'étoile alpha Orion," *Bullettino meteorologico dell'Osservatorio del Collegio Romano* 2 [1863]: 4).

5 See Janssen, "Note sur la présence du sodium dans l'étoile α Orion."

6 See Launay, *Un globe-trotter de la physique céleste*, 40–41.

7 See Secchi to Lockyer, Rome, January 27, 1873, Archives of the University of Exeter, Norman Lockyer Papers.

8 Secchi made two modifications: (1) he attached a lateral illuminated scale reflected into the instrument by one facet of the prisms' train in order to determine the relative positions of the lines; (2) in order to observe more lines, he removed the cylindrical lens as well as the eye-piece, both elements that absorbed light; unfortunately, this was obtained at the expense of their sharpness, thus losing precision in fixing their positions (see Angelo Secchi, *Le stelle: Saggio di astronomia siderale* [Milan: Fratelli Dumolard, 1877], 73). Fortunately, for inspecting the general features of stellar spectra, high precision was not required.

9 See Angelo Secchi, "Note sur les spectres prismatiques des corps célestes," *Comptes rendus de l'Académie des Sciences* 57 (1863): 71–75.

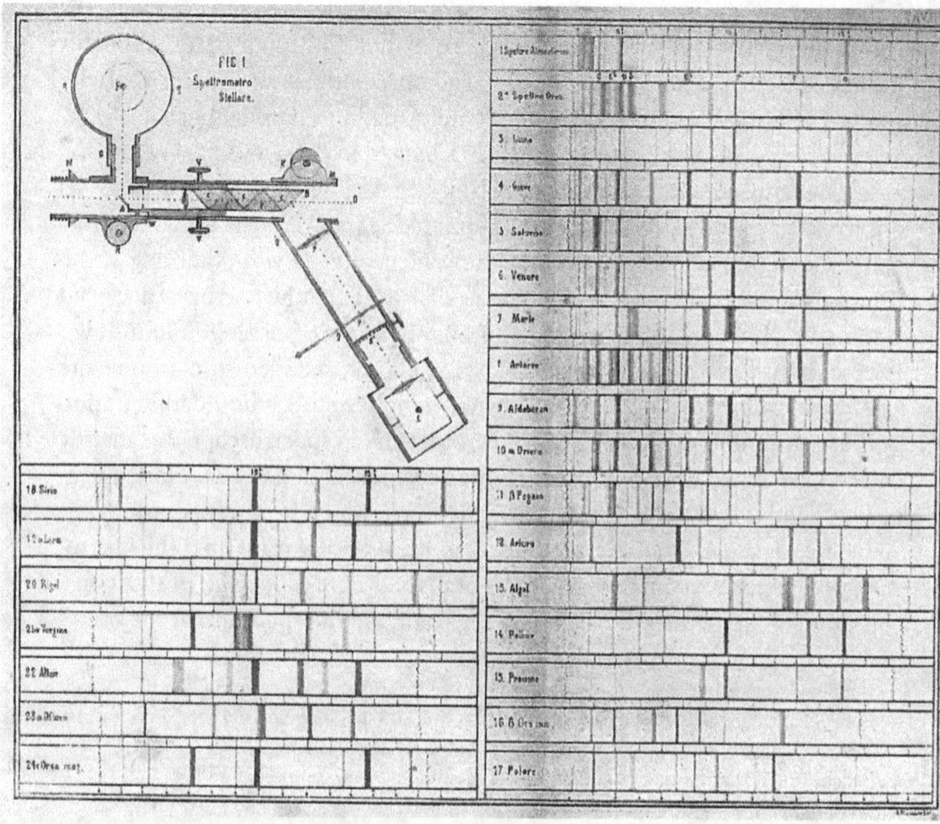

FIGURE 6.1 Plate illustrating Secchi's early spectroscopic observations of stars; on the top left: scheme of the star spectroscope. From Angelo Secchi, "Sugli spettri prismatici della luce de' corpi celesti," *Memorie dell'Osservatorio del Collegio Romano*, n.s. 2 (1863): 121–28, pl. 2.

He also announced that he was preparing a catalog containing drawings of those spectra to be published in the observatory's publications.[10]

Secchi well understood the potential impact of spectral analysis on astronomical research: "Spectral studies of celestial bodies have a double importance: first to ascertain the existence and the nature of their atmospheres, and second to find answers to some very interesting cosmic questions, especially those related to the proper motions of the stars."[11]

---

10 The catalog was published in the *Memorie*: see Angelo Secchi, "Sugli spettri prismatici della luce de' corpi celesti," *Memorie dell'Osservatorio del Collegio Romano*, n.s. 2 (1863): 121–128.

11 *L'étude des spectres prismatiques des corps célestes a une double importance: 1° celle d'établir l'existence et la nature de leur atmosphères, et 2° celle de pouvoir répondre à certaines*

In the bulletins for July and August 1863, Secchi published the results of his work, which were then assembled in the *Memoirs*. He proposed to use the Doppler shift[12] of spectral lines to study proper motions of stars and added that "spectral analysis can also be helpful for photometry"[13] by determining the peak of the emitted radiation; he then gave the description of his modified pocket spectroscope. After some comments on the spectrum of the terrestrial atmosphere, he described the spectra of the moon, Jupiter, Saturn, Venus, and Mars; the lunar spectrum was just the solar spectrum (being reflected sunlight), thus confirming the absence of an atmosphere on the moon, whereas the planetary spectra showed the solar spectrum with "evident signs of absorption, produced by their atmospheres,"[14] especially in Jupiter and Saturn, less clearly in Mars. He erroneously concluded that their atmospheres were not too different from ours, attributing the absorption effect to water vapor, as for the Earth.

Much more interesting is Secchi's analysis of stellar spectra. He divided them into two classes, colored and white stars, "though there is not a defined limit"[15] and described in detail the spectra of about twenty stars, with a plate showing the spectra of several celestial bodies he had examined. Secchi concluded:

> In general, it appears that colored stars have a different kind of spectrum in comparison with white stars. In the former, the gaps in the less refracting colors are predominant, whereas in the latter [the lines appear] in the more refracting [colors] and show some gaps in the blue. […] Many observations must be collected by many observers before anything can be determined conclusively. We consider this matter merely touched on here.[16]

---

questions d'ordre cosmique très-intéressantes, relatives surtout aux mouvements propres des étoiles. Secchi, "Note sur les spectres prismatiques des corps célestes," 71.

12 He refers to the Doppler shift of spectral lines, observed when the light source is receding from or in motion toward the observer. Huggins was the first to apply this method by measuring the radial velocity of Sirius in 1868 (see Becker, *Unravelling Starlight*, 115–120; see also Secchi, "On Stellar Spectrometry," 170).

13 *Finalmente lo studio spettrale può servire anche di mezzo fotometrico*. Secchi, "Sugli spettri prismatici della luce de' corpi celesti," 123.

14 *Manifesti segni di un assorbimento prodotto dalle loro atmosfere.* Secchi, "Sugli spettri prismatici della luce de' corpi celesti," 125.

15 *Benché non siavi limite definito.* Secchi, "Sugli spettri prismatici della luce de' corpi celesti."

16 *In generale si vede che le stelle colorate hanno altro tipo di spettro che le bianche. In quelle dominano le interruzioni nei colori meno refrangibili, e in queste nei più refrangibili, e mostrano grandi lacune nell'azzurro.* […] *molte e molte osservazioni sono da raccogliere e da più osservatori prima di nulla decidere. Noi consideriamo questa materia come appena leggermente sfiorata.* Secchi, "Sugli spettri prismatici della luce de' corpi celesti," 128.

This first work paved the way to later, ever more refined, spectral classifications by Secchi, which, together with the study of the Sun, comprised his main scientific effort for about a decade. While carrying out these studies, Secchi also improved the spectroscopic instrumentation by making adjustments and modifications.[17]

2 Cometary Spectra

The results obtained by Secchi in the field of cometary spectroscopy convinced him that "spectral analysis [was] among the most powerful means to sound out the nature of celestial bodies."[18]

Secchi was one of the first astronomers to examine the spectra of comets. He was only just preceded by Donati, the first person to make successful observations of a cometary spectrum, in Florence in 1864; he observed the spectrum of the comet Tempel (1864 I) with a wide-slit spectroscope and noted: "The spectrum of the comet looks like those produced by metals. Actually, the dark parts are larger than the luminous parts and one might say that these spectra are composed of three bright lines."[19]

Two years later, at the Collegio Romano Observatory, Secchi obtained the same results independently while carrying out spectral studies on comet Tempel-Tuttle (1866 I):

> I have found that [the spectrum of the comet] is composed of only three lines: one corresponding to the 2/5 of the distance between the Fraunhofer b and f, quite bright; its color is green [...]; the other two lines, or rays, are too short and faint to be able to measure their positions accurately. One, on the red side, is near the large main line, the other one is quite far, on the violet side.[20]

---

17 See, for instance, Angelo Secchi, *Le soleil* (Paris: Gauthier-Villars, 1875–77), 1:220–234.

18 *L'analisi spettrale* [*è*] *tra i mezzi più possenti che abbia la scienza per iscandagliare la natura de' corpi celesti.* Angelo Secchi, "Le recenti scoperte astronomiche," *Giornale Arcadico di Scienze: Lettere ed arti* 54 (1868): 168–196, here 174.

19 *Le spectre de la comète ressemble aux spectres produits par les métaux; en effet les parties noires y sont plus larges que les parties lumineuses, et on pourrait dire que ces spectres se composent de trois raies claires.* Giovan Battista Donati, "Schreiben des Herrn Professors Donati, Directors der Sternwarte in Florenza, an den Herausgeber," *Astronomische Nachrichten* 62 (1864): 375–378, here 378. Donati also indicated the approximate position of the bright bands he observed, which were included between the G and D Fraunhofer lines (λ4304–5890), in the yellow-green part of the solar spectrum.

20 *J'ai trouvé que* [*le spectre*] *est composé de trois rais seulement: une qui correspond à 2/5 de la distance entre b and f de Fraunhofer et est assez vive; sa couleur est verte* [...] *les autres*

Spectral analysis opened up interesting perspectives in cometary astronomy, though the results of the effort to measure, identify, and interpret the spectral lines of comets were sometimes unclear. By accurately comparing cometary with laboratory spectra, in 1868 Huggins (who was to become a specialist in this field[21]) identified the three bright bands observed by Donati as carbon spectral lines.[22] Secchi, however, recognized that

> the matter [was] not exhausted. [...] It is up to the chemical philosopher to resolve some of these difficulties; here, the astronomer can only walk with the lamp of chemistry. Before long, we shall know more accurately what maintains light and heat in so many bodies, scattered in the profundity of space.[23]

Later, Secchi's spectral observations of the comet Winnecke (1874 II)[24] confirmed Huggins's results. He had also remarked that cometary spectra were similar, but not identical. From there, a scientific debate on the chemical–physical nature of comets was sparked by the similarities and differences observed in their spectra.[25] The main open questions were summarized by Secchi:

> (1) Is the spectrum of the comet [when it has] just appeared identical to that developed later by the comet at its maximum brightness? (2) Do all

deux lignes ou raies sont trop petites et faibles pour en pouvoir fixer exactement la position. L'une est, du côté rouge, assez près de la large raie principale, l'autre assez éloignée du côté violet. Angelo Secchi, "Spectre de la comète de Tempel," *Comptes rendus de l'Académie des Sciences* 62 (1866): 210. The spectrum of the same comet was also described by Huggins (see below) in a more detailed article; see William Huggins, "On the Spectrum of Comet I, 1866," *Proceedings of the Royal Society* 15 (1866): 5–7.

21 See John B. Hearnshaw, *The Measurement of Starlight* (Cambridge: Cambridge University Press, 1996), 73.

22 See William Huggins, "On the Spectrum of Comet II, 1868," *Proceedings of the Royal Society* 16 (1868): 481–482. On the attribution of these bands, today known as "Swan bands," to molecular carbon (λ4380, 4740, 5165), see William Swan, "On the Prismatic Spectra of the Flames of Compounds of Carbon and Hydrogen," *Transactions of the Royal Society of Edinburgh* 21 (1857): 411–430.

23 Secchi, "On Stellar Spectrometry," 170.

24 See Angelo Secchi, "Sullo spettro della Cometa Tempel," *Memorie della Società degli Spettroscopisti Italiani* 3 (1874): 29–30. The comet 1874 II was also independently discovered by Ernst Wilhelm Tempel (1821–89) (see Kronk, *Cometography*, 2:403), and Secchi referred to it as comet Tempel.

25 Cometary spectra were initially compared with those of planetary nebulae and the existence of a background continuum spectrum, sometimes observed, was also discussed, as well as the nature—self-luminous or reflected—of the cometary light; see Polydore F.F. Swings, "Cometary Spectra," *Quarterly Journal of the Royal Astronomical Society* 6 (1965): 28–69, here 29–30.

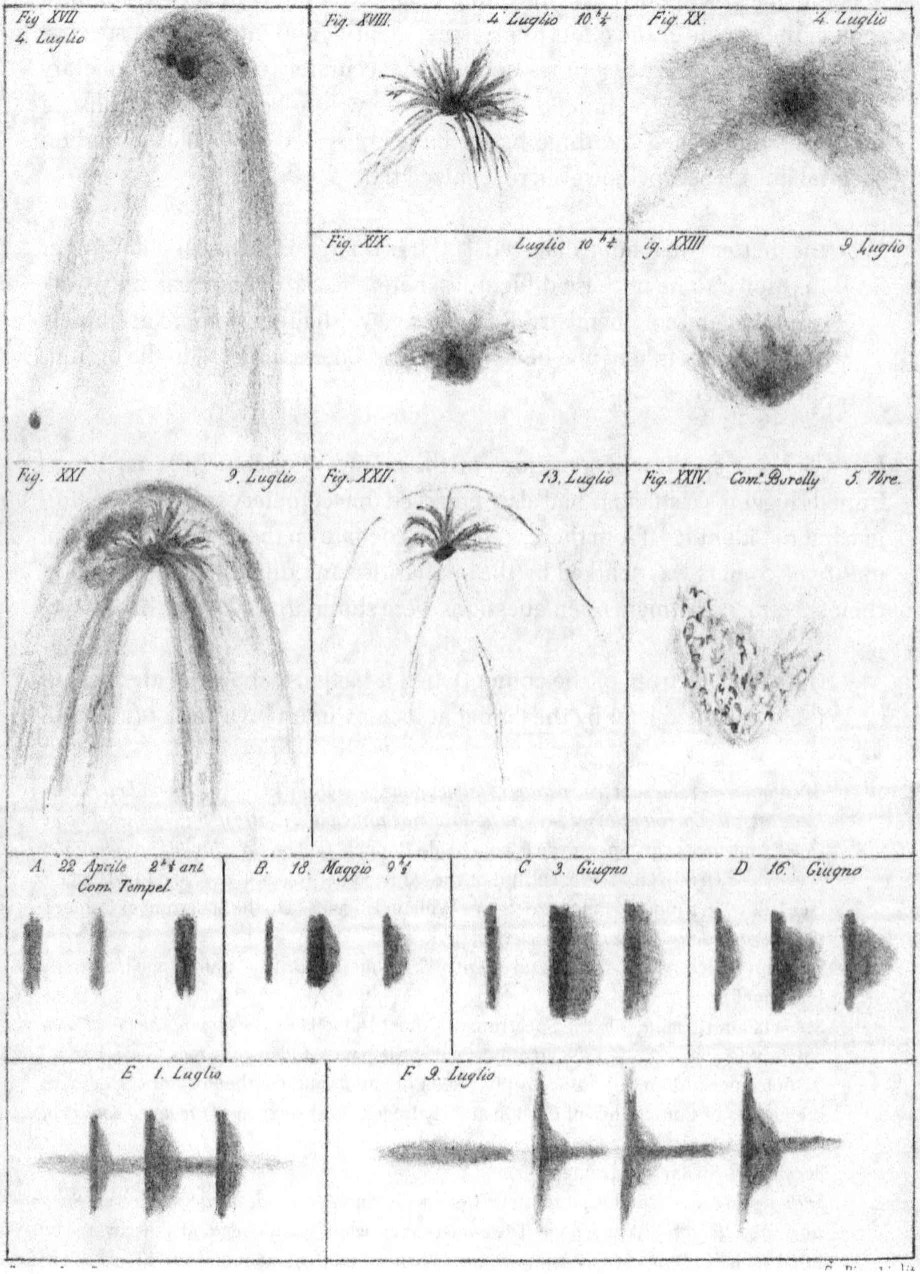

FIGURE 6.2 Visual and spectroscopic observations of comets by Secchi in 1874. From Angelo Secchi, "Studi fisici fatti all'Osservatorio del Collegio Romano sulle comete di Tempel II e Coggia III nel 1874," *Memorie della Società degli Spettroscopisti Italiani* 3 (1874): 121–26, pl. 55.

> [comets] have the same [spectral] bands, not only at the same positions, but also with the same intensity? (3) What is the influence of sunlight? (4) Is this [light] polarized?[26]

Secchi also carefully studied the spectra of the comet Coggia (1874 III), combining spectroscopic and polariscopic observations: he concluded that, in addition to the usual three-band hydrocarbon spectrum, there was a component of polarized light producing a faint continuum spectrum, due to reflected sunlight.[27]

3 Secchi's Spectral Classification

Intrigued by the results he had obtained with stellar spectra in 1863, Secchi continued this kind of research in the following years:

> Seeing that this field was so fruitful, I tried to determine what else I could glean from it; and I tried to see if there was any pattern in stellar spectra, and hence in the composition of these bodies. I basically wanted to see if the innumerable stars vary in composition proportionately with their numbers.[28]

Despite successful observations of the spectra of Betelgeuse, Sirius[29] and Antares,[30] Secchi was unhappy with the performance of his spectroscope. He continuously tried to make adjustments and improvements by interchanging several direct-vision prisms. In August 1866, it was suggested[31] that he simplify the instrument by placing the direct-vision prism between the cylindrical lens

26 *1. E' lo spettro delle comete al loro comparire identico a quello che sviluppano appresso nella massima loro luce? 2. le sono in tutte le stesse non solo nel posto, ma anche nella intensità? 3. Che influenza vi ha la luce solare? 4. E' questa polarizzata?* Angelo Secchi, "Studi fisici sulle Comete del 1874," *Memorie della Società degli Spettroscopisti Italiani* 3 (1874): 117–120, here 117.

27 See Secchi, "Sullo spettro della Cometa Tempel."

28 *Vedendo questo campo così fecondo, io cercai di spigolarvi, e mi proposi di cercare se vi fosse qualche legge negli spettri stellari, e quindi nella composizione di questi astri. Volli in sostanza vedere se, come sono innumerabili le stelle, fosse pure proporzionalmente varia la loro composizione.* Secchi, "Le recenti scoperte astronomiche," 174–175.

29 See Secchi, "Schreiben des Herrn Professors Secchi. [...] an den Herausgeber," *Astronomische Nachrichten* 66, no. 1577 (1866): 257–258.

30 See Angelo Secchi, "Schreiben des Herrn Professors Secchi [...] an den Herausgeber," *Astronomische Nachrichten* 67, no. 1608 (1866): 381–382, here 381.

31 See Secchi, "Sugli spettri prismatici delle stelle fisse," 71. Secchi does not mention who gave him this suggestion, but it is likely that it came from Merz, whose partnership was very important in the development of proper equipment for Secchi's spectral research. On Secchi's spectroscopic equipment by Merz, see Chinnici, "Maker and the Scientist,"

and the eyepiece.[32] To this end, Secchi commissioned a new spectroscope[33] with no slit and just a cylindrical lens; in addition, he replaced the spider wires of the micrometer—too difficult to see when observing faint stars—with two steel tips, overlapping the spectral line to be measured without covering it completely.[34] The performance now lived up to Secchi's expectations, and he used this "simple spectroscope" (as he called it) for his first spectral catalog of stars.

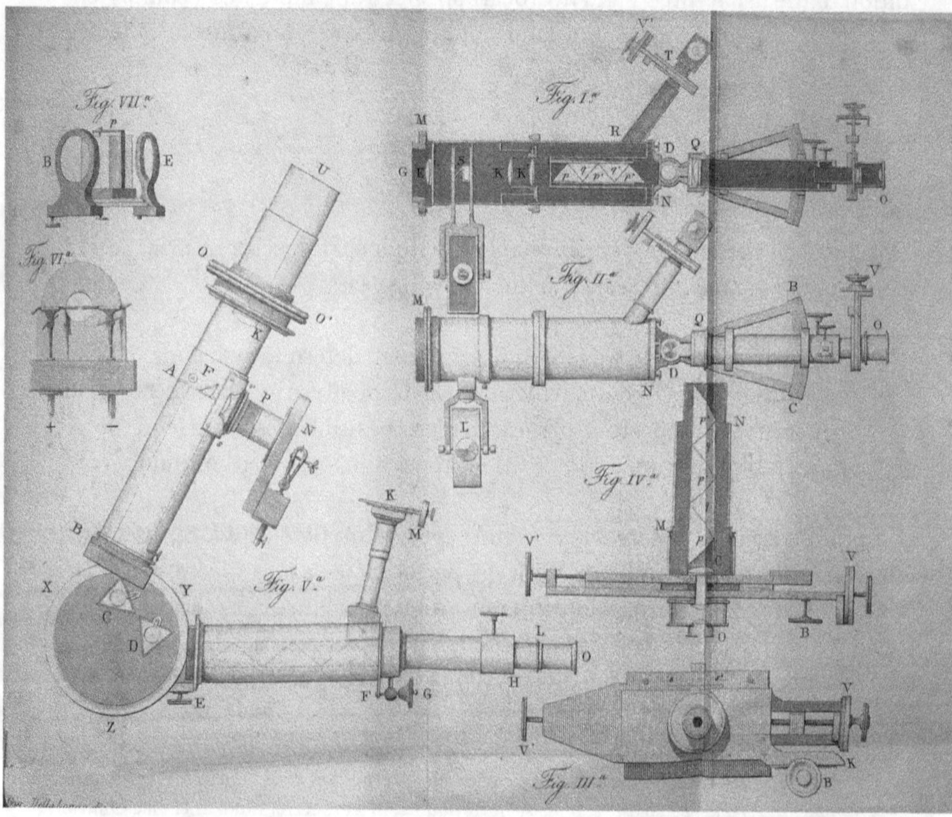

FIGURE 6.3 Diagrams of Secchi's spectroscopic instruments in 1867; on the left, the solar spectroscope and, on the right, the star spectroscope. From Angelo Secchi, "Sugli spettri prismatici delle stelle fisse. Memoria IIª," *Memorie della Società Italiana delle Scienze*, serie terza, 2 (1869): 73–133.

39–68; see also Georg Merz and Sigmund Merz, "Aus einem Schreiben der Herren G. & S. Merz an den Herausgeber," *Astronomische Nachrichten* 65, no. 1537 (1865): 13–16.

32 Secchi then changed this combination and obtained more sharpened spectral images by inverting the positions of the lens and the prism; see Secchi, "Schreiben des Herrn Professors Secchi. [...] an den Herausgeber," *Astronomische Nachrichten* 67, no. 1608 (1866): 381–382, here 382.

33 See Secchi, "Sugli spettri prismatici delle stelle fisse," 71.

34 See Secchi, "Sugli spettri prismatici delle stelle fisse," 73.

He published the catalog in 1867, together with an essay on the spectral analysis of starlight, where he proposed a classification based on three spectral types: α Lyrae-type (white stars), with few but strong dark lines; sun-type (yellow stars) with many tiny lines; and α Herculis-type (red stars), with strong absorption bands (which he called "pillars" or "columns"):[35]

> I confess that I was surprised by this unexpected result, to find so much similarity between these spectra and to see so few types among so many

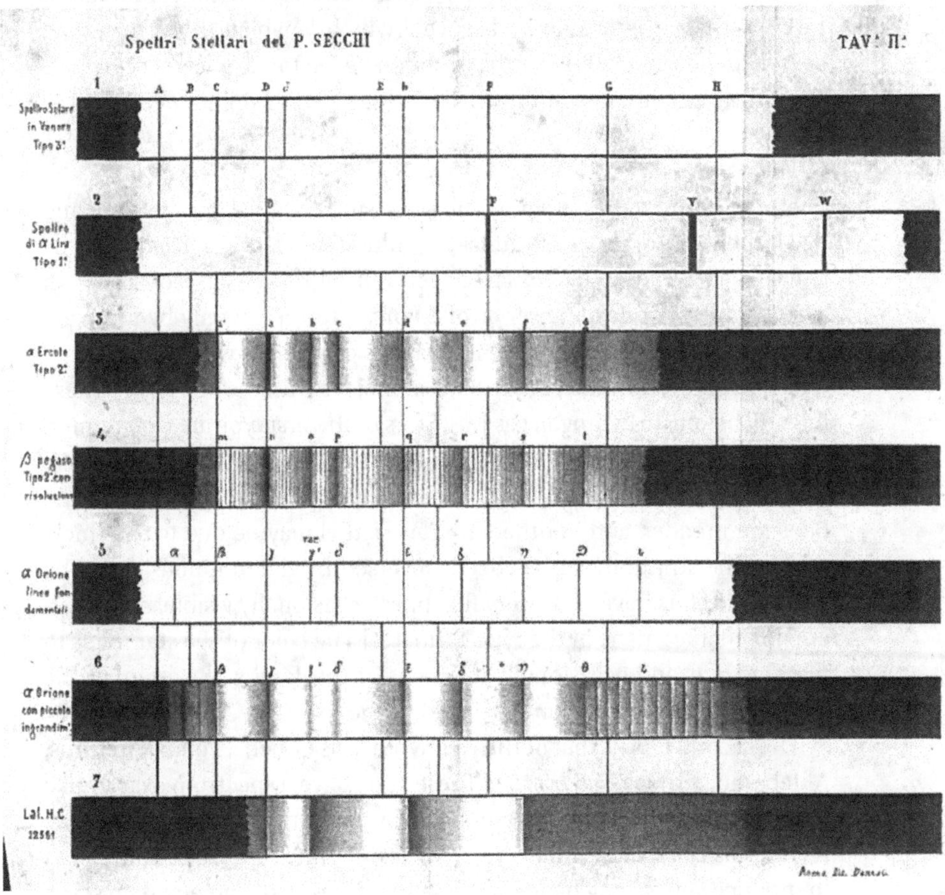

FIGURE 6.4 Secchi's three-class spectral classification. From Angelo Secchi, "Sugli spettri prismatici delle stelle fisse," *Memorie della Società Italiana delle Scienze*, serie 3, t. I-I (1867): 67–104, pl. 2.

35 See Angelo Secchi, "Catalogo delle stelle di cui si è determinato lo spettro luminoso all'Osservatorio del Collegio Romano," *Memorie della Società Italiana delle Scienze*, serie 3, t. I-I (1867): 105–152.

> bodies. Naturally, there are some transition stars, or else faint and featureless, but these are very few. One cannot believe that there are really only two types, and the third is just an exaggeration of the second, because there are many reasons that [demonstrate] the contrary [...].[36]

Secchi gave an overview of the results of his research on stellar spectrometry in a talk given to the British Association for the Advancement of Science in Norwich in 1868 (see Chapter 5). After crediting Donati and Huggins for their early studies in this field, he accurately described the "optical combination that [he] had the good fortune to discover," which "enabled [him] to extend the research to the whole of the visible stars, and even to the telescopic ones"[37] up to seventh and eighth magnitude. He also described the main spectral types as follows:

> The first type is represented by the stars Sirius, and Vega or α Lyrae and by all the white stars, as α Aquilae, Regulus, Castor, the large stars in the Great Bear, α excepted, &c. The spectra of all these stars consist of an almost uniform prismatic series of colors, interrupted only by four very strong black lines. [...] The one in the red is coincident with the solar line C of Fraunhofer; another, in the blue, coincides with the line F [...]. These lines all belong to hydrogen gas [...]. Stars of the first type are very numerous and embrace almost one-half of the visible stars of the heavens. We observe, however, some difference in individual stars; so that in some the lines are broader, and in others narrower; this may be due to the thickness of the stratum that has been traversed by the luminous rays. The more vivid stars have other very fine lines occasionally visible, but which are not characteristic of the type-form. [...] The color of the star tends to the blue hue, and occasionally to green. Of this kind is the group of the large constellation Orion and its neighborhood.
>
> The second type is that of the yellow stars, as Capella, Pollux, Arcturus, Aldebaran, α Ursae Majoris, &c. These stars have a spectrum exactly like that of our Sun—that is, distinguished by very fine and numerous lines. [...] In this class, occasionally the magnesium lines are very strong, so as to produce very strong bands, and the iron lines in the green are in some

36 *Vi confesso che io sono stato sorpreso da questo inaspettato risultato, di trovare tanta similitudine tra questi spettri e di vedere si pochi tipi fra tanti corpi. Come è naturale vi sono alcune stelle di transizione o piuttosto dubbie ed uniformi, ma queste sono pochissime. Non si può credere che i tipi realmente siano due soli, e che il terzo non sia che una esagerazione del secondo; perché vi sono varie ragioni in contrario* [...]. Secchi, "Le recenti scoperte astronomiche," 176.

37 Secchi, "On Stellar Spectrometry," 169.

> very distinct. [...] Stars of this second type are very numerous, and embrace almost the other half of the stars.
>
> The third and very remarkable type is that of orange or reddish stars. These have as a prototype the stars α Herculis, α Orionis, Antares, omicron Ceti, β Pegasi. The spectra of these stars show a row of columns at least eight in number, which are formed by strong luminous bands alternating with darker ones, so arranged as to represent apparently a series of round pillars, closely resembling a colonnade. [...] It is quite impossible to describe the beauty of the appearance that is visible in a telescope on a fine night.
>
> All the pillars are generally resolved more or less completely in different stars into smaller and finer lines, very sharp and clear. I have carefully drawn [...] the spectrum of α Orionis and α Herculis [and] those of Antares and Aldebaran [...]. In these stars, some of the divisions of the pillars correspond to some principal lines of Fraunhofer [...] but others, although very near, do not coincide with them [...]. The presence of hydrogen, however, is certain, [hydrogen lines] having been found in the principal of them.[38]

Of particular importance, from the historical point of view, was his announcement that he had examined all the stars of the Schjellerup catalog of red stars[39] to the eighth magnitude, "beyond which limit [his] instrument cannot give a good spectrum"[40] and to have added a fourth class for these stars, appearing "like drops of blood in the field of the telescope" and having carbon-like spectra consisting "of three large bands of light, which alternate with dark spaces," as later described by Secchi (see below).

Though he did not identify the corresponding chemical elements, he felt "authorized in supposing these stars to be still in a different condition from others, perhaps partly in the gaseous state, or at least surrounded by a very large atmosphere different certainly from that of the others."[41] With this statement, Secchi seemed to suggest that these stars were in a different

38 Secchi, "On Stellar Spectrometry," 166–167.

39 See Hans C.F.C. Schjellerup, "Catalog der roten, isolierten Sterne," *Astronomische Nachrichten* 67 (1866): 97–112. Hans Carl Frederik Christian Schjellerup (1827–87) was an astronomer at Copenhagen Observatory; his work on red stars opened a research field of considerable interest (see Hearnshaw, *Measurement of Starlight*, 43), especially because of the change in the observed color of some stars (see Agnes Mary Clerke, *The System of the Stars* [London: Longmans, Green and Co., 1890], 152–154).

40 Secchi, "On Stellar Spectrometry," 166.

41 Secchi, "On Stellar Spectrometry," 168.

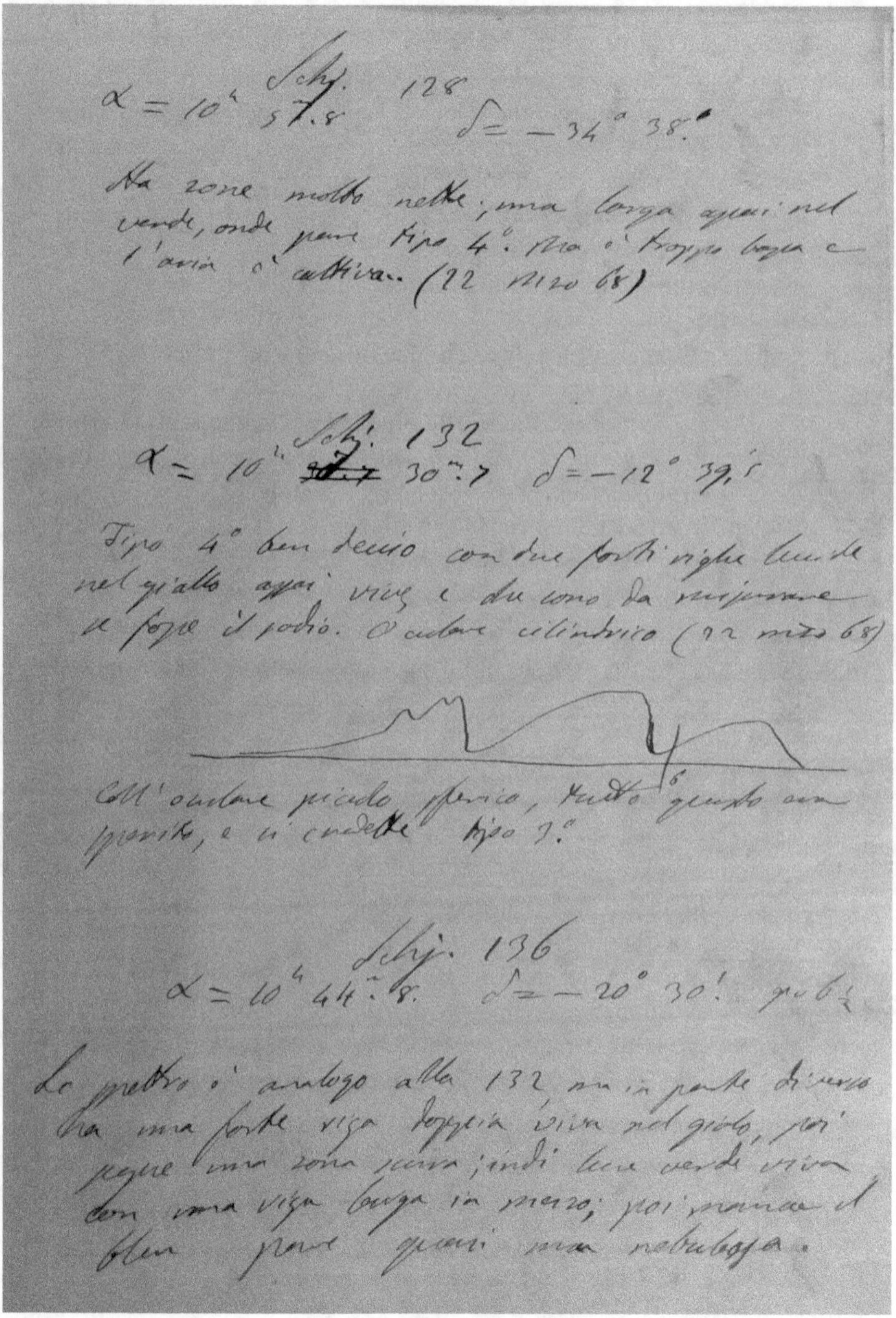

Schj. 128
α = 10h 57.8 δ = −34° 38.'

Ha zone molto nette, ma larga assai nel verde, onde pare tipo 4°. Ma è troppo bassa e l'aria è cattiva. (22 marzo 68)

Schj. 132
α = 10h 30m.7 δ = −12° 39.'5

Tipo 4° ben deciso con due forti righe lucide nel giallo assai vive e due cono da riscontrare se fosse il sodio. Oculare cilindrico (22 marzo 68)

Coll'oculare piccolo sferico, tutto questo era sparito, e si crederebbe tipo 3°.

Schj. 136
α = 10h 44m.8. δ = −20° 30'. gr 6½

Lo spettro è analogo alla 132 ma in parte diverso ha una forte riga doppia viva nel giallo, poi segue una zona scura; indi luce verde viva con una riga larga in mezzo; poi manca il bleu pare quasi una nebulosa.

FIGURE 6.5 Manuscript by Secchi with drawing of the spectrum of a type IV star. Courtesy of INAF-OAR.

evolutionary stage, in a sort of parallelism between living organisms and cosmic objects, an idea that was recurring in the years of Darwinism. Moreover, intrigued by the emission spectra of nebulae, in 1868 he wrote that he was looking for a "missing link": "I tried to see if, among small stars, there was hidden any formation that was intermediate between an ordinary star and nebula. So far, however, my observations have been unsuccessful."[42]

As he became more and more experienced in this kind of observation, Secchi refined his classification, noticing details and peculiarities. For example, he was the first to notice the presence of strong hydrogen emission lines in γ Cassiopeae[43] and, "to a very faint degree," in β Lyrae:[44]

> I was most surprised by one more thing, and it is that, among so many stars examined, I have found that two are of the first type but inverted; that is, they have bright lines where the others have black [lines]. These are γ Cassiopeia, and β Lira. The latter, indeed, is very difficult to recognize, but the other is easy. Among the curious things that deserve to be enumerated are some stars fainter than fifth and sixth magnitude that give a discontinuous spectra with bright isolated bands and look like an exaggeration of the third type; but the study of these is not yet finished and all have a vivid red color.[45]

Further investigation later led him to add a fifth class of stars with spectra of this type (see below). He also noticed that all variable stars had spectra belonging to the third and fourth type and showed changes in their light over the years, "so that it is evident that the change of these stars depends on a periodical change that happens in their atmosphere."[46]

---

42 *Ho cercato se tra le piccole stelle vi fosse nascosta qualche formazione intermedia tra la nebulosa e la stella ordinaria. Ma finora le mie osservazioni sono state infruttuose.* Secchi, "Le recenti scoperte astronomiche," 178.

43 See Secchi, "Schreiben des Herrn Professors Secchi, [...] an den Herausgeber," *Astronomische Nachrichten* 68, no. 1612 (1867): 63–64.

44 The variability of the spectrum of β Lyrae was later studied at Harvard by Antonia Maury (1866–1952); see Antonia Maury, "The Spectral Changes of Beta Lyrae," *Annals of Harvard College Observatory* 84, no. 8 (1933): 207–255.

45 *Un' altra cosa ancora mi ha più sorpreso, ed è che tra tante stelle esaminate, non ne ho trovate che due, le quali siano a rovescio del tipo primo; cioè che abbiano linee luminose, ove le altre le hanno nere. Queste sono γ Cassiopea, e β Lira. Quest'ultima però è assai difficile a riconoscere, ma l'altra è facilissima. Fra le cose curiose meritano di esser numerate alcune stelle minori di 5ª e 6ª grandezza che danno spettri discontinui e composti di fasci di luce isolati, e che sembrano una esaggerazione* [sic] *del terzo tipo; ma queste non sono ancora finite di studiare e tutte sono di color rosso carico.* Secchi, "Le recenti scoperte astronomiche," 177.

46 Secchi, "On Stellar Spectrometry," 169.

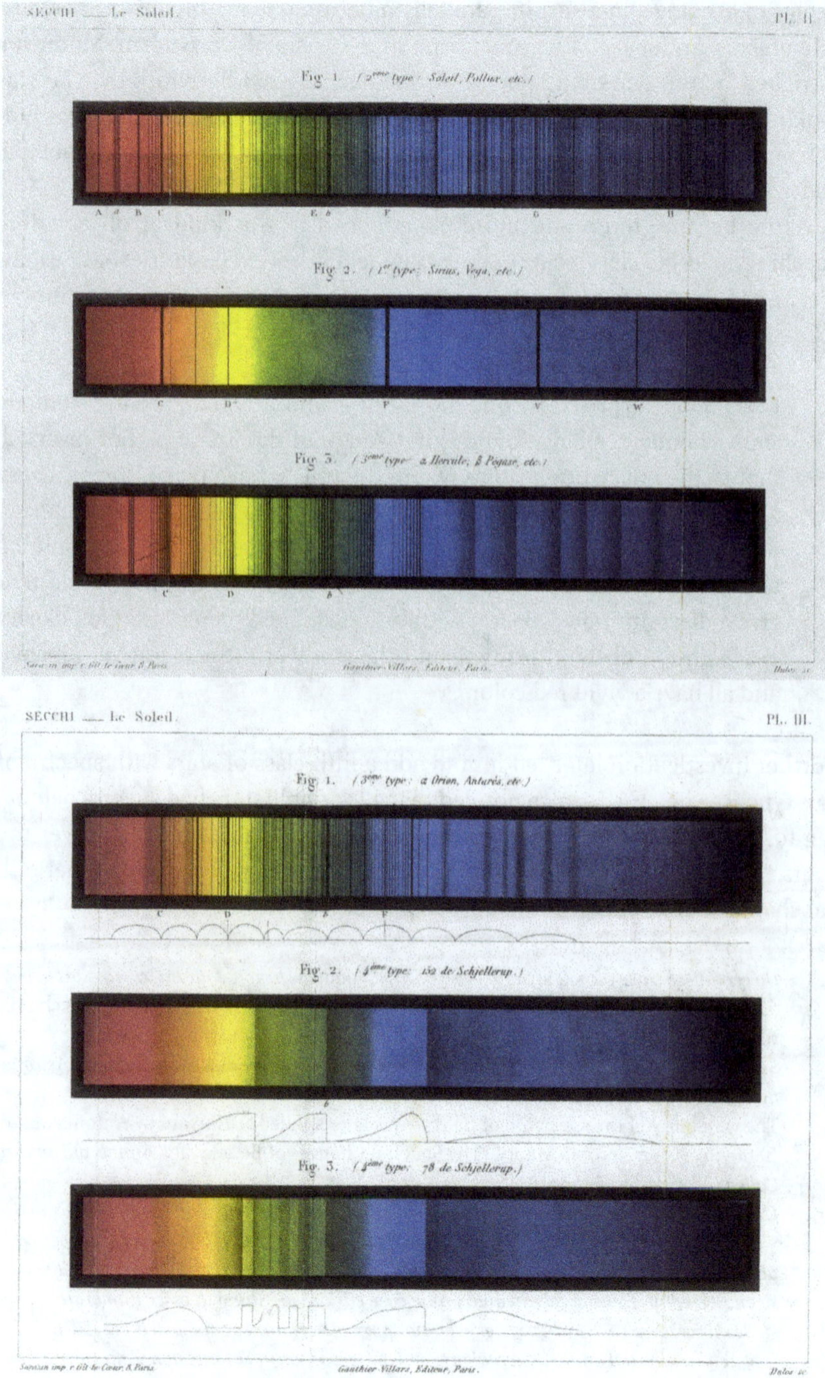

FIGURE 6.6 A-B Spectral classification of 1869, with four star types. From Angelo Secchi, *Le soleil* (Paris: Gauthier-Villars, 1870), pl. 2–3.

Thanks to his classification, Secchi arrived at some important conclusions. Beyond the fact that stars could be classified "according to a small number of types," he noticed that

> stars of the same type are occasionally crowded in the same space of the heavens. Thus the white stars are gathered in Leo, in Ursa Major, in Lyra, Pleiades, &c., while the yellow ones are very frequent in Cetus, in Eridanus, Hydra, &c. The region of Orion is very remarkable for having all around its neighborhood green stars of the first type, but with very narrow lines [...]. It seems that this particular kind of star is seen through the great mass that constitutes the great nebula of Orion, whose spectrum may contrast with the primitive spectrum of the stars.[47]

In 1869, Secchi published a second essay on the spectral classification of stars, which included the fourth class, containing faint red stars:

> In the preceding essay, we had classified stars into three fundamental classes, but current observations have compelled us to add a fourth one. This spectral type [...] includes mostly fainter stars, not exceeding sixth magnitude. [...] The spectra of these stars are the most similar to the spectra of gases, especially to that of carbon, but inverted.[48]

He chose, as the type-star of this class, no. 152 in the Schjellerup catalog of red stars; the spectrum of these stars showed many carbon bands that Secchi compared with similar lines in the laboratory spectra of nitrogen gas excited by electric sparks in Geissler tubes; Geissler tubes are early gas discharge tubes, invented by Heinrich Geissler (1814–79) in 1857. They consist of partially evacuated glass cylinders with metal electrodes at each end. The tubes contain rarefied gases, minerals or metals, that are ionized when an electric current flows through the tube, thus producing characteristic light emissions. They were ordinarily used in laboratories for spectral analysis. he described them as "shaded, bright, shutter-like bands spectra, grouped in columns, separated by graduated dark areas. Yet, some of these stellar colonnades are irresolvable at all into tiny lines, and these spectra are not easy to find in chemistry [...]."[49]

47 Secchi, "On Stellar Spectrometry," 169.

48 See Angelo Secchi, "Sugli spettri prismatici delle stelle fisse: Memoria II^a^," *Memorie della Società Italiana delle Scienze*, serie 3, t. 2 (1869): 73–133, here 79.

49 *Spettri rigati a persiane lucide sfumate, raggruppate in colonne separate da zone scure graduate. Alcuni però di questi colonnati stellari sono affatto irresolubili in linea fine, e questi spettri non sono facili a riscontrare in chimica* [...]. Secchi, *Le stelle*, 97.

Secchi also announced that he had started using a flint prism, six inches in diameter, placed in front of the object-glass of the large Merz refractor.[50] The prism was inserted in a frame that could be applied like a cap on the objective lens and could pivot to reach minimum dispersion.[51] Depending on the kinds of celestial objects to be observed, Secchi attached this device to either the large Merz refractor or to the Cauchoix telescope, which he generally used for solar observations.[52] The use of this apparatus was crucial for the development of Secchi's spectral classification of stars: the objective-prism, in fact, allowed him to observe simultaneously the spectra of all the stars in the view field of the telescope, thus collecting a high number of spectra and allowing a comparison among them. He was then in a position to extend his classification further, adding a fifth class, using γ Cassiopeae as the type-star, whose spectrum shows many emission hydrogen lines: "There are [...] a very few [stars] that do not fit into these categories, because they have isolated simple lines, and show bright instead of dark lines, though in the same positions, and can form a fifth type, with bright lines."[53]

The Jesuit astronomer paid special attention to the spectra of variable stars. He gave the first description of the spectrum of the famous variable star Mira (Omicron Ceti), which he described as continuous with absorption bands: "Strong dark flutings terminated abruptly on the violet side and shaded off gradually in the other direction."[54] In 1869, Secchi discovered bright hydrogen lines near maximum brightness in the spectrum of the variable star R Geminorum[55] and found that their presence was typical of long-period variables, all red stars of type III.[56]

---

50 See Angelo Secchi, "Studi spettrali," *Bullettino meteorologico dell'Osservatorio del Collegio Romano* 8 (1869): 81–83, here 81. Secchi stated to apply an original idea of Fraunhofer; see Secchi, *Le soleil* (1875–77), 2: 447. The priority in the use of this optical system was disputed by Respighi (see Chapter 9).

51 For a detailed description of the objective-prism, see Secchi, *Le soleil* (1875–77), 2: 447–449.

52 This instrument is preserved today at INAF-OAR; a rare photograph of it, from the time of Secchi's use, is kept in the archives of Science Museum in London, see Chinnici, "Maker and the Scientist," 63.

53 *Ve ne sono però alcune poche, le quali non entrano in queste categorie, perchè hanno righe semplici isolate, e mostrano linee lucide invece delle nere, non diverse però di posizione e possono formare un 5° tipo a righe lucide.* Secchi, *Le stelle*, 98. On the impact of Secchi's classification in the development of astrophysics, see the the conclusion to this book.

54 Helen Sawyer Hogg, "Variable Stars," in Gingerich, *General History of Astronomy, 4A*, 73–89, here 81.

55 See Angelo Secchi, "Résultats fournis par l'analyse spectrale de la lumière d'Uranus, de l'étoile R des Gémeaux, et des taches solaires," *Comptes rendus de l'Académie des Sciences* 68 (1869): 761–765, here 763–764.

56 See David Leverington, *A History of Astronomy: From 1890 to the Present* (London: Springer-Verlag, 1995), 162.

Finally, it is interesting to note that Secchi was open to the possible existence of planets in other star systems (i.e., "exoplanets"). While examining the spectra of variable stars, he noticed that Algol's spectrum was different from the others and did not show any special change during the periods of decreasing brightness:

> Having seen that variables stars had spectra of the third type, I looked to see if Algol (β Persei) were of this type, instead it is of the first, and remains the same [type] as its brightness decreases. Hence I think its variability is not due, like in the other cases, to atmospheric absorption or variable physical activity; but only to an opaque body orbiting around it. If so, we might have evidence of the existence of an opaque planet among the stars![57]

Secchi's spectroscopic observations therefore reinforced the eighteenth-century hypothesis that Algol's variability "was an example of eclipse of a fixed star by its own dark planet."[58]

Secchi also observed the spectrum of Nova Cygni 1876[59] and made other interesting contributions on variable stars, which he divided into eclipsing binaries, intrinsic variables, and temporary variables, like novae, "burning suddenly and rarely, in some cases only one time,[60] to form a true new world in this way."[61]

4 Planetary Spectra

In the field of planetary spectroscopy, Secchi's most remarkable achievement was the first spectroscopic observation of the planet Uranus in 1869. He

57 *Avendo veduto che le variabili appartenevano al tipo 3°, ho cercato se Algol (β Perseo) fosse di questo tipo; ma essa invece è del primo, e lo conserva nelle sue diminuzioni di luce. Quindi io credo che la sua variabilità non avvenga come nelle altre da assorbimento atmosferico o fisica attività variabile; ma solo da un corpo opaco che le gira intorno. Così avremmo la prova dell'esistenza di un pianeta opaco tra le stelle!* Secchi, "Le recenti scoperte astronomiche," 177.

58 Secchi, "On Stellar Spectrometry," 169. The English amateur astronomer John Goodricke (1764–86) was the first to study the variability of Algol and to propose the eclipse hypothesis, which was strongly contested at that time.

59 See Angelo Secchi, "La nuova stella del Cigno," *Bullettino meteorologico dell'Osservatorio del Collegio Romano* 16 (1877): 1–3.

60 This is the case of supernovae.

61 *Per veri incendii che accadono in esse a periodi incerti e lontani, e forse in alcune per una volta sola, per formare così un vero mondo novello.* Secchi, "La nuova stella del Cigno," 2.

announced this unexpected discovery to the Académie des Sciences: "The spectroscope continues to reveal some very curious things, but the one that I have just found out will probably appear to you, like to me, quite singular. This is the spectrum of Uranus, which presents a completely unexpected composition."[62]

The discovery was made while Secchi was using the spectroscope to observe the faint variable star R *Geminorum.* He reported:

> I turned the spectroscope to Uranus, which was in the field of the finderscope, with no expectation of seeing anything special, because the planet seemed barely a sixth-magnitude star, and the moon, which was not far away, was quite bright. What a surprise, seeing a very bright spectrum and with a very sharp band near its end![63]

Secchi sought to pinpoint the position of the dark band, roughly located in the green region of the visible spectrum; after a more careful observation, he was able to see another dark band toward the violet, a considerable gap in yellow and a faint emission in the red.

He concluded:

> We thus see that the spectrum of this planet has no resemblance to the solar spectrum.[64] The huge gap is quite unique; there is no example in

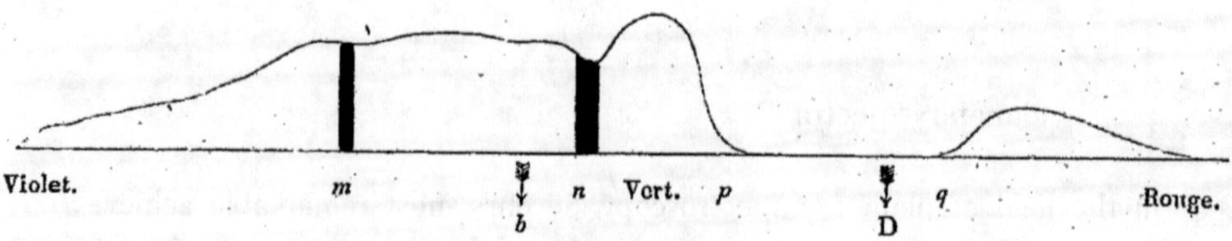

FIGURE 6.7 First observation of the spectrum of Uranus. From Angelo Secchi, "Résultats fournis par l'analyse spectrale de la lumière d'Uranus, de l'étoile R des Gémeaux, et des taches solaires," *Comptes rendus de l'Académie des Sciences* 68 (1869): 761–65, here 762.

62 *Le spectroscope continue à nous révéler des choses bien curieuses, mais celle que je viens de découvrir vous paraîtra sans doute, comme a moi, assez singulière. Il s'agit du spectre d'Uranus, qui présente une constitution tout à fait inattendue.* Secchi, "Résultats fournis par l'analyse spectrale de la lumière d'Uranus," 761.

63 *Je tournai le spectroscope vers Uranus, qui était dans le champ du chercheur, sans espérer y rien voir de particulier, car la planète semblait à peine une étoile de 6e grandeur, et la Lune, qui n'était pas éloignée, était assez grande. Quelle fut ma surprise, en voyant un spectre très-vif et avec une bande très-prononcée près de son extrémité!* Secchi, "Résultats fournis par l'analyse spectrale de la lumière d'Uranus," 761–762.

64 The Fraunhofer lines in Uranus' spectrum were so faint that they were not easily detected; they were only observed in 1889, thanks to the application of long-exposure photographic

> the other planets; and though Jupiter and Saturn show a few bands of absorption, they are far from having the same gap, as if the yellow color was entirely lacking. If this spectrum is purely due to reflected sunlight (which might perhaps be questioned), it must undergo a considerable modification in [the planet's] atmosphere.[65]

Two years later, Huggins confirmed Secchi's observations; he also found additional bands and measured their positions, identifying only a line of hydrogen (H_β). The nature of these broad dark bands remained unknown until the 1930s, when they were identified as belonging to gaseous methane.

5 Dark Nebulae

In the nineteenth century, some astronomers detected dark clouds, or dark "nebulae," in the sky, which were eventually to be one of the early starting points for the discussion of what we now call dark matter.[66] (These clouds are now known to be made of baryonic mass, too cold to be luminous in visible light, and thus very different from the non-baryonic "dark matter" that is responsible for the shapes and orbits of galaxies.) William Herschel considered them just as "holes" in the sky, containing no celestial objects. Astronomer Giorgio Abetti (1882–1982) attributed the discovery of dark nebulae to Secchi, in the sense of dark masses covering the starry background and absorbing its radiation.[67] Indeed, in 1877, twenty years after his first research on nebulae (see Chapter 3), Secchi stated:

> Among these studies, there is the interesting probable discovery of dark masses scattered in space, whose existence was revealed thanks to the

techniques (see David Leverington, *Babylon to Voyager and Beyond: A History of Planetary Astronomy* [Cambridge: Cambridge University Press, 2003], 265).

65 *On voit donc que le spectre de cette planète n'a aucune ressemblance avec le spectre solaire. L'énorme lacune est tout à fait singulière; on n'en trouve pas d'exemple dans les autres planètes; et quoique Jupiter et Saturne présentent quelques bandes d'absorption, ils sont loin d'offrir un pareil intervalle c'est comme si toute la couleur jaune était supprimée. Si ce spectre est purement dû à la lumière solaire réfléchie (ce qu'on pourrait peut-être mettre en question), elle doit subir une modification considérable dans son atmosphère.* [...] *Cette découverte donnerait lieu à plusieurs réflexions que je passe sous silence quant à présent.* Secchi, "Résultats fournis par l'analyse spectrale de la lumière d'Uranus," 763.

66 See Gianfranco Bertone and Dan Hooper, "A History of Dark Matter" (US Department of Energy, High Energy Physics Division, 2016), 8.

67 See Giorgio Abetti, "Le nebulose oscure: Una scoperta del p. Angelo Secchi," *L'Universo* 9 (1928): 315–319.

> bright background on which they are projected. Until now, they were classified as black cavities, but this explanation is highly improbable, especially after the discovery of the gaseous nature of the nebular masses,[68] and it is more likely that their blackness is the result of a dark nebula, projected onto the bright background and blocking the rays of the latter.[69]

Secchi added that he had found them in the constellation Sagittarius and in the Andromeda nebula: "In these regions, we found those spaces absolutely dark, well defined, and determined, which made us believe that, instead of being real holes, they were rather shadows of dark masses."[70]

Though intrigued by the subject, Secchi could not carry out more extensive research in this field because of his premature death, which left many of his other contributions incomplete.

6 Solar Physics Studies

Secchi had a particular interest in the Sun and greatly contributed to the development of solar physics. He is considered an example of "the prevalent viewpoint among the founders of the solar physics community, who assigned a particular significance to solar research [...] because [the Sun's] proximity made it relatively knowable to scientists and profoundly important to humanity."[71] One of his early interests was the study of solar radiation, which led him to create the thermoheliometer, an instrument to measure solar heat radiation. This interest actually originated from his meteorological studies, because it was Secchi's intention to investigate the solar influence on the Earth's atmosphere. In 1863, he devoted a number of interesting articles to solar radiation: its chemical action, its absorption by the atmosphere, and so on. Its measurements were generally made with a simple blackened thermometer, exposed

---

68 See William Huggins and William Allen Miller, "On the Spectra of some Nebulae," *Philosophical Transactions of the Royal Society* 154 (1864): 437–444.

69 *Tra questi studi vi è il fatto interessante della scoperta probabilità di masse oscure disperse nello spazio, delle quali si rilevò l'esistenza dal fondo chiaro del cielo su cui sono proiettate. Finora si classificarono queste masse come fori neri, ma è assai improbabile questa spiegazione, specialmente dopo la scoperta della natura gassosa delle masse nebulari ed è invece più probabile che quella nerezza risulti da una nebulosità oscura proiettata sul fondo lucido, che intercetta i raggi di questo.* Secchi, "L'astronomia in Roma nel pontificato di Pio IX," 58.

70 *Fu in queste regioni che trovammo quegli spazi assolutamente oscuri, che ben definiti e determinati ci fecero credere, che invece di essere veri fori fossero piuttosto ombre di masse oscure.* Secchi, "L'astronomia in Roma nel pontificato di Pio IX," 59.

71 Hufbauer, *Exploring the Sun*, 64.

to the Sun; Secchi noted that this measurement, however, was influenced by the environment and could therefore be questioned because of "the influence exerted on the thermometer by (1) the temperature of the surrounding air; (2) the radiation from surrounding objects; (3) the nature of the surface covering the thermometer."[72] After making several attempts to isolate the various components and measure the extent of their influence, he designed a device, "free from such influences [*libero da tali inconvenienti*]," which he called a thermoheliometer, whose operation he explained in the *Bullettino meteorologico*.[73]

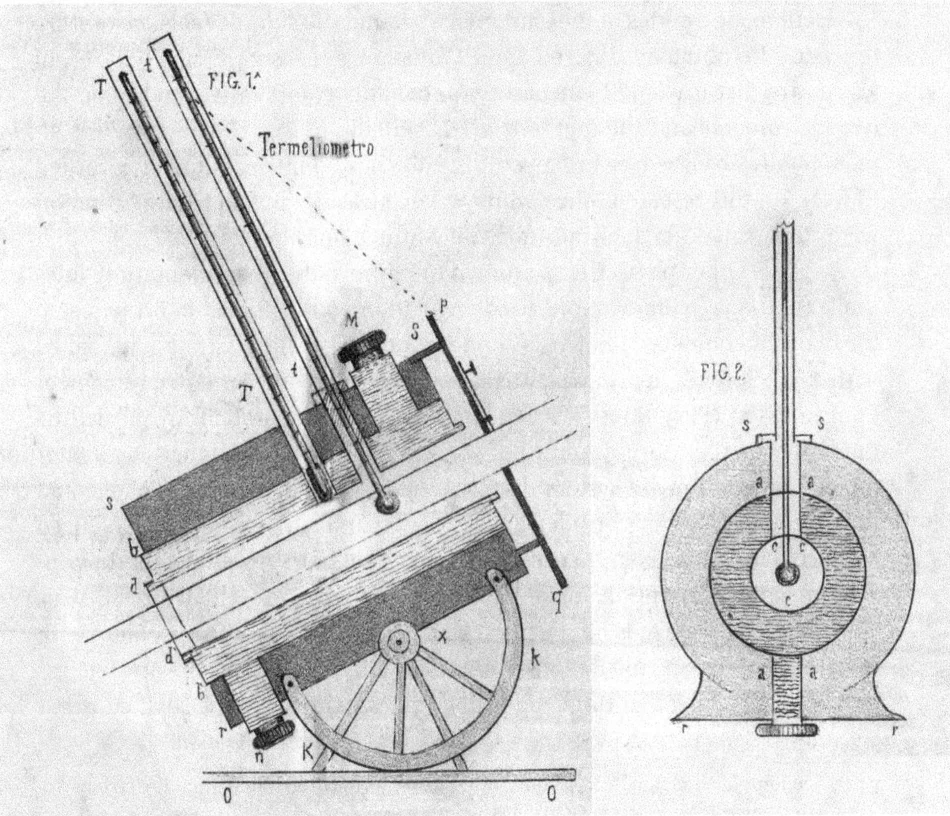

FIGURE 6.8 Diagram of Secchi's thermoheliometer. From Angelo Secchi, "Intorno all'influenza olare sull'atmosfera terrestre: Parte III," *Bullettino Meteorologico dell'Osservatorio del Collegio Romano* 2 (1863): 105–8, pl. 2, figg. 1–2.

72 *Per l'influenza che esercitano sul termometro 1º. la temperatura dell'aria circostante; 2º. la radiazione degli oggetti circonvicini; 3º. la natura della superficie che riveste il termometro.* Angelo Secchi, "Intorno all'influenza solare sull'atmosfera terrestre: Parte III. Radiazione solare," *Bullettino meteorologico dell'Osservatorio del Collegio Romano* 2 (1863): 105–108, here 105.

73 See Secchi, "Intorno all'influenza solare sull'atmosfera terrestre: Parte III. Radiazione solare".

Secchi described his thermoheliometer as a modified version of a device invented by Scottish physicist John James Waterston (1811–83).[74] It consisted of a sort of metal cylindrical boiler, filled with water, with a wooden coaxial finish; the bulb of a thermometer was placed in an empty axis channel, while a second thermometer was placed in contact with the metal surface of the boiler. The difference between the temperatures of the two thermometers gave an estimate of the solar constant. The instrument was supported by an old parallactic mounting, which allowed him to make small altazimuth and equatorial movements to optimize the positioning of the instrument.

Secchi made a series of measurements around the summer solstice of 1863; the value he obtained differed from Waterston's measurements, but Secchi was aware that it would be necessary to conduct comparative studies on the two instruments and the methods used. Actually, at that time Secchi was almost totally absorbed by his spectral studies, and only occasionally returned to work on this research. Meanwhile, other measurements of solar radiation were being made at high altitude, on Mont Blanc, by Jacques-Louis Soret (1827–90).[75] Other physicists continued to improve the instrumentation: Jules Violle (1841–1923), for example, used an actinometer—similar, in principle, to the thermoheliometer—having a copper sphere blackened with soot and filled with ice.[76] Secchi's instrument was later criticized by the American engineer John Ericsson (1803–89).[77] Ericsson proposed a better system for measuring the boiler temperature, where the water circulated in the boiler at a determined temperature—a system that was also adopted by Soret and Violle and that Secchi, while not considering it to be completely necessary, recognized as "excellent." Moreover, "[Ericsson] claims that I am grossly wrong in considering solar absorption made by its atmosphere"—Secchi wrote in 1875: "However, the mistake is his, because he thinks he can treat the problem with certain rough mean calculations while I treated it with Laplace's and Plana's formulae."[78] During that time, the various measurements of the solar constant

---

74 See John J. Waterston, "An Account of Observations on Solar Radiation. Radiazione solare," *Monthly Notices of the Royal Astronomical Society* 22, no. 2 (1862): 60–68.

75 See Jacques-Louis Soret, "On the Intensity of the Solar Radiation," *Philosophical Magazine* 34, no. 231 (1867): 404–407.

76 See Jules Violle, "Sur la température du soleil," *Annales scientifiques de l'ENS* 4 (1875): 363–370.

77 See John Ericsson, "The Difference of Thermal Energy Transmitted to the Earth by Radiation from Different Parts of the Solar Surface," *Nature* 12 (1875): 517–520. Secchi summed up his responses to these and other criticisms in the second volume of *Le soleil* (see Secchi, *Le soleil* [1875–77], 2:235–240).

78 [*Egli*] *pretende che io mi sia sbagliato grossolanamente nel valutare l'assorbimento solare, fatto dalla sua atmosfera. Ma lo sbaglio è suo, perché crede di poter trattare il problema con certe medie maccaroniche, mentre io l'ho trattato colle formule di Laplace e di Plana.*

gave discrepant results, until it was determined correctly in 1881 by Samuel P. Langley (1834–1906), inventor of the bolometer, which can be considered to be a development of the thermoheliometer.

Secchi was also a pioneer in the application of photography during the observations of solar eclipses. As we saw in Chapter 3, he carried out the first attempts in 1851, and achieved his most significant results during the eclipse of 1860; unfortunately, the photographic attempts during the eclipse of 1870 were partially unsuccessful (see Chapter 5).

Secchi chose to use both optical and photographic instruments to observe eclipses because he considered the two methods to be complementary and was aware that the use of only one of the two types of instruments gave incomplete results:

> There were two classes of instruments to be used, that is, optical and photographic. The first ones naturally have precedence, being those that supply surer and complete data, but unhappily, given the transience of the phenomena, they are far from being committed out of any danger [risk of error], and also, leaving no trace of itself, do not allow us to return to the phenomenon except by means of reminiscences, not always sure. The photographic ones, though incomplete by their nature because they are incapable of fixing colors, are hard to handle in such critical moments, and subject to many misconceptions unless they are assisted by the eye and intellect, have the great advantage of permanently fixing the phenomenon on which to return with a cool head; thus, as the two systems complement each other, the use of both ensures complete success.[79]

Secchi's results played a crucial role in settling the dispute over the nature of solar prominences, a subject that had been debated for many years—that is,

---

Secchi to Tacchini, Rome, October 29, 1875, quoted in Chinnici and Gasperini, *Alle origini dell'astrofisica italiana*, 396.

79 *Due erano le classi degli strumenti da usarsi, cioè gli ottici e i fotografici. I primi hanno naturalmente la preminenza, come quelli che forniscono i dati più sicuri e complete, ma sventuratamente per la fugacità de' fenomeni sono lungi dal poter essere impegnati fuori da ogni pericolo, e inoltre non lasciando traccia di sé, non permettono di ritornare sul fenomeno che per mezzo delle reminiscenze sempre mal sicure. I fotografici, benchè di lor natura incompleti perchè incapaci di fissare i colori, di difficile maneggio in momenti sì critici, e soggetti a molti equivoci, ove non siano assistiti dall'occhio ed all'intelletto, hanno però il grande vantaggio di fissar permanentemente il fenomeno su cui poter ritornare a mente fredda, e così l'uno dei due sistemi compensando l'altro, fu risoluto diimpiegarli ambedue per assicurare un completo successo.* Secchi, "Relazione delle osservazioni fatte in Spagna durante l'eclisse," 59, quoted in Gasparini, "Padre Angelo Secchi e l'applicazione della fotografia nelle osservazioni astronomiche," 42.

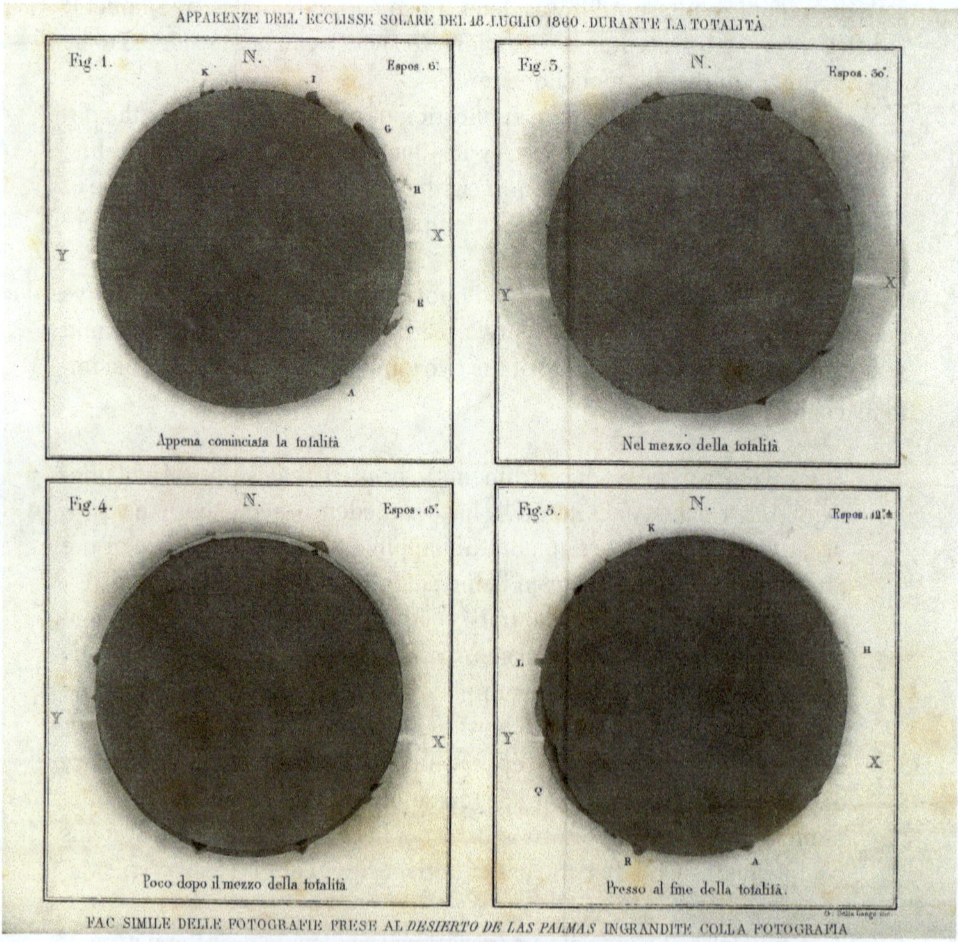

FIGURE 6.9 Photographs of the total solar eclipse of 1860, taken by Secchi in Spain. Facsimile, from Angelo Secchi, "Sull'ecclisse solare totale osservato in Spagna nel 18 luglio 1860," *Memorie dell'Osservatorio del Collegio Romano* n.s. 2 (1863): 33–52, pl. [4].

if they were actually structures belonging to the Sun or just diffraction effects produced by the lunar limb, the Earth's atmosphere, or even the optics of the instruments. On the occasion of the 1860 eclipse, the photographs taken by Secchi at Desierto de Las Palmas were compared with those obtained about five hundred kilometers away, in the Ebro Valley at Miranda, in the outskirts of Rivabellosa (near Bilbao), by De la Rue.

Secchi, who had the first prints, made a detailed comparison of the two sets of photographs and found that they coincided perfectly.[80] Unfortunately, the

80 See Angelo Secchi, "Relazione delle osservazioni fatte in Spagna durante l'eclisse totale del 18 luglio 1860," *Giornale Arcadico di Scienze: Lettere ed arti*, n.s. 20 (1860): 53–121, here 107–115.

photographs that were printed by Aguilar y Vela for general distribution were not as detailed as the first prints, and some details were almost invisible in the distributed copies. De la Rue wrote to Secchi:

> It's very interesting to hear that your own observations accord so well with mine; results obtained at two stations apart so far as ours would completely set at rest the question of the protuberances being possibly produced by diffraction of the Sun's light, and show that they are realities existing beyond the Sun's photosphere.[81]

The English astronomer made a comparison of the two sets of photographs in November 1862 when he visited the Collegio Romano Observatory; he examined the first prints and found that the details matched with those of his photographs:

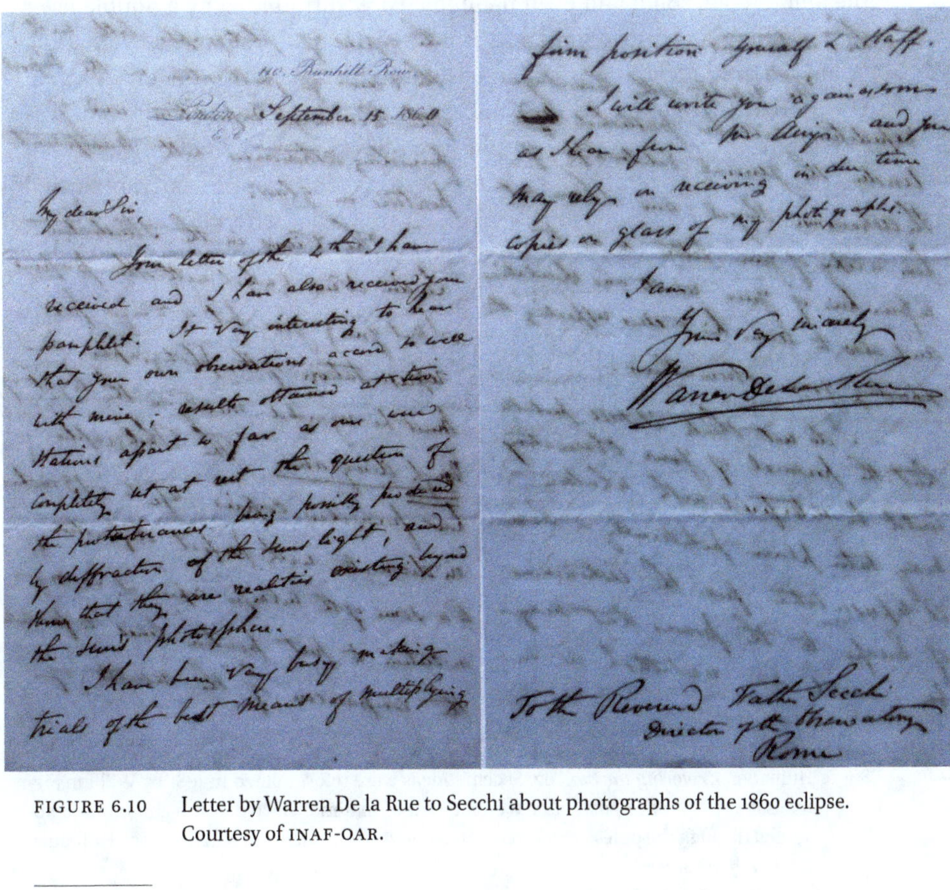

110 Bunhill Row
London September 15 1860
E.C.

My dear Sir,

Your letter of the 4th I have received and I have also received your pamphlet. It's very interesting to hear that your own observations accord so well with mine; results obtained at two stations apart so far as ours would completely set at rest the question of the protuberances being possibly produced by diffraction of the Sun's light, and show that they are realities existing beyond the Sun's photosphere.

I have been very busy making trials of the best means of multiplying

firm position yourself & staff.

I will write you again as soon as I hear from Mr Aguilar and you may rely on receiving in due time copies on glass of my photographs.

I am
Yours very sincerely
Warren De la Rue

To the Reverend Father Secchi
Director of the Observatory
Rome

FIGURE 6.10 Letter by Warren De la Rue to Secchi about photographs of the 1860 eclipse. Courtesy of INAF-OAR.

81 De la Rue to Secchi, London, September 15, 1860; AOAR, Fondo Secchi, I.E.2.1, 48

> The photographic images of the prominences, so far as they are common to the two photographs taken at Miranda and Desierto de Las Palmas, accord in their most minute details. The photographs must, from the difference of position of the two stations, have been made at an absolute interval of about seven minutes; and this fact, while it strongly supports the conclusion that the protuberances belong to the Sun, at the same time shows that there is no change in their form during an interval much greater than the whole duration of an eclipse.[82]

The comparison between the two sets of photographs showed a full concordance in the position of solar prominences, whose actual presence on the solar limb was thus recognized. The results that Secchi and De la Rue obtained demonstrated "the desirability of equipping all future eclipse parties with photographic apparatus."[83]

In the meantime, it became more and more clear that Herschel's model of the Sun (a cold, solid, and even habitable core, surrounded by a double layer of hot luminous atmosphere) was irreconcilable with the laws of radiation. Thanks to the contributions of physicists and chemists, who worked to explain the process that generated the Sun's energy,[84] a new model of the Sun was developing "based on the new insights from solar thermodynamics. [...] Secchi led the way in January 1864 by suggesting that the trend of research indicated that the Sun was a gaseous heat engine."[85] In fact, in 1864, after commenting on the eighteenth-century hypothesis that sunspots were clouds and rejecting the idea of the habitability of the Sun, he concluded that:

> Without contradicting the laws of physics, (1) the photospheric layer can have a brightness greater than the inner core; (2) what we call core is not supposed to be solid nor liquid, but it can be gaseous too, though denser; (3) despite the closeness of the photospheric layer, it can have not only

---

82 Warren De la Rue, "Comparison of Mr. De la Rue's and Padre Secchi's Eclipse Photographs," *Proceedings of the Royal Society* 13 (1863): 442–444, here 442, quoted in Hughes, *Catchers of the Light*, 776.

83 Royal Astronomical Society Council, "Report [...] to the Forty-First Annual General Meeting of the Society," *Monthly Notices of the RAS* 12 (1861): 73–123, here 117, quoted in Hufbauer, *Exploring the Sun*, 52.

84 See Meadows, *Early Solar Physics*, 81–82.

85 Hufbauer, *Exploring the Sun*, 62. Secchi's ideas were reformulated in 1881 by William Siemens (1823–83), who postulated a circulation of material ejected at the equator and absorbed through the poles, which generated energy from chemical reactions; see Hufbauer, *Exploring the Sun*, 62.

> different light, but also different temperature; (4) the appearance of the sunspot shapes absolutely exclude the structure of clouds […].[86]

In the early 1860s, Secchi participated in the international debate about the appearance of the solar surface, today called solar granulation. The main protagonists of this controversy were Nasmyth, who invented the term "willow-leaf" to indicate the shape of the elementary photospheric structures; Dawes, who questioned this description, proposing the term "granule"; and Secchi, who described them as "rice grains" though recognizing their shapes were quite irregular.[87] The controversy inevitably raised a general debate over the nature of granulation.[88] In Nasmyth's opinion, the solar surface was covered by willow-leaf-shaped clouds, whereas Dawes considered the granules as irregular solid masses of chemical precipitates.

Secchi remarked that, since they were continuously in motion and undergoing modifications, the bright granules were very difficult to observe; however, he shared the theory expressed in 1865 by Faye, who considered them as the top of hot ascending currents, coming from the Sun's interior, "whereas the dark interstices represented regions where cooler currents were descending."[89] In reply to Secchi, Dawes wrote a long letter[90] with the aim of clarifying the different names used to describe the same solar features. Ultimately, the dispute "had the beneficial effect of standardizing the terminology used in describing the solar surface."[91]

Sunspots were another major topic of Secchi's research. In May 1855, he accurately observed the evolution of a sunspot[92] and was further convinced that

86 *Senza contraddire le leggi della fisica, 1º lo strato fotosferico può avere una luminosità maggiore del nucleo interno, 2º che questo che diciamo nucleo non è mestieri punto supporlo solido, né liquido, ma può esser anch'esso gassoso solo però più denso, 3º che malgrado la vicinanza dello strato fotosferico esso può avere non solo diversa luce, ma anche diversa temperature, 4º che le apparenze delle forme delle macchie escludono assolutamente la struttura di nubi* […]. Angelo Secchi, "Sulla teoria delle macchie solari proposta dal sig. Kirchhoff," *Bullettino meteorologico dell'Osservatorio del Collegio Romano* 3 (1864): 1–4, here 4.

87 See Secchi, *Le soleil* (1870), 33.

88 See Meadows, *Early Solar Physics*, 28–45.

89 Meadows, *Early Solar Physics*, 71.

90 See Angelo Secchi, "Sulla struttura della fotosfera solare: Lettera del Rev. Sig. Dawes," *Bullettino meteorologico dell'Osservatorio del Collegio Romano* 4 (1865): 93–99.

91 Meadows, *Early Solar Physics*, 30.

92 See Angelo Secchi, "Sulle macchie e sulla temperatura del Sole," *Memorie dell'Osservatorio del Collegio Romano 1852–55* (1856): 130–133, here 130–131.

they originated from internal gas explosions or vortices generated within the photosphere and disrupting it.

Secchi's conviction was reinforced by his studies on the Wilson effect[93]—the apparent contraction and disappearance of the sunspot penumbra when it is observed near the solar limb—an effect that was associated with depressions in the solar photosphere. Indeed, in the 1850s, the Jesuit astronomer determined geometrically the depth of spot cavities, finding a value not exceeding one Earth radius.[94]

Along with Tacchini, at the beginning of the 1870s Secchi entered into a controversy with Faye on the cyclonic nature of sunspots. Their two "rival theories" were summarized by American astrophysicist Charles Augustus Young (1834–1908) as follows: "Faye conceives the sun-spots to be the effect of solar storms; Secchi believes them to be dense clouds of eruption-products settling down into the photosphere near, but not at, the points where they were ejected."[95] Secchi started from the observation that only a very small percentage of sunspots appear as whirls, showing a vortex motion, whereas the sunspots penumbra generally appeared with radial structures that looked like materials flowing into a depression; moreover, together with Tacchini, he noticed that sunspots were always surrounded by faculae and prominences and deduced the existence of a ring of gaseous eruptions all around the sunspot: "Secchi's [...] theory is based essentially upon the idea [...] that eruptions are continually breaking through the photosphere, and carrying up metallic vapors from the regions beneath. He imagines that these vapors, after becoming considerably cooled, descend upon the photosphere and form depressions in it [...]."[96]

In 1868, Secchi was one the first astronomers to apply the method devised by Janssen and Lockyer to observe the prominences of the uneclipsed Sun by placing the slit of the spectroscope tangentially to the solar limb. The observations of the total solar eclipse of that year revealed the gaseous nature of

---

93 It was observed and studied in 1774 by Alexander Wilson (1714–86), who proved that it was due to perspective effects that could be explained by the association of sunspots with photospheric cavities.

94 See Secchi, *Le soleil* (1870), 46–51; see also Angelo Secchi, "Ricerche fisiche intorno ai corpi celesti. I. Sulle macchie solari e sul modo di determinarne la profondità," *Memorie dell'Osservatorio del Collegio Romano*, n.s. 1 (1859): 9–12 (for figure, see 10).

95 Charles A. Young, *The Sun* (New York: Dr. Appleton & Company, 1896), 182–183.

96 Young, *Sun*, 185. The main objections to Secchi's theory were how the eruption was long maintained and why the cloud should continue to descend in the same place (see Young, *Sun*, 185–186). Later, after the discovery of the Zeeman effect, in 1896, accurate observations of sunspots' spectra revealed that they are regions of intense magnetic fields and, in the twentieth century, thanks to the ionization theory, their interaction with erupted plasma explained the permanence of the coronal loops.

the prominences and their chemical composition, made of hydrogen and an unknown element (helium), whose main line was denominated D^3; [97] Secchi commented on these results:

> The abundance of hydrogen [...] is not surprising, since from our research it was shown that half of the stars show no other sharpened lines than those of this gas; [...] the bright solar line next to D [i.e., the double Fraunhofer line of sodium] [...] does not correspond to any notable dark line.[98]

From then on, spectroscopic observations of prominences were part of Secchi's regular daily work program, especially after the establishment of the Società degli Spettroscopisti in 1871 (see Chapter 8). He investigated their correlation with sunspots and faculae, studying their statistics and looking for a general law connecting all solar features; his analysis, however, remained mainly phenomenological—he suggested hypotheses, but without developing theories.

Secchi also encouraged Tacchini to try to observe the solar corona when the Sun was not in eclipse, in order to better determine the position of the mysterious green emission line at 1474K (in Kirchhoff's scale, today 5303 Å), observed in the coronal spectrum by William Harkness (1837–1903) in 1869. The corona was only observable during total solar eclipses, in the few minutes of totality, and the exact determination of the line position was quite difficult with ordinary means, so that it was sometimes confused with some known iron lines and also with a spectral line of polar aurorae.[99] Tacchini proposed building an astronomical station on Mount Etna in order to take advantage of its high elevation and reduce the atmospheric diffraction effects, and Secchi's intervention was crucial in supporting the project on a scientific basis (see Chapter 8).

Secchi was convinced of the existence of continuous magneto-electric influences of the Sun on the Earth (what is now called the "Sun–Earth connection"). In this field, he shared the same view as Tacchini, who expressed this idea well by asserting that "Sun and Earth are just an electric machine and

---

97 Helium was later isolated in a laboratory, while dealing with radioactive minerals, by William Ramsay (1852–1916), in 1896 ; see Meadows, *Early Solar Physics*, 24–25.

98 *La copia dell'idrogeno* [...] *non deve sorprendere, poichè dalle nostre ricerche fu dimostrato che la metà delle stelle non mostrano altre righe distinte che quelle di questo gas;* [...] *la lucida solare accanto a D* [...] *non corrisponde a nessuna linea nera notabile.* Angelo Secchi, "Sullo spettro delle protuberanze solari osservato in pieno sole," *Bullettino meteorologico dell'Osservatorio del Collegio Romano* 7 (1868): 89–91, here 90.

99 See Meadows, *Early Solar Physics*, 133.

an electroscope in small distance inside the cabinet of a physics professor."[100] The connection between the variation in solar activity and that of terrestrial magnetism could be explained by these effects: "The variation in solar activity could be transmitted to the Earth through the heat radiation, or through some other still unknown means, for example, the electrodynamic induction, producing some meteorological and electric phenomena on our globe."[101] The concomitant appearance of polar aurorae and geomagnetic disturbances, often observed during the maximum of solar activity, further confirmed this connection.[102] However, the physical causes of these connections were still unclear. The propagation of electromagnetic waves through ether was theorized by James Clerk Maxwell (1831–79) in 1864, and Secchi mentioned it in his essays,[103] but in the face of new theories he often adopted a prudent attitude, declaring that "for the moment, we have to wait [*pour le moment, il faut attendre*]."[104]

7 Redaction of Scientific Treatises

Secchi also made an important contribution to science by publishing major treatises[105] that helped to circulate the main theories and ideas of that time in the field of physics, astrophysics, and geophysics. Secchi was not afraid of the challenge posed by new ideas: rather, he considered them as a stimulus toward a deeper knowledge of creation. Moreover, he probably hoped to promote a renewal of the Catholic tradition, superseding the Thomistic–Aristotelian view that was still in vogue on doctrinal more than scientific grounds (see Chapter 9). Secchi's role in editing scientific treatises further reflects his efforts to bring

100 *Sole e terra non sono né più né meno che una macchina elettrica e un elettroscopio a piccola distanza nel gabinetto di un professore di fisica.* Pietro Tacchini, "Il Sole e le aurore boreali osservate a Palermo nell'aprile 1871," *Rivista Sicula* 5 (1871): 415–427, here 420.

101 *Les variations de cette activité pourraient bien se communiquer à la Terre, soit par le moyen de la chaleur, soit par quelque autre moyen encore inconnu, par exemple l'induction électrodynamique, produisant ainsi sur notre globe des phénomènes météorologiques ou électriques.* Secchi, *Le soleil* (1870), 330.

102 Initially, Secchi was also convinced of the existence of a reciprocal relationship between geomagnetic and meteorological perturbations; see Angelo Secchi, "Intorno alla corrispondenza che passa fra le variazioni meteorologiche e quelle del magnetismo terrestre," *Memorie dell'Osservatorio del Collegio Romano*, n.s. 2, (1863): 81–120.

103 See Secchi, *Le soleil* (1875–77), 2:110.

104 Secchi, *Le soleil* (1875–77), 2:112.

105 See Fioravanti, "Per una bibliografia secchiana," 277–290.

science out of the inner circles of the academies and to educate members of the general public, who were often ignorant of new scientific discoveries.

Secchi intensified this activity during the last years of his career: his health, by then in decline, no longer allowed him to work at the telescope with the same earnestness and energy of his early years, nor did he have sufficient financial resources to maintain the observatory (see Chapter 10). Instead, writing books was the best way to do a useful service for science and, at the same time, scrape together revenue for the observatory.

7.1 *L'unità delle forze fisiche* (On the Unity of Physical Forces)

L'unità delle forze fisiche, the first edition of which was published in 1864, is a summation of Secchi's vision of nature. The book was "specially aimed at the education of zealous students,"[106] but Secchi hoped that "it would be useful for scholars as well":[107] its aim was both to educate and to popularize science. In the book, Secchi expressed the unified vision of nature that continuously drove him to look for correlations between different physical phenomena.[108] Moreover, he also introduced his readers to modern theories, including the concept of the atom and the kinetic theory of gases, which drew virulent criticism from the supporters of neo-Thomism (see Chapter 9).

The book's introduction deserves to be quoted extensively. Secchi started by presenting the purpose of the work:

> The great discovery that currently preoccupies all learned people, and illustrates our age, is that of the mechanical theory of heat, in which this agent is reduced to a simple mode of motion. To reveal the foundations of this theory and extend its applications to the imponderable [matter] and other physical forces is the purpose of this work.[109]

---

106 *Dirette specialmente all'istruzione della gioventù studiosa.* Angelo Secchi, *Sull'unità delle forze fisiche* (Rome: Tipografia Forense, 1864), 2.

107 *Alcun poco di utile troveranno pure quelli che già conoscono la scienza. Secchi, Sull'unità delle forze fisiche,* 2. On March 6th, 1864, Secchi delivered a sealed envelope, containing a copy of the proofs of the book - slightly different from the published version - to the Pontificia Accademia dei Lincei. The reasons of this decision are unclear. An accurate comparison between the proofs (today kept at the Accademia Nazionale dei Lincei) and the published text would deserve further attention.

108 See Aldo Altamore, "La nascita della Nuova Astronomia," in Altamore and Maffeo, *Angelo Secchi*, 109–144, here 111–112.

109 *La grande scoperta che presentemente preoccupa tutti i dotti, e illustra la nostra epoca, è quella della teoria meccanica del calore, per la quale questo agente viene ridotto ad un semplice modo di movimento. L'esporre le basi di questa teoria ed estenderne le applicazioni agl'imponderabili e alle altre forze fisiche, è lo scopo dell'opera presente.* Secchi, *Sull'unità delle forze fisiche*, iii.

Secchi then justified why he, an astronomer, was publishing such a physics book:

> It is no wonder that an astronomer should treat a subject that is commonly thought to be the domain of physicists, if you reflect that the question about the nature of forces is especially interesting in astronomy. Indeed, you can consider celestial motions in two ways: either as mere corollaries of an abstract force, gravity, [...] or as effects from a proximate cause of the highest order, whose consequence would be gravity itself. [...] Recent observations have shown that within interplanetary space, even activities like repulsion, resistance of the heating medium, magnetic and electric actions, are operating; so that there is no doubt that these forces are common to all creation and bind together, like light, all the bodies of the universe.
>
> In the face of such a seductive idea, anyone who wandered into an astronomical career after many years spent in the study of physics could not but feel greatly encouraged to study the mutual relationship of these forces, and investigate their principles, and see whether it is possible to reduce them all to a common source. The great progress that science has made in recent times has removed the main difficulties one meets in dealing with such a problem, and one no longer considers these physical forces as independent of those supporting the motion of celestial bodies. The existence of an absolute vacuum in planetary space is no longer accepted by anyone, and these forces are no longer considered as occult properties instilled to matter, but as the results of simple motion, which opens a new and unexplored field [...].[110]

---

110 *Non farà meraviglia a veruno che un astronomo entri a trattare una materia creduta communemente del dominio de' fisici, se si rifletta che la questione sulla natura delle forze interessa in modo speciale l'Astronomia. In due modi infatti si possono considerare i moti celesti; o come semplici corollari di una forza astratta, la gravità,* [...] *ovvero come effetti provenienti da una causa immediata di ordine più elevato, di cui la gravità stessa non sarebbe che una conseguenza.* [...] *Le recenti osservazioni han dimostrato che entrano in attività anche nello spazio planetario la ripulsione calorifica, la resistenza del mezzo, l'azione magnetica e l'elettrica, talché non v'è dubbio che queste forze non siano comuni a tutta la creazione e leghino insieme, come la luce, tutti i più remoti corpi dell' universo.*

In faccia a un'idea sì seducente, chi si avviò per la carriera astronomica dopo molti anni spesi nello studio della fisica, non può a meno di non sentirsi grandemente incitato a studiare la mutua relazione di queste forze, e investigarne i principi, e vedere se sia possibile ridurle tutte a una origine commune. I grandi progressi che ha fatto la scienza in questi ultimi tempi han tolto le principali difficoltà che incontravano i dotti a trattare un tale problema, e non si riguardano più come indipendenti le forze fisiche e quelle che reggono il moto de' corpi

Secchi had already touched on this subject in 1858[111] and in 1862[112] and was persuaded to write a book on this matter, "in the short spare time that was left to me by other duties,"[113] by other physicists, and by the awareness that "many of the ideas that were already integral to science were not yet sufficiently widespread and well known."[114]

Secchi then clarified his position and his challenge:

> In this work, I don't pretend to create a new philosophy of nature, but only to reveal what is prevailing today, after the study of the phenomena. [...] I will not conceal that the challenge of dealing with this matter in a way that would render the work accessible to the greatest number of readers was greater than I had originally believed it would be.[115]

It was necessary to minimize formulas and calculations, without weakening the demonstrations, and synthesize the results of experiments, avoiding the details; otherwise, the book "would have been boring to some of the readers and unintelligible to others."[116]

Secchi warned his readers that the book was intended for a well-educated audience and required a solid background in physics:

---

celesti. L'esistenza di un vuoto assoluto nello spazio planetario non è più ammessa da nessuno, e le forze non si stimano più come qualità occulte infuse alla materia, ma come risultati di semplice movimento, e ciò apre un nuovo campo inesplorato [...]. Secchi, *Sull'unità delle forze fisiche*, iii–iv.

111 He presented a short communication to the Accademia Tiberina, a pontifical academy established in 1813 to promote sciences and literature in Rome; the talk was addressed to non-scientists and is an example of Secchi's efforts to disseminate scientific findings to the broader public (see Angelo Secchi, "Sulla correlazione delle forze fisiche e sulla sua influenza nel concetto dell'universo," *Giornale arcadico*, n.s. 155, no. 9 [1858]: 1–27).

112 He published a letter on this subject (see Angelo Secchi, *Intorno alla soluzione di un problema fisico-cosmologico. Lettera al R. P. Nardini* [Rome: Tipografia Marini, 1862]), in the context of the debate between neo-Thomism and atomism (see Chapter 9).

113 *Nei brevi ritagli di tempo che mi permettevano gli altri doveri* [...]. Secchi, *Sull'unità delle forze fisiche*, v.

114 *Molte delle idee già invalse nella scienza non erano ancora diffuse e conosciute sufficientemente tra di noi.* Secchi, *Sull'unità delle forze fisiche*, iv–v.

115 *In quest'opera io non ho la pretensione di creare una novella filosofia della natura, ma solo di esporre quella che oggidì va prevalendo dietro lo studio de' fenomeni.* [...] *Non voglio dissimulare che la difficoltà di trattare questa materia, in modo da render l'opera accessibile al maggior numero de' lettori, l'ho trovata maggiore di quella che credeva da principio.* Secchi, *Sull'unità delle forze fisiche*, v.

116 *Sarebbe riuscito noioso ad alcuni de' lettori e inintelligibile ad altri.* Secchi, *Sull'unità delle forze fisiche*, v.

> [It] demands serious reading and mindful attention because of the multiplicity and complexity of subjects. It is very different from those pleasant popular readings, half-scientific and half-literary, which have been recently introduced to spread science and, though being useful to spread it, are far from making it advance. My purpose was more the latter one [...].[117]

Secchi chose to end each chapter with a summary of its contents,[118] so that "even without reading the whole work, anyone can have an idea"[119] of the subjects it discusses, and concluded the introduction by saying that he would have been satisfied to write this book just for himself, for the pleasure of examining in depth all the forces of nature: "A pleasure that is really supreme and inexpressible, compensating a thousand-fold the considerable effort it costs."[120]

The general idea guiding Secchi's thought is that "heat, light, electricity, magnetism, molecular attractions, chemical affinities, universal gravitation, are the various known principles that result in all the phenomena of matter. Their nature, for a long time mysterious or poorly known, now appears to be unfolding under a novel and simpler form, namely that of motion."[121]

According to Secchi, everything is matter and motion: in fact, all physical forces are attributable to the motion of the ether, an "imponderable matter" (a sort of hyper-dissociated matter) that pervades the entire cosmos and acts as operator of physical phenomena, through purely mechanical action.[122]

---

117 *Esige una seria lettura e una ponderata attenzione per la molteplicità delle materie e per la loro difficoltà. Esso è molto diverso in ciò da quelle amene letture popolari, mezzo scientifiche, mezzo letterarie, introdotte in questi ultimi tempi per diffondere la scienza, le quali benché utilissime a spargerla, sono però ben lungi dal farla progredire. Il mio scopo era principalmente questo secondo* [...].Secchi, *Sull'unità delle forze fisiche*, vi.

118 Secchi intended to insert additional notes, but the delay in publication dissuaded him from doing so; see Secchi, *Sull'unità delle forze fisiche*, vi–vii.

119 *Potrà chi vorrà, anche senza leggere tutta l'opera, averne un'idea.* Secchi, *Sull'unità delle forze fisiche*, vi.

120 *Piacere veramente sommo ed inesprimibile, che compensa a mille doppi la non poca fatica che costa.* Secchi, *Sull'unità delle forze fisiche*, vii.

121 *Il calore, la luce, l'elettricità, il magnetismo, le attrazioni molecolari, le affinità chimiche, la gravitazione universale, sono i vari principi a noi cogniti che epilogano tutti i fenomeni della materia. La loro indole per gran tempo misteriosa, o mal conosciuta, sembra ora svelarsi sotto una forma novella e più semplice, cioè quella del movimento.* Secchi, *Sull'unità delle forze fisiche*, 2.

122 Secchi resumed Pianciani's ideas; in fact, in his treatise *Istituzioni fisico–chimiche* [Physical-chemical principles], Pianciani had adopted the waves-in-ether theory as the basis of his physics course.

Starting with the relationship between heat and the kinetic energy of molecules, in the first chapter, Secchi introduces some of the concepts of thermodynamics. In the second chapter, he deals with the nature of light and matter–radiation interactions; and, in the third chapter, with electromagnetism. The book closes in the fourth chapter with an overview of the structure of matter, including physical chemistry (atomic and molecular theory).[123]

The first edition, printed in haste, lacked illustrations; these were inserted later, in the first French edition of 1869. Secchi was persuaded to publish a French edition because the book, in his words, "met so favorable a welcome from the scientific public that the edition was quickly sold out. Although written in Italian, it obtained such a wide success as to move me to accept the proposal of Doctor Deleschamps[124] to make a French edition."[125] This edition was not a simple translation of the first book but contained significant improvements, being illustrated, extended, and updated, with further information, corrections, and clarifications.

The book was widely successful, as can be seen in the number of editions, reprints, and translations. Beyond the three Italian editions (1864, 1874, and 1885), there were two editions in French (1869 and 1874), three German (1876, 1884–85, 1892), and three Russian editions (1872, 1873, and 1880). This certainly contributed to the consolidation of Secchi's international reputation and popularity.

7.2 *Le soleil* (The Sun)

Le soleil was the main astronomical treatise written by Secchi. It collected all the most advanced theories on the Sun that were in circulation at that time, presenting them clearly and comprehensively to a non-specialist educated audience. The treatise also included various elements of stellar and planetary astrophysics, in a way prefiguring the next treatise, *Le stelle* (The stars). As in

123 The book also contained some elements of biophysics, about energy in living organisms, and astrophysical topics; the latter would be revisited by Secchi in his later treatises.

124 Secchi greatly appreciated the translation made by medical doctor Albert Deleschamps, who sacrificed the elegance of the French language and privileged the accuracy of the concepts to be transmitted. Deleschamps was a friend (more probably, a pupil) of Ganot (see Josep Simon, *Communicating Physics: The Production, Circulation and Appropriation of Ganot's Textbooks in France and England* [London: Routledge, 2011], 245n107) and, in the same year 1869, presented and published - by the same publisher - his doctoral thesis on physiology of the human voice (see Albert Deleschamps, *Etude physique du son de la parole* [Paris: F. Savy, 1869]).

125 *Ce livre* [...] *rencontra dans le public scientifique un si favorable accueil que l'édition se trouva promptement épuisée. Bien qu'écrit en italien, il obtint en France un succès assez grand pour me déterminer à accepter la proposition du docteur Deleschamps d'en faire une édition française.* Angelo Secchi, *L'unité des forces physiques* (Paris: F. Savy, 1869), ix.

the case of the previous book on the physical forces, Secchi again avoided the use of complex mathematical formulas, thereby ensuring that the treatise was accessible to a wider, non-specialist audience.

The first edition of the book appeared in French in 1870. While Secchi claimed to have chosen to publish it in French as "a tribute of gratitude toward France, for its sympathetic welcome on the occasion of the Exposition Universelle,"[126] it also meant that the book would reach a wider audience than would have been the case if it had been published in Italian. The first edition of the book contained, in essence, the themes of all the subsequent treatises by Secchi. It was divided into three parts, preceded by a general introduction: the first part, bearing the title "Structure du soleil" (Structure of the Sun), contained nine chapters and described the current state of scientific knowledge about the Sun. The remaining two parts are more like appendixes: the second part, entitled "Activité extérieure du soleil" (Exterior activity of the Sun), contains two chapters on radiation and gravitation and describes solar emissions in the different regions of the electromagnetic spectrum as well as the structure of the solar system. In the third part, consisting of only one chapter, entitled "Les soleils ou les étoiles" (The suns, that is, the stars), Secchi presents his spectral classification of stars into five classes and makes some general remarks of interest. He states, for instance, that stars are distant suns and that they are not fixed in the sky, as they appear; that it was "absurd to pretend to fix the center of the universe"; that the Sun "is not placed in the middle of the shell forming the Milky Way" but is located "in a very off-centered position"; and that the distribution of the other stars in the Milky Way is not uniform.[127]

From these considerations, Secchi concluded that

> the depth of heaven is really unfathomable and we will never know its boundaries. It is likely that the association of brilliant stars surrounding our Sun is one of the clusters that form the Milky Way, and, from a distance, this cluster would appear to us as a whiter spot in the Milky Way itself.[128]

---

126 *En publiant cet Ouvrage en français, l'auteur est heureux de pouvoir payer à la France une faible partic du tribut de reconnaissance qu'il lui doit pour l'accueil plein de sympathie qu'il a reçu pendant l'Exposition universelle.* Secchi, *Le soleil* (1870), xii.

127 See Secchi, *Le soleil* (1870), 388.

128 *Que la profondeur des cieux est réellement insondable et que nous n'en connaîtrons jamais les bornes. Il est probable que la réunion des grandes étoiles qui environnent notre Soleil n'est qu'un des amas qui forment la voie lactée, et que, vue d'une certaine distance, cet amas nous apparaîtrait comme une tache plus blanche dans la voie lactée elle-même.* Secchi, *Le soleil* (1870), 417.

Moreover, if the stars are other suns, "it appears absurd to look at these wide spaces as uninhabited deserts."[129] In Secchi's opinion, in fact, it is reasonable to admit that the other stars, like the Sun, are "destined to support the life of a crowd of creatures of all sorts."[130]

More than a year passed between the submission of the manuscript to the publisher and the book finally appearing in print, and in 1871 Secchi, replying to a critical review by Newcomb,[131] admitted that during the intervening time "great progress has been made in this branch of science" and he had found "so many things to add," recognizing that "in the rapid progress of discovery, a book becomes old very soon."[132]

The first edition of *Le soleil* was followed two years later by a German translation (*Die Sonne* [1872]), with some additions; it was edited by Heinrich Schellen (1818–84), physics teacher and director of the Realschule in Cologne, who paid special attention to teaching and disseminating the theory and practice of spectroscopy.[133] Secchi soon saw the need to have an extended version in French, and he thus prepared a huge two-volume second edition (the two parts were published in 1875 and 1877), enriched with many illustrations and furnished with an atlas of spectral lines. In this updated edition, Secchi included advances in the field of solar physics, in both theory and observation. The first part (*tome premier*) contained books 1–4 (each divided into several chapters), which, respectively, dealt with general concepts of solar phenomena, an examination of the solar surface, the solar atmosphere, and eclipses.

The second volume (*tome deuxième*) included books 5–8, concerning solar prominences, the temperature of the Sun, exterior activity of the Sun (i.e., interaction with the bodies of the solar system), and stars as suns—the last two books being quite short, as with the first edition. The atlas, included in the first volume, contained plates reproducing the solar spectrum as observed by many scientists: the one by Volkert Simon Maarten van der Willigen (1822–78) and the telluric lines observed by Janssen and by David Brewster (1781–1868); a more extended and detailed solar spectrum by Anders Jonas Ångström (1814–74); the solar diffraction spectrum, photographed by Henry Draper (1837–82); and the extension of the spectrum into the ultraviolet region, observed by Alfred Cornu (1841–1902).

---

129 *Il nous semblerait absurd de regarder ces vastes régions comme des déserts inhabités.* Secchi, *Le soleil* (1870), 418.

130 *Destinés à entretenir la vie d'une foule de créatures de toute espèce.* Secchi, *Le soleil* (1870), 418.

131 Simon Newcomb, "The Sun," *Nature* 4 (1871), 41–43, here 41–42.

132 Angelo Secchi, "The Sun," *Nature* 4 (1871): 82–83, here 82.

133 This important German edition contained the first lithographic reproductions of spectrograms; see Hentschel, *Mapping the Spectrum*, 210–211.

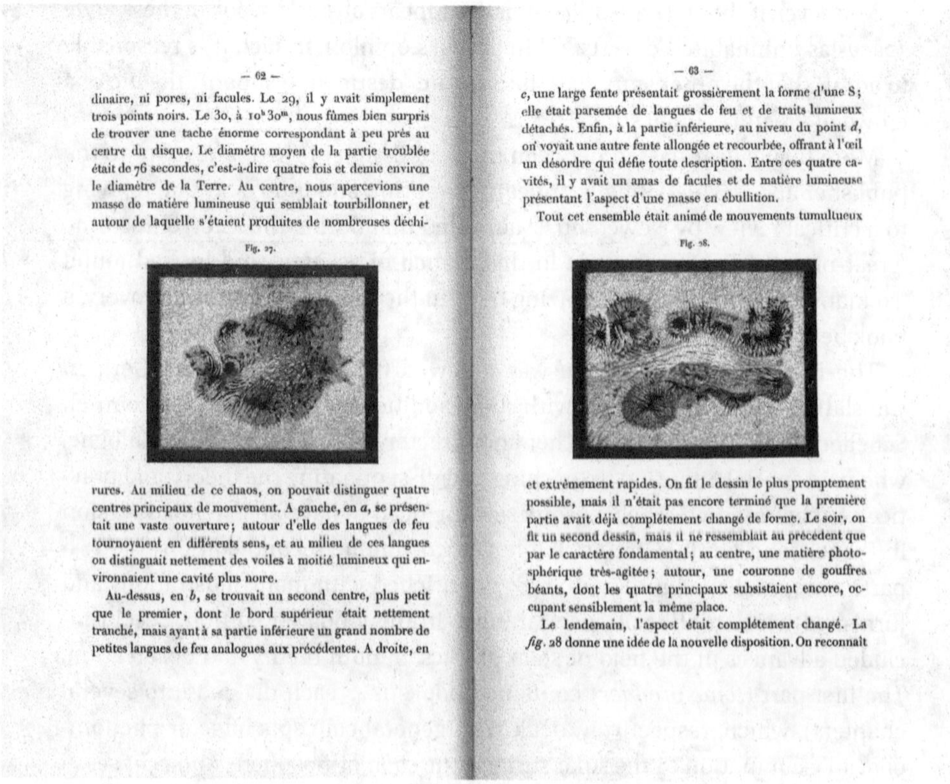

— 62 —

dinaire, ni pores, ni facules. Le 29, il y avait simplement trois points noirs. Le 30, à 10ʰ30ᵐ, nous fûmes bien surpris de trouver une tache énorme correspondant à peu près au centre du disque. Le diamètre moyen de la partie troublée était de 76 secondes, c'est-à-dire quatre fois et demie environ le diamètre de la Terre. Au centre, nous apercevions une masse de matière lumineuse qui semblait tourbillonner, et autour de laquelle s'étaient produites de nombreuses déchi-

Fig. 27.

rures. Au milieu de ce chaos, on pouvait distinguer quatre centres principaux de mouvement. A gauche, en *a*, se présentait une vaste ouverture; autour d'elle des langues de feu tournoyaient en différents sens, et au milieu de ces langues on distinguait nettement des voiles à moitié lumineux qui environnaient une cavité plus noire.

Au-dessus, en *b*, se trouvait un second centre, plus petit que le premier, dont le bord supérieur était nettement tranché, mais ayant à sa partie inférieure un grand nombre de petites langues de feu analogues aux précédentes. A droite, en

— 63 —

c, une large fente présentait grossièrement la forme d'une S; elle était parsemée de langues de feu et de traits lumineux détachés. Enfin, à la partie inférieure, au niveau du point *d*, on voyait une autre fente allongée et recourbée, offrant à l'œil un désordre qui défie toute description. Entre ces quatre cavités, il y avait un amas de facules et de matière lumineuse présentant l'aspect d'une masse en ébullition.

Tout cet ensemble était animé de mouvements tumultueux

Fig. 28.

et extrêmement rapides. On fit le dessin le plus promptement possible, mais il n'était pas encore terminé que la première partie avait déjà complétement changé de forme. Le soir, on fit un second dessin, mais il ne ressemblait au précédent que par le caractère fondamental; au centre, une matière photosphérique très-agitée; autour, une couronne de gouffres béants, dont les quatre principaux subsistaient encore, occupant sensiblement la même place.

Le lendemain, l'aspect était complétement changé. La *fig*. 28 donne une idée de la nouvelle disposition. On reconnait

FIGURE 6.11 Pages from the second edition of *Le Soleil*, with illustrations of solar sunspots and granulation. From Angelo Secchi, *Le soleil* (Paris: Gauthier-Villars, 1875–77), 1:62–63.

This second edition of *Le soleil* was an editorial masterpiece and a milestone among treatises on solar physics. It was translated into Spanish in 1879 and into Italian in 1884 and was considered "the most complete and beautifully illustrated treatise on the Sun which has yet appeared."[134]

7.3 *Le stelle: Trattato di astronomia siderale* (The stars: A Treatise on Sidereal Astronomy)

Le stelle: Trattato di astronomia siderale was the last book that Secchi would see published. It appeared in 1877, and Secchi dedicated it to Schiaparelli, who had revised the drafts. The structure was that of an astronomy textbook, but the style was, as usual, fluent and accessible, with minimal mathematical apparatus, like a sort of handbook. The book was illustrated with figures and plates, though there were fewer of them than in *Le soleil,* which had been

134 Newcomb, *Popular Astronomy*, 547.

produced to a much higher standard. *Le stelle* was divided into five chapters, preceded by an introduction and followed by a conclusion, and contained as appendixes various catalogs of variable stars, binaries, colored stars (with spectral types), and peculiar objects. The first chapter, "Aspetto generale del cielo" (On the general appearance of the sky), was followed by a catalog of fundamental stars for the year 1870 (extracted from the *Nautical Almanac*); the other chapters were entitled "Fisica stellare" (Stellar physics), which included Secchi's spectral classification of stars and other spectroscopic studies on nebulae and galaxies; "Movimenti stellari" (Stars' motions); "Vastità dello spazio stellato" (Vastness of the starry space), on astronomical distances; and "Struttura dell'universo" (Structure of the universe), with a discussion of stellar statistics. In the concluding chapter, bearing the title "Idee sulla grandezza e la struttura dell'universo" (Ideas about the vastness and structure of the universe), he again expressed his conviction about the existence of life in the universe, as in *Le soleil*.

The treatise was well received, and after Secchi's death, it was considered a book that "will [...] prove a work of great importance, and likely to procure for its author a lasting reputation."[135] A second Italian edition was published in 1878; the German and French translations, which appeared posthumously (in 1878 and in 1880, 1884, 1895 respectively), confirmed the success of the book.

7.4 *Lezioni elementari di fisica terrestre* (Elementary Lessons of Earth Physics)

Lezioni elementari di fisica terrestre, published in 1879, one year after the author's death, was certainly the least-known treatise by Secchi. It had a limited circulation (there was only one translation, into Spanish, which appeared seven years later, in 1886) and was probably considered an unusual book in comparison with Secchi's astrophysics books, which continued to be republished until the 1890s. Indeed, it is a sort of scientific testament from Secchi, summarizing his views about a unified universe. The text is accompanied by an appendix with two *Discorsi sulla grandezza del creato* (Discourses about the greatness of creation), an interesting synthesis of science and faith.

In the introduction, Secchi explained that the book was not an original work but resulted from assembling the thirteen lessons he gave to college students "to complete the study of physics, showing its applications in a large scale in Earth physics."[136] He declared that he had taken advantage of many published

135 Philip M.A. Kelland, "Father Secchi," *Proceedings of the Royal Society of Edinburgh* 10 (1880): 11–12, here 12.

136 *Per completare lo studio della fisica, facendone vedere le applicazioni in vasta scala nella fisica mondiale.* Angelo Secchi, *Lezioni elementari di fisica terrestre, con l'aggiunta di due discorsi sulla grandezza del creato* (Turin: Ermanno Loescher, 1879), 1.

works, especially the three-volume *Corso di geologia* (Geology course) by Antonio Stoppani (1824–91), priest and patriot, who is considered the founder of Italian geology and paleontology. "The study of this application of physics is supremely important," Secchi remarked:

> Our own home is the first thing that every living being should try to know [...]; the Earth is our home, where we live and that we inhabit, and from which we draw our livelihood. [...] From its particulars, we can draw the data needed for choosing the most advantageous residence, a healthy climate, an abundance of food, the comforts of life, the leisure of traveling, etc. This study is therefore of practical utility, more than anything else among the studies of this part of nature.[137]

The book, well illustrated and furnished with nine colored plates, further highlights Secchi's efforts to provide students with quality texts: according to Ferrari, it was a further "proof of his tireless hard work, especially when it was done to benefit the young scholars, whose profit was the only [reason why] he had composed these lessons."[138] The contents of the lessons, corresponding to the chapters of the book, contained the latest scientific discoveries: they included a general description of the terrestrial globe, the action of the water in molding the Earth's surface, the water cycle, the circulation of the oceanic currents, phreatic eruptions and artesian wells, volcanism, elements of stratigraphic geology, the Paleozoic, Mesozoic, and Neozoic eras, and the Ice and Anthropic Ages.

8 Contributions to Other Sciences

As we have seen in previous chapters, Secchi adopted a multidisciplinary approach to science, and this is reflected in the number of important contributions he made to sciences other than astronomy. The following focuses on Secchi's

---

137 *Lo studio di questa applicazione della fisica è sommamente importante. La propria casa è la prima cosa che deve cercare di conoscere ogni vivente* [...] *la Terra è questa nostra casa, dove abitiamo e viviamo, e da cui tiriamo il nostro sostentamento.* [...] *dalle sue particolarità noi possiamo trarre informazione per la scelta più vantaggiosa nell'abitazione, per la salubrità del clima, l'abbondanza degli alimenti, la commodità degli agi della vita, la distrazione dei viaggi, ecc. Tale studio è dunque d'utile pratico, più che altro mai, tra gli studi parziali della natura.* Secchi, *Lezioni elementari di fisica terrestre*, 1.

138 *Prova novella della sua infaticabile operosità, specialmente allorchè si trattava di giovare alla gioventù studiosa, in pro della quale aveva unicamente composte queste lezioni.* Gaspare Stanislao Ferrari, "Introduzione" in Secchi, *Lezioni elementari di fisica terrestre*, 2.

contributions to meteorology and oceanography (particularly, the study of coast waters): the former in consideration of the competence he acquired in the field thanks to his multidisciplinary approach to natural sciences; the latter, though relatively minor, because of its current applications, especially in the field of limnology (the study of inland waters, such as lakes, ponds, and wetlands).

8.1 *Meteorology*

Secchi's interest in meteorology was largely stimulated by Commodore Maury's studies on, dynamic meteorology (see Chapter 4). For Secchi, the study of meteorology could be reduced to a purely physical problem: "This problem, though vast, is basically an application of the known laws of physics; the difficulties of the solution are due rather to the large number of disturbing causes and to the incalculable reactions of the effects on the causes than to a real gap in the general theory."[139] It was therefore important, for Secchi, to collect a large amount of data and to order the data in such a way that the information obtained would be easy to interpret.[140]

In January 1857, Secchi announced to the Académie des Sciences in Paris the creation of a balance barometer, allowing the value of atmospheric pressure to be determined by weighting a mercury column, instead of measuring its height, as was usually done.[141] This system also meant it was possible to record the instantaneous value of the atmospheric pressure on a diagram and, for this reason, Secchi called his instrument a "barometrograph."[142] Thanks to the encouraging results obtained with this instrument, he decided to install recording devices in other meteorological instruments, and in 1859, he described his first "meteorograph":[143]

---

139 *Ce problème, si vaste qu'il soit, n'est au fond qu'une application des lois les plus connues de la physique; les difficultés de la solution tiennent plutôt au grand nombre des causes perturbatrices et aux réactions incalculables des effets sur les causes qu'à une véritable lacune dans la théorie générale.* Angelo Secchi, *La météorologie et le météorographe du P. Secchi S.J. à l'Exposition Universelle* (Paris: Gauthier-Villars, 1867), 8, quoted in Brenni, "Il meteorografo di Padre Angelo Secchi," 201.

140 Secchi strongly supported the introduction of recording instruments, as they could record the phenomena in a continuous way, reduce the errors, and guarantee the reliability of the recorded data; see Angelo Secchi, "Intorno alla corrispondenza che passa fra le variazioni meteorologiche e quelle del magnetismo terrestre," 82.

141 Secchi believed he had invented something new, ignoring that rare balance barometers were built in the eighteenth century (see W.E. Knowles Middleton, *The History of the Barometer* [Baltimore: Johns Hopkins University Press, 1964]; and Brenni, "Il meteorografo di Padre Angelo Secchi," 203–204).

142 See Secchi, "Sur un nouveau baromètre à balance."

143 Other meteorographs were developed in those years, especially by British makers (see Robert Bud and Deborah Jean Warner, eds., *Instruments of Science: An Historical Encyclopedia* [New York: Garland, 1998], 375).

> The purpose of this machine is to record simultaneously all weather events with uninterrupted and mutually compared graphical indications. The use of recording instruments in meteorology is going to be extended more and more every day, and this is the only method of observation from which we may, hopefully, advance the science. Those that are now in use (as far as I know, and that I saw in the visits to the main observatories last year, in October 1858) do not have the benefit of bringing together on one machine and on a single sheet of paper the diagrams of multiple instruments, and thus directly comparing the various phenomena in order to deduce the relationships and influences. Such an ensemble was the main purpose of this recorder, specially designed to this comparative study, which for brevity I will call a meteorograph.[144]

To the initial barometrograph, Secchi added, progressively, a telegraphic transmission anemograph; a mechanic thermometrograph; and a rain gauge. The instrument could then register simultaneously, using purely mechanical systems, pressure, temperature, and the duration of rainfall; the wind speed and direction were recorded with electromechanical systems, while the amount of rainfall was still recorded manually, after reading the rain gauge. Later, Secchi also introduced an electromechanical system connected to a psychrometer for recording humidity, and finally an electromagnet system to record rainfall. Secchi designed the meteorograph so that its clock was connected to a cam that drove an arm with a pen point; in turn, the arm was connected to a frame with platinum wires that regularly dipped into small columns of mercury in the instruments,[145] which completed an electrical circuit composed

---

144 *Lo scopo di questa macchina è di registrare simultaneamente tutte le vicende atmosferiche con indicazioni grafiche non interrotte e messe a reciproco confronto. L'uso degli strumenti grafici nella meteorologia va estendendosi ogni giorno più, ed è questo il solo metodo di osservazione da cui possa sperarsi un avanzamento nella scienza. Quelli però che sono ora in uso (per quanto io conosco, e per ciò che ho veduto nella visita fatta ai principali Osservatorii nell'ottobre dello scorso anno 1858) non hanno il vantaggio di riunire su una sola macchina e su di un solo foglio, le indicazioni di più strumenti relativi e di mettere così direttamente a confronto i vari fenomeni, onde dedurre le reciproche relazioni ed influenze. Una tale unione è stata lo scopo principale del presente registratore, che per brevità chiamerò meteorografo, il quale è destinato specialmente a questo studio comparativo.* Angelo Secchi, "Descrizione di un meteorografo ossia registratore meteorologico universale eretto all'osservatorio," *Memorie dell'Osservatorio del Collegio Romano 1857–1859* (Rome: Tipografia delle Belle Arti, 1859), 1:1–8, here 1.

145 The idea of using electricity for reading the meteorological instruments had been introduced by Wheatstone, who had proposed the use of platinum wires for contacts with mercury.

FIGURE 6.12 The Secchi meteorograph at the Jesuit College of Belén in Cuba, with Benito Viñez, S.J. (1837–93), the Jesuit meteorologist who discovered the laws of Caribbean anticyclones; https://es.slideshare.net/Ofranca/contribucion-de-igl-catolica-a-la-ciencia (accessed June 6, 2018).

of a battery and an electromagnet on the mobile arm. Over the years, Secchi improved the instrument,[146] which became more and more efficient, up to the final version, which was presented at the Exposition Universelle in Paris in 1867 (see Chapter 5). Among early recording meteorological instruments, the meteorograph stands out for its efficiency and can be considered the prototype

146 For a detailed description of the meteorograph and its evolution, see Brenni, "Il meteorografo di Padre Angelo Secchi."

of a modern meteorological unit. For this reason, it was acquired and installed in other astronomical and meteorological observatories, mainly in Jesuit colleges, as in the cases of Paris, Manila, Havana, Shanghai, and Washington.[147]

Secchi's interests in meteorology were wide-ranging, including volcanic eruptions, floods of the Tiber, sandstorms, hail,[148] and haze, among myriad other things. His ideas pioneered many different aspects of modern atmospheric physics and environmental sciences. Some of his ideas, for example about climate change, are surprisingly topical:

> Has our climate changed since some time in the past up to now? A complaint of this kind is not uncommon, and if we believed in certain people, every year would always be the hottest or coldest or most intemperate that we have ever had! The thermometer, however, refutes these impressions. Consulting ancient texts, we realize that episodes of freezing that occurred in the past no longer occur and hence definitely we are facing a climate change, due to human activity such as deforestation and the introduction of artificial sources of heat. In large industrialized centers, like London, the average temperature has increased by 2° C over the last few years [...]. Though in Italy these variations have not occurred, there is, however, the evidence that the glaciers of the Alps have decreased as a result of increasing temperatures [...].[149]

---

147 See Brenni, "Il meteorografo di Padre Angelo Secchi," 235–237. In Italy, the only meteorograph operating at the same time as the one at the Collegio Romano was installed at Palermo Observatory, but it had been completely destroyed by the middle of the twentieth century. The meteorograph of the Collegio Romano is still extant: it was restored in the 1990s and is on display at INAF-OAR (site of Monte Porzio Catone). Because of its worldwide distribution, the meteorograph has been regarded as one more example of "Secchi's truly catholic vision of scientific research" (Massimo Mazzotti, "The Jesuit on the Roof: Observatory Sciences, Metaphysics, and Nation Building," in *The Heavens on Earth: Observatories and Astronomy in Nineteenth-Century Science and Culture*, ed. David Aubin, Charlotte Bigg, and H. Otto Sibum [Durham, NC: Duke University Press, 2010], 58–85, here 67).

148 Secchi's hypothesis on the hail formation was inspired by the account by Rosalie Goujon (1819–1908), also known as Madame Poitevin, a pioneer of aeronautics. In 1859, while in a hot air baloon over Rome, she encountered a cold air vortex and lost control of the baloon for a number of minutes. She later said that, in those few minutes, the water in the bottles on board became frozen. From his observatory, Secchi observed the ascent and the effect of the vortex on the baloon with the Merz telescope, and mentioned the report by Madame Poitevin as evidence supporting his theory that hail is produced by cold air vortex at some height in the atmosphere. See Angelo Secchi, "Di alcuni fatti relativi all'origine della grandine", *Atti dell'Accademia Pontificia dei Nuovi Lincei* 29 (1876): 1–7.

149 *È il nostro clima cambiato da qualche tempo in qua? Un lamento di questo genere non è raro, e se credessimo a certe persone, ogni anno sarebbe sempre il più caldo o il più freddo e il più stemperato che siasi mai avuto! Il termometro però smentisce queste sensazioni popolari.*

8.2 *Oceanography*

Secchi also made an important contribution to the study of oceanography, in particular the development of a method, still in use in limnology, for measuring the transparency of water. The method involves the use of the "Secchi disk," a device that is immersed in water with a graduated tape until it is no longer perceptible to the eye. This limit, now called the "Secchi depth," is an indicator of the transparency of the water. In oceanography, the measurement of this parameter, along with the color of the water, had always been recommended to sailors for safe navigation, particularly to assess the visibility in the water and discern cliffs, sandbanks, shoals of fish, proximity to the coast and currents.

Systems similar to the Secchi disk were already known and in use among mariners, especially those who were interested in the study of the seabed.[150] Alessandro Cialdi (1807–82), engineer and commander general of the Pontifical Navy, was among them. Cialdi was especially interested in the correlation between transparency, waves, and currents; and he took into account the recommendations given by renowned physicist and astronomer François Arago (1786–1853) to include optical experiments in these kinds of studies. For this reason, in 1865 Cialdi invited Secchi to join him in a measurement campaign aboard the pontifical steam corvette *Immacolata Concezione*,[151] off the port of Civitavecchia.

Attracted as usual by the possibility of new experiences and new discoveries, Secchi gladly accepted the invitation and remained aboard from April 20 to June 1, a total of forty-three days. The research program included an analysis of the transparency and color of the water, relative to the altitude of the Sun, weather and sea conditions, measurements of sea depth, height from the surface of the water and the observer's eyesight. Secchi was interested primarily in the effects of light refraction and absorption, in order "to reduce the reflected light of the sea, and hence to better see the bottom, and help our view to

---

Consultando i testi dell'antichità ci rendiamo conto che fenomeni di gelate avvenute in passato non si verificano più e quindi sicuramente siamo di fronte ad un cambiamento climatico dovuto all'azione dell'uomo, quale il disboscamento ed introduzione di sorgenti artificiali di calore. Nei grandi centri industrializzati come Londra la media della temperatura è aumentata di 2°C negli ultimi anni, come riportato dal sig. Glaisher. Pur non essendosi verificate in Italia tali variazioni è da evidenziare che i ghiacciai delle Alpi si sono ridotti con conseguente aumento della temperatura [...]. Angelo Secchi, "Sul clima di Roma," *Bullettino meteorologico dell'Osservatorio del Collegio Romano* 5 (1866): 44–48; 57–58, here 57.

150 See Marcel R. Wernand, "On the History of the Secchi Disc," *Journal of the European Optical Society: Rapid Publications* 5 (2010): 1–6 ; see also Natalia Ptitsyna, "Il disco di Secchi," in Altamore and Maffeo, *Angelo Secchi*, 241–253, here 242–243.

151 In operation since 1859, the *Immacolata Concezione* was supposed to be at the pope's personal disposal during his travels; Pius IX, however, never used it given the increasingly complex political situation, which made it inopportune to depart from Rome.

FIGURE 6.13 The papal steamship *Immacolata Concezione* in an illustration of 1859. From *The Illustrated London News* (August 10, 1859), 179.

penetrate inside the waves,"[152] as indicated by Arago.[153] To measure the transparency of the seawater, he used some disks of different sizes and materials:

> The objects destined to be plunged in the water were disks of different colors and sizes: a system of ropes and weights were attached to keep them horizontal in the water, and graduated lanyards were used to determine the depth.
>
> The first one was a large disk of 3.73 meters in diameter, made of an iron ring, covered with a ceruse painted veil canvas. The others were small disks of 0.40-meter diameter, but of different materials [...] and different colors [...].[154]

152 *Utili a diminuire la luce riflessa del mare e quindi a scorgere meglio il fondo, e aiutare la penetrazione della visuale dentro l'onda.* Angelo Secchi, "Relazione delle esperienze fatte a bordo della pontificia pirocorvetta L'Immacolata Concezione per determinare la trasparenza del mare," in Alessandro Cialdi, *Sul moto ondoso del mare e su le correnti di esso*, (Rome: Tipografia delle Belle Arti, 1866) 260.

153 See Alessandro Cialdi and Angelo Secchi, "Sur la transparence de la mer," *Comptes rendus de l'Académie des Sciences* 61 (1865): 100–104, here 100.

154 *Les corps destinés à être plongés dans la mer étaient des disques de couleurs et de grandeurs différents: un système de cordes et de poids était disposé pour les tenir horizontaux dans l'eau, et*

Through these experiments, Secchi could deduce experimental laws for determining the transparency and the color of the water. Secchi and Cialdi communicated the results of their meticulous analysis in a report to the Académie des Sciences,[155] which, almost a century later, was summarized in the following way:[156]

> Secchi's experiments showed that the depth of visibility increased with the size and whiteness of the disc and with the altitude of the Sun. He noted that image dissection by surface refraction caused the visibility of the disc to decrease and that the ship's shadow underwater also influenced its visibility. He demonstrated the detrimental effect of surface reflections on the measurement and recommended a wide shadow over the place where the observations were being made. He also made observations on the effect of the clearness of the sky, the color of the water and of the disc, and the height of the observer above the water. His work established the experimental procedure for obtaining transparency with a Secchi disc, as it is now called. In the years following Secchi's work, the Secchi disc became a standard piece of equipment.[157]

At the end of the nineteenth century, in order to measure the transparency (and hence the quality) of drinking water in reservoirs, the original all-white disk was modified by engineer George C. Whipple (1866–1924), an expert in sanitary microbiology, into a well-known version with alternately black-and-white quadrants. Widely used even today, in different versions, to determine the turbidity and transparency of seawater and freshwater, it was named the Secchi disk in honor of the Jesuit scientist by François-Alphonse Forel (1841–1912), the founder of limnology.[158]

---

des lignes graduées servaient à déterminer la profondeur. Le premier était un grand disque de 3^{m},73 de diamètre, formé d'un cercle en fer recouvert de toile à voile vernie au blanc de céruse. Les autres étaient des petits disques de 0^{m},40 de diamètre seulement, mais de diverses substances [...] *et de différentes couleurs* [...]. Cialdi and Secchi, "Sur la transparence de la mer," 101.

155 Secchi published the full report in *Il nuovo cimento*, and Cialdi included it in his book on the motion of seawaves; see Cialdi, *Sul moto ondoso del mare e su le correnti di esso*, 258–287.

156 Theoretical studies on the Secchi disk came quite late (see John E. Tyler, "The Secchi Disc," *Limnology and Oceanography* 13, no. 1 [1968]: 1–6, here 2).

157 Tyler, "Secchi Disc," 1–2. It is important to stress that, rather than the disk itself, Secchi invented a method of using it, which was later standardized and is still in use today.

158 See François-Alphonse Forel, *Le Léman, monographie limnologique* (Lousanne: F. Rouge Editeur, 1895), 433.

CHAPTER 7

A Crucial Political Change: New Risks and Missed Opportunities

The year 1870 was an extremely difficult one for Angelo Secchi. In March, a famous Florentine newspaper, *La Nazione*, published a report from Rome containing the following news:

> The Jesuit astronomers Rosa and Secchi to leave the Society [of Jesus]. It is unknown what were the causes for such a decision from Count Rosa [...], who so loved his order as to donate to it several thousand scudi at the expense of his family [...]. Secchi believed that many of the instruments of the Roman Observatory belonged to him as his private property, and wanted to convert them into cash. The provincial [superior] opposed him; and, naturally, the general superior of the Society asserted that the provincial was right. Secchi, a very irascible man, retorted to the general a fortnight ago: for which he was invited to seek accommodation elsewhere.[1]

The Catholic world was shocked and confused by this news, as was Secchi's family, who were unable to imagine Secchi ever making such a decision. The attitude within the family is well described in a letter Secchi received from his audacious eleven-year-old grandniece, Gabriella (see Chapter 1), which deserves to be reported almost in its entirety:

> My illustrious and dear uncle,
>
> I have already reached the age of eleven years and the respectable most Reverend Your Lordship may think that, after many cares and expenses incurred by Mom to educate me, at least I learned to pen a line. Rereading

1 *I gesuiti astronomi Rosa e Secchi escono dalla compagnia. Ignoro le cagioni che per una tal risoluzione può avere il conte Rosa* [...], *il quale tanto amò il suo ordine da donargli parecchie migliaia di scudi a scapito della propria famiglia* [...]. *Il Secchi credeva che non poche macchine dell'Osservatorio romano gli appartenessero come proprietà private, e voleva convertirle in denaro sonante. Il provinciale gli si oppose; e come è ben naturale il preposto generale della compagnia diede ragione al provinciale. Secchi, uomo irascibilissimo, rispose per le rime al generale, un quindici giorni fa: per la qual cosa è stato invitato di cercare alloggio altrove* (*La Nazione*, no. 72 [March 13, 1870], 1).

 | DOI:10.1163/9789004387331_009

now your precious letters written to my poor Daddy, where in several of them I find greetings also for me, I was already set to address to you my first letter, and if today I have the honor and luck to do it, nonetheless my hand trembles for the awe inspired by a distinct personality like Your Lordship. On the other hand, I thought that indulgence is always to be found in a Venerable Priest, so I find the courage to express the cause which moved me to write you.

Three days ago, we read in the newspaper the news of your sudden retirement from the Society of Jesus, and this unexpected news, that is considered certain, has naturally aroused utmost concern to our whole family, which is unaware of what cause would have led Y.L. to that resolution and doubts that you can find now in the midst of terrible sorrows.

Every day, my excellent Uncle Don Alessandro says to write to you, but until now he has not had the courage, not even Mom, for fear that this may be displeasing for you. And I ask for forgiveness if today, seeing my good Grandmother more and more worried about you, and also my Aunt, all in short, and being in the dark if the Uncle has written, I begged my Mom to allow me to write to you, trusting again in your kindness, and thinking about the joy that I would feel in being the first who will be able to inform and ensure your dear mother and all about your perfect health, as I hope.

And while, I believe to be forgiven by Y.R.L. if you would be pleased to address a single line to the family for its peace of mind, as it also, as you can imagine, is required at any time to know news about you.[2]

---

2 *Illustre e caro mio zio.*
Ho già raggiunta l'età di 11 anni e la Rispettabile S.V. Rev.ª penserà che dopo tante cure e spese sostenute da Mammà per educarmi, io abbia almeno imparata a vergare una riga. Rileggendo ora le preziose di Lei lettere scritte al povero mio Babbo, dove trovo in parecchie anche un saluto per me, mi ero già prefissa di dirigere a V.S. la mia prima lettera, e se oggi compio tale onore e fortuna, mi trema d'altra parte la mano per la soggezione che inspira un distinto Personaggio qual è V. Signoria. Il pensiero d'altronde che in un Venerando Sacerdote si rinviene sempre l'indulgenza, mi faccio coraggio esternandole la causa per cui mossi a scriverle.

Tre giorni or sono abbiamo letta nel Giornale la notizia del di Lei improvviso ritiro dalla Compagnia di Gesù, e tale inaspettata nuova che si vorrebbe positiva ha naturalmente suscitata la massima inquietudine a tutta la nostra famiglia, la quale ignara della causa che avrebbe indotta V.S. a tale risoluzione, dubbita [sic] *ancora che possa essere ora in mezzo a gravissimi dispiaceri.*

Ogni giorno l'ottimo mio Zio Don Alessandro dice di scrivere, ma poi sin'ora non ne ha avuto il coraggio, e neppure Mammà nel timore che ciò possa disgustarla. Ed io domando perdono se oggi vedendo vie più in pena la mia buona Nonna, la Zia, tutti insomma, ed altresì nell'incertezza che lo Zio abbia scritto, ho pregato la mia Madre di permettermi di scrivere, confidando ripeto

As was to be expected, Secchi's reaction to the newspaper's report was prompt: in a letter to the director of *La Nazione*, he wrote that "all of that story, and the reasons attached thereto, [were] a pure invention"[3] and that the supposed controversies with his superiors were absurd and mere inventions as well. At Secchi's request, his denial was published in the same newspaper on March 19, 1870, a week after the fake announcement. However, the idea of a conflict between Secchi and the Society had now entered into the public imagination, and when the political scenario changed after the annexation of Rome by the Kingdom of Italy, the controversy was reignited.

The politically oriented newspapers of that time gave voice to three main currents of thought: liberals (the ruling party); republicans, who were the opposition; and Catholics, mostly supporters of the pope's temporal authority. The liberal–Freemason government needed to justify its conquest of Rome in the eyes of Catholics, while many republicans were openly anti-clericals. Catholic supporters of the pope's temporal power opposed both sides in the name of safeguarding what they considered the divine rights of the pope; they were animated by a spirit of militant re-conquest of the conscience to Catholicism, which they saw as being threatened on two fronts.[4] In this context, because of his popularity, Secchi was often used (and attacked) by one side or the other to engage public opinion for their own political ends.

1 The Chair at La Sapienza University: Acceptance and Resignation

Secchi's acceptance and resignation from the chair that was offered to him at La Sapienza University in Rome was one of the most painful events in his life. It also had serious consequences for his scientific career and, more generally, for the development of astrophysics in Italy. With it was lost the opportunity to create the very first school of astrophysics, even earlier than those established in Germany and the United States at the end of the century. It is interesting

---

nella di Lei bontà, e pensando alla gioia che proverei nel potere per la prima partecipare e assicurare la Sua cara mamma e tutti della di Lei perfetta salute siccome spero.

Intanto crederò di essere perdonata da V.S.R. se si compiacerà dirigere una sola righa [sic] *alla famiglia per tranquillità, la quale inoltre come Ella può ben immaginare, è ad ogni momento richiesta per sapere di Lei nuove.* Gabriella Secchi to Angelo Secchi, Reggio Emilia, March 15, 1870; APUG, FS 19.

3 *Tutto quel racconto, e le ragioni allegate ivi, sono una pura invenzione.* APUG, FS 9.V.

4 See Feliciano Gallon, "La stampa romana e la vita religiosa," in *La vita religiosa a Roma intorno al 1870: Ricerche di storia e sociologia*, ed. Paul Droulers, Giacomo Martina, and Paolo Tufari (Rome: Università Gregoriana Editore, 1971), 47–96, here 50.

to read an account of this incident from a diary that Secchi himself wrote to defend himself from various accusations:

> Since many people now write about me in the newspapers, and some believe I've dishonored the Society [of Jesus] by agreeing to stay at the observatory while others believe they have done me a great favor by letting me stay, I think it well to write down how the matter came about.[5]

Secchi reported that, after the entry of the Piedmontese army in Rome,[6] he was immediately approached by Domenico Cipolletti (1840–74), an emissary of the general secretary for public education, Giovanni Cantoni (1818–97).

Cantoni hastened to inform Secchi that the government intended to offer him a chair. In Secchi's words:

> He told me that the minister wished that I stay on at the observatory and that I would receive from them all possible respect and the salary of a professor, etc. I replied that this was a serious matter, and that for the moment I could not give a definitive answer, but I would prefer by far to provide my services in my homeland rather than elsewhere, considering my own convenience. I spoke only in broad terms, although in the previous days I had searched the will of the provincial [superior] [Fr. Gaetano Tedeschi (1820–91)], […] who recommended that I do everything possible to defend my personal rights at the observatory, and be compensated for the money I had spent on this establishment […].[7]

Two days later, a second letter from Cantoni was delivered to Secchi by Diamilla Müller (see Chapter 5), while Francesco Brioschi (1824–97), member of the Superior Council of Public Instruction, visited Secchi in order to explore

5 *Siccome molti si occupano adesso nei fogli di me, e mentre alcuni credono che io abbia disonorato la Compagnia accettando di restare all'Osservatorio; altri, credono avermi fatto un gran favore lasciandomici: io credo bene di notar le cose come sono andate.* APUG, FS 23.II.C.

6 An interesting report of that difficult day at the Collegio Romano was written by the rector; see Giacomo Martina, "Al Collegio Romano il 20 settembre 1870: Dalla relazione del p. Pietro Ragazzini," *Archivium historiae pontificiae* 8 (1970): 332–347.

7 *Questi mi disse che il ministro desiderava che io restassi all'osservatorio e che io avrei avuto da loro ogni riguardo possibile e soldo da professore, ecc* […] *Risposi che la cosa era grave, e che nel momento non poteva dare risposta definitiva, che però io avrei preferito sempre di prestare i miei servizi alla mia patria che altrove, salve le mie convenienze. Mi tenni così sulle generali, benché avessi giorni prima scandagliato l'animo del Provinciale* […] *che mi avea detto di fare il possibile di sostenere i miei diritti personali all'osservatorio, e farmi ricompensare i capitali che avea speso in questo stabilimento* […] APUG, FS 23.II.C.

FIGURE 7.1 The Battle of Porta Pia in a painting by Carlo Ademollo (1824–1911); https://commons.wikimedia.org/wiki/File:Breccia_di_Porta_Pia_Ademollo.jpg (accessed June 8, 2018).

his intentions. He also showed him a letter from the minister of finance, Quintino Sella (1827–84), asking him to mediate between the Society of Jesus and the government—which Secchi refused to do, as the conditions set by the government were considered unacceptable by the Jesuits. The minister, however, pushed for an answer from Secchi, who, after a second visit from Diamilla Müller, at last transmitted his personal conditions to Cantoni:

> First, I have certain duties and obligations: these were the boundaries of my actions. It is my obligation to do nothing against my conscience, so I would not make an oath to anyone. Second, I have duties toward the pope and the Society [of Jesus], and I would never make any improper act toward them, or act without their approval. But if I knew that to stay at the observatory under such terms was acceptable to them, I would stay. Among my other considerations, I then added that I wanted to have freedom in the choice of the assistants and staff at the observatory, because I did not want to be put into the hands of people who did not have my trust, nor to dismiss those who had responded to my every gesture with such care up to now.
>
> I believed I could also ask for those conditions with a certain right as I was not receiving an observatory that was in any way unconnected and unimportant to me. It was the fruit of my labor for a total of twenty-one years, and of the financing obtained through my efforts. […] To keep me

> at the observatory was as if to allow me stay in my own home […]; this was not a favor, but an act of justice […].[8]

Secchi's requests were accepted: "Then I received a letter from Cantoni in which he said that my convictions of conscience would be respected and I would have no obligation to make any oaths and that I could make the observatory what I wanted."[9]

The situation seemed to be turning out for the best: Secchi could remain director of the Collegio Romano Observatory, under conditions set by him and with maximum freedom of action. However, a problem soon arose that was connected to the Jesuits and their freedom to teach in colleges. Secchi was sent by his superiors to make contact with Brioschi, who was in charge of the reorganization of the education system in Rome, in order "to fathom the intentions" of the government in this matter:

> He greeted me with every courtesy, really […] the conversation turned to teaching and he asked me if I would accept a teaching role at La Sapienza. I said yes, but not knowing if I would continue my teaching at the Collegio [Romano] next year or not, I wished to know this before accepting, for my usual reasons. He replied that there was no problem with continuing to teach at the college and the Jesuits could have a private and free school as the Scolopi [Piarists] did in Florence, because the government wanted to please everyone.[10]

---

8 *Primieramente avea io dei doveri e degli obblighi: questi erano regola del mio agire. È mio obbligo di non far nulla contro la mia coscienza, quindi io non avrei fatto giuramento a chicchessia. Secondo avea dei doveri verso il Papa e la compagnia, e io non avrei mai fatto atto alcuno sconveniente verso di loro, o che non fosse di loro piacere. Siccome però sapeva che il restare all'osservatorio dietro tali patti non era loro discaro, io sarei restato. Tra le convenienze poi io allegava che volevo avere libertà sulla scelta del personale di mio aiuto e servizio all'osservatorio, perché non volevo restare in mano di persone che non fossero di mia fiducia, né cacciare quelli che si erano finora prestati ad ogni mio cenno con tanta premura.*

Credeva io di poter domandare tali condizioni anche pel titolo che io non riceveva un osservatorio qualunque a cui fossi estraneo ed affatto indifferente. Esser l'osservatorio frutto totale di 21 anni di mio lavoro, e del credito che mi era saputo procacciare colle mie fatiche. […] *Il lasciarmi all'osservatorio valeva quanto lasciarmi in casa mia* […] *non* [*era*] *questo un favore, ma un atto di giustizia* […] APUG, FS 23.II.C.

9 *In seguito giunge lettera di Cantoni in cui diceva che io sarei rispettato nelle mie convinzioni di coscienza, e non avrei obbligo di giuramento e che avrei fatto all'osservatorio ciò che avessi voluto.* APUG, FS 23.II.C.

10 *Egli mi accolse veramente con ogni cortesia* […] *cadde il discorso sulle scuole, domandandomi esso se io avrei accettato una scuola in Sapienza. Dissi di sì, ma che non sapendo se quest'altr'anno sarebbero continuate o no le scuole in collegio, avrei avuto prima desiderio di sapere questo per mia norma.* APUG, FS 23.II.C.

Secchi transmitted this favorable response to his superiors; but, a few days later, Brioschi's attitude had changed. During their next visit, the rector, Fr. Pietro Ragazzini (1827–77), and the above-mentioned father provincial, Tedeschi, found that Brioschi gave them no guarantee that Jesuit teaching in public colleges would be respected:

> The result of this meeting was that Ours saw all hope lost, and the kindest thing I heard said about me at home was that I was a sucker who had been hoodwinked; others said that I was already cooked [i.e., subjugated] and half liberal, as if I had reported what was not [true].
>
> This was very annoying to me, and since I had to go see Brioschi because of my departure [to Sicily (see Chapter 5)], I wanted to tell him what was happening. I asked him to tell me the future of the C[ollegio] R[omano] and if, upon my return from Sicily, I would still find my [religious] family there. He told me that the thing was very difficult and that he was studying whether the Collegio Romano was or was not owned by the Society [of Jesus], if it was convenient to leave it as an educational institution, and by which way one could stay there. I replied that I could save him the study, because it was uncontested that the Collegio Romano was founded by St. Ignatius and Francis Borgia […]; it is a building for hosting religious people and not a governmental or municipal institute for teaching, it is ecclesiastic, and it could not be taken away without doing a clear act of injustice to the lay boarding students […]. Well, he said, these are things of another order that need further study and decision. Meanwhile, go and let me know if you want to give lessons at La Sapienza on your beautiful discoveries. […] The next day, I actually received a letter in which I was asked if I wanted to accept the chair of physical astronomy. After consulting the provincial about everything, I replied in writing, yes. Those who might be scandalized, let them think about what the provincial was telling me: accept this chair so that you will be free to refuse any teaching in any school that they would open here, and thus you would not be obligated to act in competition with our schools. Therefore, I accepted.[11]

Esso mi rispose, non esservi veruna difficoltà a continuare le scuole in collegio, e che i Gesuiti potevano fare una scuola privata e libera come gli Scolopii a Firenze, perché il governo volea contentar tutti. APUG, FS 23.II.C.

11 *Il risultato di questa conferenza fu che i nostri videro perduta ogni speranza, e di me parlavasi in casa dai più benevoli come di un balordo che era stato gabbato; da altri che io era già cotto e mezzo liberale, quasichè avessi riferito ciò che non era.*

Questo mi seccava assai, e dovendo io andare dal Brioschi per la mia partenza, volli dirgli il fatto mio. Lo chiesi adunque che mi dicesse che cosa si sarebbe fatto del C.R. e se al mio

Meanwhile, Brioschi sent a letter to the rector of the Collegio Romano, as reported by Secchi:

> The content of this letter was that the registration of students was not prohibited, but it was under the penalty of not being recognized. [...] This letter, as well as an exhortation from the pope that begged us not to dismiss the external students from the schools, meant that Ours began to get much too busy accepting as many boys as they could [...] inde irae [then came the wrath], and in the evening [the anti-Jesuit party] staged a demonstration in the streets,[12] and thus everything was well played out, and it was forbidden for Italian boys to come to us, while freedom of education was shouted throughout the streets from those who were expelling us. Long live logic.[13]

These events aggravated the conflicts between the Jesuits and the government, and tensions increased. On November 5, Secchi left Rome to go to Sicily and, the day before leaving, having been urged to give an official reply, he wrote a letter to Brioschi where he formally accepted the chair of physical astronomy at the prestigious Roman University.

He also suggested changing its title:

---

ritorno dalla Sicilia avrei potuto ritrovare la mia famiglia. Egli mi disse che la cosa era assai grave che stava studiando se il CR fosse o no proprietà della compagnia, se fosse convenuto a luogo d'istruzione e con che titoli noi ci stavamo. Risposi che gli studi glieli avrei risparmiato io, perché era fuori di dubbio che il CR fu fondato da S. Ignazio e Francesco Borgia [...] *esso è convento e non istituto governativo né municipale di insegnamento, ma ecclesiastico e pei collegi esteri non potrebbe togliersi senza chiara ingiustizia* [...] *Sta bene, disse esso, ma queste sono cose di altro ordine e che bisogna decidere e studiare. Intanto vada , e mi risponda se vuole dare lezioni in Sapienza sulle sue belle scoperte.* [...] *Il giorno seguente venne di fatto una lettera a me in cui mi chiedeva se io voleva accettare la cattedra di astronomia fisica. Io dopo aver di tutto consultato il provinciale risposi di sì per iscritto. Quelli che ne fossero scandalizzati pensino ciò che mi diceva il Provinciale: Accetti questa cattedra che così Ella sarà libero a non accettare nessun insegnamento nel liceo che si aprirà qui, e così non sarà obbligato di fare un contro altare alle nostre scuole. Perciò io accettai.* APUG, FS 23.II.C.

12 The high number of enrollments at the Collegio Romano provoked an anti-Jesuit protest.

13 *Il tenore di questa lettera era tale che non inibiva il pigliar scolari, ma solo poneva la penale del non riconoscimento* [...] *Questa lettera, e una esortazione del Papa che scongiurava a non respingere gli esterni dalle scuole fece sì che i nostri si mettessero con troppa alacrità ad accettare ragazzi quanti più potevano* [...] *Ma inde irae, e la sera si fece la commedia della dimostrazione in piazza, e tutto fu così accomodato, e proibito ai ragazzi italiani di venire da noi mentre si gridava libertà di istruzione per le vie da quelli che ci cacciavano. Viva la logica.* APUG, FS 23.II.C.

FIGURE 7.2 Seventeenth-century engraving of the University La Sapienza. From Giuseppe Vasi, *Delle magnificenze di Roma antica e moderna, libro nono* (Rome: Nella Stamperia di Niccolò e Marco Pagliarini, 1759), 161, pl. 9.

> Realizing that there is another chair of astronomy, and that this simple title could excite some apprehension in the other professor [Respighi], to avoid this perhaps it would not be a bad idea to also add meteorology; so that the title [of the chair] would be of physical astronomy and meteorology.[14]

A couple of comments are necessary here. The first is that Secchi took care to see that his educational curriculum did not overlap with Respighi's, probably because he felt that the relationship with him was becoming more and more difficult; in a couple of years, it actually came to a head when Respighi accused Secchi of plagiarizing his spectroscopic research and incited an underground campaign against him (see Chapter 9). The second is that this proposal was quite innovative, because no chair of physical astronomy existed at that time,

14 *Solamente avendo io riflettuto che vi è un'altra cattedra di Astronomia, e che questo semplice titolo potrebbe eccitare qualche suscettibilità nell'altro professore, ad evitare questa, forse non sarebbe mal fatto di aggiungere anche di Meteorologia; onde il titolo sarebbe di Astronomia Fisica e Meteorologia* (*La Nazione*, no. 191 [July 10, 1871]).

either in Italy or abroad, and there were only a few chairs of meteorology, in Germany. Secchi stayed true to his original background in considering himself an astrophysicist and a meteorologist (ultimately, a physicist), rather than an astronomer.

In the meantime, the controversy between the Jesuits and the government over the freedom of teaching continued, and Secchi was placed in an embarrassing position. He had accepted a charge from a government that was in conflict with the religious order to which he belonged. His most intransigent confrères criticized his decision and demanded that he resign.

The days Secchi spent in Palermo were intense and full of anxiety. He was in a extremely difficult situation, pressured to resign to keep true to his Society but wishing to maintain an obligation to be loyal to the government. He was probably aware that he was a strategic pawn in the hands of both actors. Even though he certainly wanted to keep both promises, there was no way out: he had to choose between his personal scientific career and his religious duties. While in Palermo, he did not miss any chance to meet with his local confrères and other possible advisors who might help him find a way to save face. The only other churchman taking part in the scientific expedition to Sicily was Fr. Denza, who was among his confidants during this period. Secchi had been offered accommodation at Palermo Observatory, and its anti-clerical director Gaetano Cacciatore was disturbed by the frequent visits that Secchi received, as he angrily wrote to his colleague Donati in Florence after Secchi's departure:

> I start with running over some facts that happened during the residence of those most evil men in Palermo. The ugly mug [Secchi] was accommodated at my home, the little tramp [Denza] at Tacchini's; he only slept there. The whole day until after midnight the couple in love were always alone together. The normal visits resumed, and this time with greater fervor: the most horrible mob of the country came into my immaculate apartments. A kind of Catalog in White circulated through the city: there was an Oremus pro Pius IX at the head, and a signature of Fr. Secchi at the end. The newspapers became indignant and wrote this and that; a certain anonymous letter was also sent:[15] the Jesuit got the wind up and set off at a breakneck pace.
>
> Before leaving, he was invited by the Academy of Sciences and Letters (a reactionary and clerical body) to say something about the eclipse: a two-for-one deal! He went there. All the barbers and grocers of the

15 It was a threatening letter; see Chapter 9.

city were invited. The immense crowd turned into a bacchanalia, which forced Fr. Secchi to give his lecture in the Great Hall of the university. Our Society of Natural Sciences, indignant at the proceeding of the academy, sent the same invitation to me as vice-president of the commission, and to Tacchini, and we made our presentations [...]

He [Secchi] was oh so genteel and cloying while he was guest in my house. Once outside of it, he caused me rancor and grief. Back at my home, he started up again with his cloying conduct, and now that he has gone away I expect something uglier. But I laugh and hold in contempt him and his peers, and while I worship science, on the other hand I have contempt for it when it can serve to mask such abhorrent, filthy, obscene, and lascivious men. I hope that our government will finally open its eyes to discern the true nature of these things, to appreciate real merit, stripping away the coarse exaggeration owed largely to tonsure and cocottes. [...] Rumor has it that the government will give honors to Fr. Secchi. I do not believe it. But then, what of us! It would be really weird! I see that I can say no more.[16]

16 *Comincio col cennarvi qualche fatto durante la dimora di quei tristissimi in Palermo. Il brutto muso riprese stanza presso di me, la squaldrinella presso Tacchini solo per dormire. Il giorno sin dopo la mezzanotte l'amorosa coppia stava chiusa sempre insieme. Le consuete visite ritornarono, e questa volta con maggior fervore: la più orrenda genìa del Paese frequentò i miei immaculati appartamenti. Girò per la Città una specie d'Album in bianco: vi era in capo un Oremus pro Pio IX, e ai piedi una firma di P. Secchi. I giornali se ne adontarono, e scrissero qualche cosa: qualche lettera anonima fu anche spedita: e il Gesuita si ebbe timore e partì a rotta di collo. Prima di partire ebbe invito dall'Accademia di Scienze e Lettere (corpo retrivo e clericale) di dire qualche cosa sull'ecclisse: affare combinato! Ei vi andò. Furono invitati tutti i barbieri e i pizzicagnoli della Città. Folla immensa che fece un baccano, ed obbligò il P. Secchi a far la sua conferenza nell'Aula Grande dell'Università. La Società nostra di scienze naturali, indignata dal procedere dell'Accademia, mandava contemporaneamente lo stesso invito a me qual Vice Presidente della Commissione ed al Tacchini, e noi fecimo la nostra relazione* [...] *Fu gentilissimo e sdolcinato mentre fu ospitato in mia casa. Appena uscitone mi fu causa di rancori, e di dispiacere. Ritornato presso di me riprese le prime sdolcinature, ed ora che ne è uscito e lontano mi attendo qualche cosa di più brutto. Ma mi rido e sprezzo lui ed i suoi pari, e se da un lato venero la scienza, dall'altro la disprezzo quando può servire a mascherare uomini turpi, osceni, luridi e lascivi. Io voglio sperare che il nostro governo avrà finalmente occhi a discernere il vero valore delle cose, ad apprezzare il merito reale, spogliandolo dalle esagerazioni da trivio dovute in gran parte alla cherica e alle cocottes.* [...] *Corre voce che il governo vorrà dare degli onori al P. Secchi. Io non lo credo. Ma che cosa siamo adunque noi! Sarebbe veramente strano! Vedo bene che non la finirei più.* Gaetano Cacciatore to Giovan Battista Donati, Palermo, January 23, 1871; Archives of INAF-Osservatorio Astrofisico di Arcetri.

The letter continues in this tone before launching into a general invective against priests and churchmen. It highlights the anti-clerical climate of the time and the rivalry between the various political currents. It also confirms Secchi's popularity, whom even the simplest people went to hear. On the one hand, it betrays a deep sense of envy at a man being "courted" by the government and preferred to others for the recognition of his scientific merits. At the same time, it also demonstrates that Secchi's decision to travel to Palermo had political repercussions and, as was to be expected, provoked various reactions.

In the meantime, in Rome, the superior general, Peter Jan Beckx (1795–1887, in office 1853–87), was in an embarrassing position as well, as Secchi was seen to be a Jesuit who was collaborating with the "enemy" government. Did Secchi really make this decision in agreement with his provincial superior? If so, to refuse the appointment might raise a problem of obedience. On the other hand, Secchi's resignation of the chair could provoke further reactions from the government against the Jesuits, and they even risked being expelled from the college.

Secchi wrote the superior general a letter of explanation to try to defend his decision and justify his acceptance of the chair. He wrote to his friend and confidant, Marchetti, assistant at the observatory, to reassure him; he also asked him to show the superior general his correspondence with Cantoni:

> Here enclosed you will find the letter in response to our father [general]. The acceptance was made with full awareness of the superiors […]. So we're quite sure all of us. […] Father general tells me to declare that I accept, under conditions that respect my duties of conscience and my membership in the Society [of Jesus]: this is fine. I had already put in these conditions as essential even before, ever since the beginning of the negotiations. […]
>
> The question is now reduced to these terms, either to answer immediately by renouncing, or to accept after these statements. To answer immediately by renouncing could perhaps give some an opportunity to further persecute Ours, and would be interpreted with suspicion, as if I did so under threat of violence or moral pressure. However, if father general tells me to do it, I shall do it. […] I only prayed father general that, if I give it up, he sends me to another destination, because I could not return to Rome without exposing myself to the danger of receiving a letter of immediate dismissal from the observatory, and I could not bear this affront, that they could treat me as someone who failed to keep his promise. If the Jesuits are expelled from the college, questio finita est [the issue is over], I will go with them […]

> […] The observatory is my work, not only for my efforts there, but also for the money that I have sacrificed there from the university, and other benefits […]. Therefore, to leave me at the observatory, and to give me a chair to survive in case of diaspora [of the Jesuits], that is not a gift but an act of justice. […] I have been so stupid as to invest everything into the observatory, who would cry if I accept a reward from the government in order to survive in case of a diaspora? I am ready to give up even this reward, but then father general would certainly take this sacrifice into account, and I hope that he would not leave me wandering, but he would send me to some house where I could end my days in peace, no matter if in studies or not; it is enough for me to live my vocation, and I am happy now to be able to make some real sacrifice for it.[17]

While waiting for Beckx's reply, Secchi prepared for the worst. He wrote to Marchetti that he had consulted some confrères in Palermo and that he feared not being able to return to Rome. He therefore asked him to secure what little money that was left in order to pay for the instruments that had been ordered and "to show that [his resignation was] not due to financial chicanery" but for

---

17 *Qui acclusa troverete la lettera di risposta che dirigo a nostro padre. L'accettazione fu fatta con piena intelligenza de' superiori* […]. *Quindi siamo abbastanza sicuri tutti quanti.* […]

Il P. Generale mi dice di dichiarare che accetto salvi i miei doveri di coscienza e l'essere della compagnia: sta benissimo. Io già avea messe queste condizioni come essenziali anche prima, fino dagli esordii delle trattative […].

La questione adesso è ridotta a questi termini, o di risponder subito rinunziando, o di accettare con queste dichiarazioni. Il risponder subito rinunziando forse potrebbe dare occasione di perseguitare sempre più i nostri, e si presterebbe a sinistre interpretazioni, e anche si direbbe che mi fu fatta violenza o pressione morale. Tuttavia lo farò se il P. Generale mi dice di farlo. […] *Solo ho pregato il P. Generale che se io devo rinunziare, mi destini ad altra residenza, perché non potrei ritornare a Roma senza espormi al pericolo di ricevere una lettera di congedo immediato dall'osservatorio, e io non gradirei questo sfregio, che potrebbero farmi come a mancatore di parola. Se i gesuiti sono cacciati dal collegio, questio finita est io andrò con loro* […].

[…] *l'osservatorio è non solo opera mia per le premure, ma anche pei danari sagrificati in esso provenienti dall'Università, e altre propine* […] *perciò il lasciarmi all'osservatorio, e darmi una cattedra per vivere in caso di dispersione non è un regalo ma un atto di giustizia.*

[…] *io sono stato minchione tanto da metter tutto all'osservatorio e si griderà se io accetto dal governo una ricompensa per poter vivere in caso di dispersione? Io però non voglio nemmeno questa, ma allora il P. Generale terrà a calcolo questo sagrifizio certamente, e spero che non mi lascierà ramingo, ma mi invierà in qualche casa ove possa finire in pace i miei giorni, poco importa poi se studiando o no, a me basta di vivere colla mia vocazione, e ora sono lieto di poter fare ad essa qualche vero sagrifizio.* Secchi to Marchetti, Rome (*sic*; Palermo), November 20 1870; APUG, FS 9.II.

other reasons. He also asked him to pack and secure his personal belongings in case things got complicated.[18]

At last, the reply arrived, and Secchi, reluctantly, resigned—a decision that caused him great stress and even led to physical illness. Secchi wrote from Palermo to Marchetti:

> A letter is enclosed for our holy father [general], where is included my pure and simple disclaimer. He then will do what he considers worthwhile, but first I would wish that the holy father[19] be informed. Since it is possible that father general is not in Rome, send my letter to whomever is taking his place, or else directly to Cardinal Antonelli. Tomorrow after midday we leave for Augusta, henceforth you will send my mail there. The ship is good but the captain belongs to the Italian Royal Navy; therefore, he got lost on the road coming in from Naples! Let's hope that he will lead us towards Scylla or Charybdis. At the moment, in either case, I would not feel sorry. For the last three nights, I have slept very little. Yesterday, I think I had a fever, but today I am fine. My sorrow for what is said about me is at its peak. However, in the face of God, I know that I have not done wrong, and this is my consolation. The rest will be what God wants.[20]

Secchi's self-regard had been deeply injured, and he asked Marchetti to preserve his reputation with the other Jesuits and among the broader public, making known in the *Bollettino meteorologico* the reasons that had previously led him to accept the chair; but then he changed his mind and recommended that nothing be published, in order to avoid any provocation that could give

---

18 See Secchi to Marchetti, Palermo, November 22, 1870; APUG, FS 9.II.

19 The role of Pius IX in this story is unclear. He always supported the Jesuit astronomer and Secchi was bound to him with feelings of deep gratitude. If the final decision was taken by Pius IX, Secchi would not have resisted the pope's will, in consideration of both his personal friendship and his vow of obedience.

20 *Vi accludo una lettera per N. Padre Santo, ove è acclusa la mia rinunzia pura e semplice. Esso poi farà quel che crede, ma prima desidero che ne informi il S.P. Essendo possibile che il P. Generale non si trovi in Roma, mandate la mia a chi fa le sue veci, ovvero al Card. Antonelli direttamente. Domani dopo mezzodì partiremo per Augusta, e voi d'ora innanzi dirigete colà le lettere. Il bastimento è buono ma il Capitano è della R. Marina Italiana. Perciò sbagliò la strada nel venire da Napoli! Speriamo che ci porterà o contro Scilla o dentro Cariddi. Nel momento presente non mi dispiacerebbe né l'uno né l'altro. Sono tre notti che dormo pochissimo. Ieri credo ebbi la febbre a oggi sto bene. La mia afflizione per ciò che si dice di me è al colmo. Ma in faccia a Dio so che non ho mancato, e questo mi consola. Del resto sarà ciò che Dio vorrà.* Secchi to Marchetti, Palermo, November 26, 1870; APUG, FS 9.II.

occasion for the government to expel the Jesuits. However, he asked Marchetti to prepare himself to pack instruments, books, and all his manuscripts.[21]

The journey from Palermo to Augusta "would have been good"—Secchi says—"if I hadn't been in circumstances so sad and weighed down by a real melancholy […]."[22] The rumors about him in Rome were hard to bear for the Jesuit:

> You finish your letter with a phrase that really pierced me. You say that in sum they want me to break with the revolution! […] I am fully persuaded that many have a bad opinion of me, even among Ours, but that I could ever have gone along with the revolution so that I must now break it, that is too much. […] I know that nothing is forgiven of us, and […] now I must be sacrificed to the hatred of our hypocritical enemies. I hope that I will not have to bear this persecution, which is beyond my strength, and therefore please recommend me to the prayers of all, because [otherwise] such a stain would be too painful for me to bear.[23]

Among some hardline conservatives, the acceptance of the chair at La Sapienza at a time of conflict between the papacy and the Italian government was interpreted as an act of betrayal: collusion with a government that had put an end to the Papal States, expropriating their assets, and warring against the rights of the pope. It was difficult in that period to understand that Secchi's decision, far from being a political move, was instead taken in the interest of science and for the sake of his personal reputation and career. However, the entire affair again serves to highlight the frequency with which Secchi was used as a pawn in the difficult post-unity relations between the papacy and the Kingdom of Italy.

The government viewed Secchi's resignation as an offensive act. He was aware of this interpretation and feared that he would suffer the consequences

---

21 See Secchi to Marchetti, Palermo, November 27, 1870; APUG, FS 9.II.

22 *Tutto sarebbe stato buono se non fossi io stato in circostanze così tristi e oppresso da una vera malinconia* […]. Secchi to Marchetti, Augusta, November 30, 1870; APUG, FS 9.II.

23 *Voi concludete la lettera vostra con una frase che veramente mi trafisse. Dite che in somma si vuole che io la rompa colla rivoluzione!* […] *Che moltissimi abbiano di me una pessima opinione anche tra i nostri io ne sono persuasissimo, ma che io abbia potuto mai transigere colla rivoluzione in modo che ora debba romperla, questo è troppo.* […] *so bene che a noi nulla si perdona, e* […] *potrei esser sagrificato anch'io all'odio de' nostri nemici ipocriti. Io spero che non avrò tale persecuzione che superi la mia virtù, e perciò vi prego a raccomandarmi alle orazioni di tutti, perché l'aver io a dover portare in fronte tal macchia sarebbe troppo doloroso.* Secchi to Marchetti, Augusta, November 30, 1870; APUG, FS 9.II.

of his decision—primarily, his possible expulsion from the observatory, a risk that provoked much anxiety. How could he justify his right to remain there? How was he to defend the observatory from being confiscated? This became a real obsession for Secchi, who was afraid of losing the results of his lifetime's work. He wrote to both his assistants Marchetti and Egidi:

> Now my position is such that I cannot stay at the observatory without absolute independence from the government [...]. If [this] cannot be accomplished, begin to remove as much stuff as you can; and, if Count Rosa can do it, pack up the equatorial too. It would be a good opportunity to send it to Ecuador.[24]
>
> Even without an official expulsion, it has become opportune for me now to withdraw from the observatory, in case that absolute independence from the government does not occur. It is better to move these [things] before and in time. [...] My renunciation of the chair puts me in a position that if these [people] are still in Rome I would not be tolerated there.[25]

The day after the eclipse, Secchi sought consolation by meeting the members of the American expedition of the US Naval Observatory, who had sited their astronomical station near Augusta, in Syracuse. As we saw in Chapter 5, it was during this period that, thinking about his future, in case of a departure from Rome, Secchi entertained the idea of returning to the United States.[26]

2 Defending the Collegio Romano

In the meantime, however, an international outcry prevented the Italian government from taking any action against the Collegio Romano. On November 11,

24 *Ora la mia posizione è tale che non posso restare all'osservatorio senza una assoluta indipendenza dal governo [...] Se [ciò] non può concludersi, si cominci a mandar via più roba che si può e se il conte Rosa può farlo, faccia ritirare anche l'Equatoriale. Sarebbe una buona occasione per spedirlo all'Equador.* Secchi to Marchetti, Augusta, December 18, 1870; APUG, FS 9.II.

25 *Anche senza pensare ad una espulsione ufficiale, ora per me è divenuta una cosa di convenienza anche il ritiro dall'osservatorio, qualora non si verifichi un mezzo di una assoluta indipendenza dal Governo. E' bene prendere questi pezzi prima e a tempo. [...] la mia rinunzia alla cattedra, mi mette in una posizione tale che se questi restano ancora a Roma io non vi sarò tollerato.* Secchi to Egidi, Augusta, December 18, 1870; APUG, FS 9.II.

26 *If I am removed, I go straight to America, in California, not to look for gold but to find a bit of peace, leaving this old crazy Europe (Se mi cacciano vado diritto in America in California, non mica a cercar l'oro ma per trovare un po' di pace lasciando questa pazza e vecchia Europa).* Secchi to Denza, n.d. [1873] quoted in Maffeo, "Padre Angelo Secchi e la meteorologia," 29.

1870, the rectors of the German–Hungarian, English, Scottish, Belgian, French, Latin American, and Polish colleges in Rome sent an official protest against the rumored confiscation of the Collegio Romano to General Alfonso La Marmora (1804–78), lieutenant of the Italian government in Rome; included in the message were the ambassadors of those nations to the Holy See and the bishops of the home dioceses of their students. They emphasized that the college was not Jesuit property but a Catholic institution. After recalling the history of the college, set up by the popes to improve the instruction of the young people destined for the priesthood, who came from about fifteen countries, they concluded:

> Therefore, the Roman College is not an institution belonging to the city and municipality of Rome nor to any particular government: it is entirely an international and Catholic institution. [...] The popes and Catholic princes who gave financial support and promoted its advancement intended it for the religious and scientific advantages of all nations [...]; the royal government will not deprive the Catholic nations of what for several centuries has been their due for their own legitimate use and purpose. [...] Foreign nations, therefore, have a real right over the schools of the Roman College: there is no way that they could not take offense, should an institution like this one, which is by its nature Catholic, be arbitrarily reduced to the status of a royal government high school.[27]

The Italian government soon realized that any action against the Collegio Romano would be inappropriate and constitute an unpopular measure. The issue was not only international but also felt locally, since the best Roman families would often send their children to study at the Collegio Romano to ensure they received a high-quality education.[28] As a result, Secchi was allowed to maintain his position at the Collegio Romano Observatory. Nonetheless, he remained

27 *Non è adunque il collegio romano una istituzione spettante alla città e municipio di Roma o propria di un qualsiasi particolare governo: è una istituzione al tutto cattolica e internazionale.* [...] *i Sommi Pontefici e i Principi cattolici nel beneficarlo e nel promuoverne gli avanzamenti ebbero in mira i vantaggi e religiosi e scientifici di tutte le nazioni* [...] *non potrassi al certo ritogliere dal regio governo alle nazioni cattoliche ciò che per sua propria destinazione ed uso parimenti legittimo di più secoli spetta.* [...] *Le nazioni estere adunque hanno un vero diritto sopra le scuole del collegio romano: né può senza offesa manifesta di un tal diritto, una istituzione qual è questa di sua natura cattolica, essere arbitrariamente ridotta alla condizione di liceo regio ovvero governativo* (*Brevi memorie intorno al Collegio Romano* [Rome: Stamperia della S.C. De Propaganda Fide, 1870], 12–13; 20–21). Secchi's name is explicitly mentioned in the anonymous pamphlet publicizing the protest as well as in the rectors' letter itself: in fact, his international reputation was an argument in support of the Collegio Romano.

28 See notes 12 and 13.

FIGURE 7.3 Bust of Pius IX at the Collegio Romano Observatory. Courtesy of Aldo Altamore.

profoundly scarred by this experience. He sensed that things would inevitably become more difficult, and that he was entering a very trying phase of his life: from that point on, every act he made would have political implications.

3 Criticisms and Invectives

Secchi's concerns were evident a few months later, when Abbé Moigno, editor of the popular scientific magazine *Les Mondes*, published a letter by Secchi[29] expressing his anxiety for the future on account of the current political situation, fearing that it would evolve in a similar way to what had occurred with the Commune[30] in Paris. He complained that he was leading a life "much more

29 See *Les Mondes* 25, no. 8 (June 22, 1871): 307.

30 The Commune was the popular and revolutionary government of workers set up in the French capital after the revolution that broke out in March 1871; it was violently repressed a few months later, with the establishment of the Third Republic.

pitiful than death." He found solace from his "cruel suffering" in his solar physics studies, even if he could not properly carry out his research because "the resources of the observatory were diminished." The letter gave rise to an immediate reaction from Brioschi, who considered Secchi's complaints as a form of provocation and did not hesitate to say so in the newspapers. After citing the letter in question, Brioschi added some vitriolic comments:

> This is not the first time that Father Secchi, whose childish vanity is proverbial among men of science, deals with the public, posing himself as an implacable enemy of the Italian government, though, he said, he was begged by the same government to accept a chair at the University of Rome. Father Secchi is a clever man: being a witness, for many months, of a new direction in history, that of a government and a people who endure all kinds of provocation with the greatest restraint, he believed he could count on this. [...] He actually has not failed, and even if his provocation had continued to appear in the newspapers [...] of the faction to which he belongs, I probably wouldn't have decided to break the silence. But Father Secchi [...] took the opportunity to launch false accusations in a communication to a scientific journal, and with surely an unenviable courage to wish upon his own country those tragic events that recently happened in a noble nation [France], which horrified the whole civilized world. [...] I will instead bear it upon myself to publish the letter of November 4, where he claimed to accept that offer [of the chair].[31]

After quoting Secchi's entire letter, Brioschi concluded: "Whoever compares even superficially this letter with the one addressed to Abbé Moigno has

31 *Non è la prima volta che il Padre Secchi, di cui la puerile vanità è ormai proverbiale fra gli uomini di scienza, occupa di sè il pubblico, atteggiandosi ad implacabile nemico del Governo italiano, per quanto, a suo dire, egli fosse stato pregato dal Governo stesso di accettare una cattedra nell'Università romana. Il Padre Secchi è un uomo abile: assistendo da mesi ad uno spettacolo nuovo nella storia, di un governo ed un popolo che a provocazioni di ogni natura contrappongono la più grande moderazione, ha creduto di poter contare su di essa* [...] *Egli non si è infatti ingannato, e se le sue provocazioni avessero continuato a comparire nei giornali* [...] *del sodalizio al quale egli appartiene, non mi sarei probabilmente deciso a rompere il silenzio, Ma il Padre Secchi* [...] *ha colto l'occasione di una comunicazione ad un giornale scientifico per lanciare accuse false, e, con un coraggio al certo non invidiabile, augurare al proprio paese sciagure che toccata ad una nobile nazione fecero non ha guari inorridire tutto il mondo civile. Io non seguirò l'esempio dato dal Padre Secchi; non accennerò ad alcun colloquio* [...] *Mi limiterò invece a pubblicare la lettera del quattro novembre colla quale dichiarava di accettare quella offerta* ("Il Padre Secchi," *La Nazione* [July 10, 1871]).

enough to judge the man who wrote it; we wish him that, at least, his fame as a scientist should always remain intact."[32]

After this incident, it was abundantly clear to Secchi that he could no longer count upon the support of the Italian government to finance his observatory. On the other hand, the papal government could hardly support an observatory that was always at risk of being confiscated by the Italian government. Indeed, it was only to be expected that the laws for the expropriation of ecclesiastic properties, applied in 1866 and 1867 in the Kingdom of Italy, would be extended to the urban territory of Rome following its annexation.

The political situation was very uncertain, and Secchi saw dark clouds on the horizon. He wrote the following anecdote to Denza, revealing the spirit of the time:

> The other night, while I was observing [a polar] aurora, a rogue shouted with a boorish stentorian voice from one of the nearby lodges: "Ah, Secchi! Ah Secchi! Look at the aurora borealis [...]. This is the Paris Commune arriving. Rome will have the Commune!" You see, even the boors move in retrograde [backward] motion![33]

Despite the changing Italian political landscape during this period, Secchi could count upon the support of Sella, scientist and statesman, who often visited Secchi at the Collegio Romano and, together with Tacchini, did his best to help the Jesuit astronomer defend his observatory (see Chapter 10).

---

32 *Chi raffronta anche superficialmente questa lettera con quella diretta all'abate Moigno, ha quanto basta per caratterizzare l'uomo che la scriveva; auguriamo a lui che la sua fama di scienziato rimanga almeno sempre intatta* ("Il Padre Secchi," *La Nazione* [July 10, 1871]).

33 *L'altra sera, durante l'aurora [boreale], un briccone da una delle logge vicine gridava a squarcia gola con voce stentorea buzzurresca queste parole: «Ah Secchi, Ah Secchi ! Guarda l'aurora boreale. È l'aurora boreale. È la commune di Parigi che viene. Roma dovrà avere la commune!» Ella vede che anche i buzzurri sono retrogradi abbastanza.* Secchi to Denza, February 7, 1872, quoted in Maffeo, "Padre Angelo Secchi e la meteorologia," 27.

CHAPTER 8

A Society for a New Astronomy: Building a Network for Spectroscopic Research

In the 1870s, astronomy was changing: a new perspective was being opened up by the application of spectroscopy to astronomical research. But the astronomical community resisted the change. Unlike many of his colleagues, Secchi clearly understood that, in future, successful astronomical research would require the use of spectroscopy and the formation of a collaborative network of scientists. The astronomer was no longer a lonely scientist working in an ivory tower, isolated from the rest of the community; on the contrary, science could only advance via cooperation. However, if you believe in collaboration, you need to find reliable collaborators.

Over the years, Secchi developed an important scientific partnership with Tacchini, adjunct astronomer at the Palermo Observatory.

The two scientists shared a modern vision of astronomy, a keen interest in spectroscopic research, and, most importantly, they possessed two almost identical telescopes, which meant that the results of their observations could be easily compared, as in 1865 Tacchini had installed a twenty-five-centimeter aperture Merz equatorial refractor, which he began to use to observe the Sun.

1 Secchi and Tacchini: One and the Same Goal

Despite their deeply different convictions and lifestyles—though Tacchini also had Emilian roots, having been born in Modena to a wealthy middle-class family, he was twenty years younger than Secchi, anti-clerical, and something of a playboy—Secchi and Tacchini shared the same views about the overall aims of astronomical research. As a result, Tacchini would often ask for Secchi's advice and collaboration, even though he sometimes disagreed with Secchi's theories: he was ready to confront Secchi, but he was also willing to consult the Jesuit in astronomical matters.

Secchi, in turn, found in Tacchini an ideal scientific partner, intelligent and efficient, someone who was equally committed to opening up the field of astronomical research to new perspectives. Despite his older age, and being

 | DOI:10.1163/9789004387331_010

FIGURE 8.1 Pietro Tacchini. Courtesy of INAF-OAPa.

a priest, Secchi felt free to pour out his heart to his young colleague, confiding his worries and disappointments in him. This attitude of mutual respect led to a solid and fruitful partnership, a sort of "scientific friendship," which lasted until Secchi's death. The two men's common love for science and its advancement thus overcame their divergent ideological, political, and moral beliefs.

In 1865, once he had installed the Palermo Merz telescope, Tacchini proposed to Secchi that they compare their drawings of features in the solar photosphere. The proposal was made in the context of a controversy that had arisen in 1862 over the description of the solar surface and the use of the terminology to describe granulation and sunspot structures (see Chapter 6). Secchi was directly involved in these debates, and Tacchini proposed to him that they carry out simultaneous observations of the same structure (sunspots or facula) from Rome and Palermo in order to standardize the terms used to describe them:

> We will have to run on the same spot and on different points on the solar disk over short intervals of time […]. For this purpose, it would be necessary that the means employed by the various observers were almost equal […]. We, having an instrument similar to yours, could volunteer ourselves for this kind of work, and on an appropriate occasion, we could make drawings of certain spots at the same time and then record the changes and add detailed drawings if necessary […].[1]

Secchi replied in general terms: "About the invitation to make pictures of the sunspots, let me just say you that I am not capable at drawing […]. Fr. Ferrari could occasionally do them."[2]

This was the beginning of a series of exchanges that led to a long scientific partnership: both Secchi and Tacchini sensed the revolutionary impact of the new spectral studies in astronomy. This was in contrast with the traditional mindset, which restricted astronomy to the domain of celestial mechanics.

Significantly, both astronomers regretted that, due to a lack of funds, the Italian government had missed the chance to organize an expedition to observe the total solar eclipse of 1868 in India.[3] As it happened, this turned out to be one of the most important eclipses in the history of solar physics, as it revealed the gaseous nature of the prominences and the existence of helium. Two years later, the Italian government finally organized an expedition to observe the total solar eclipse visible from Sicily (see Chapter 5); on that occasion,

---

1 *Si dovranno eseguire osservazioni su di una stessa macchia e a punti differenti sul disco solare a piccoli intervalli di tempo, cosa che non conosco se sia finora stata fatta da alcuno. A questo scopo sarebbe anche necessario che i mezzi di osservazioni impiegati dai diversi osservatori fossero pressoché eguali* […] *Noi che ci troviamo ad avere un istrumento del tutto simile al suo potremmo prestarci benissimo a questo genere di lavoro, ed alla opportuna occasione si potrebbero eseguire contemporaneamente i disegni di talune macchie e poscia registrarne le variazioni ed aggiungere disegni particolari di dettagli quando occorra* […]. Tacchini to Secchi, Palermo, December 24, 1865, quoted in Chinnici and Gasperini, *Alle origini dell'astrofisica italiana*, 60–61.

2 *In quanto all'invito di fare le figure delle macchie, le posso dire soltanto che io non so disegnare* […] *Il P. Ferrari cercherà di occuparsene quando potrà* […]. Secchi to Tacchini, Rome, December 31, 1865, quoted in Chinnici and Gasperini, *Alle origini dell'astrofisica italiana*, 63.

3 See Chinnici, *L'eclisse totale di sole del 1870 in Sicilia*, 24–26.

Tacchini negotiated with the government to invite Secchi to take part in the expedition,[4] and being in charge of the spectroscopic observations of the phenomenon, he often visited the Jesuit to receive training under his guidance.[5] In June 1871, once he was in possession of his own spectroscopic instruments, Tacchini proposed that he and Secchi carry out simultaneous observations of the prominences and to invite his friend Giuseppe Lorenzoni, astronomer at Padua Observatory, to take part in their research:

> Since you [Secchi] and I have instruments of equal capacity, I would propose the following experiment on the shapes of the prominences, which are so important for explaining these phenomena. Therefore, if you wish to acquiesce to my proposal, I suggest we make drawings of the whole limb for three or four days, as you prefer, at the same times [...]. Today, I shall write to Lorenzoni, who has a smaller equatorial, inviting him to do the same, and if he is able to contribute to this work, then we shall have three series, made with different instruments, and the resulting comparison will be even more interesting.[6]

This preliminary program was successfully carried out in July 1871.[7] The comparison showed that the general appearance of each observed prominence was the same, that different details were due simply to different weather

4 See Chinnici, *L'eclisse totale di sole del 1870 in Sicilia*, 27–30.

5 See Tacchini to Secchi, Palermo, August 28, 1869, quoted in Chinnici and Gasperini, *Alle origini dell'astrofisica italiana*, 104. Until 1871, Italian astronomical observatories lacked spectroscopic instruments (see Ileana Chinnici, "19th-Century Spectroscopic Instruments in Italian Astronomical Observatories," *Nuncius* 15, no. 2 [2000]: 671–680); the total solar eclipse of 1870 was crucial for the development of spectroscopic research in Italy, because on that occasion spectroscopic instruments were acquired for the observatories in Palermo, Padua, and Naples, thus constituting, in principle at least, the basic equipment for a potential Italian solar physics research network (see Chinnici, *L'eclisse totale di sole del 1870 in Sicilia*).

6 *Siccome io e lei abbiamo istrumenti della stessa portata, così proporrei il seguente esperimento relativo alla forma delle protuberanze, che tanta importanza ha sul modo di spiegare questi fenomeni. Se adunque a lei piace di aderire alla mia proposta, io direi di fare il disegno dell'intiero bordo per tre o 4 giorni conforme le piacerà meglio, e nelle stesse ore* [...] *Oggi stesso scriverò a Lorenzoni che ha un equatoriale più piccolo invitandolo a fare la stessa cosa e se potrà anche lui concorrer al lavoro, allora si avrebbero 3 serie con instrumenti diversi e il confronto diverrebbe anche più interessante*. Tacchini to Secchi, Palermo, June 19, 1871, quoted in Chinnici and Gasperini, *Alle origini dell'astrofisica italiana*, 118.

7 See Pietro Tacchini, "Fisica solare," *Bullettino meteorologico del R. Osservatorio di Palermo* 7 (1871): 93–95, and Tacchini, "Protuberanze solari osservate contemporaneamente a Palermo, Roma e Padova nel luglio ed agosto 1871," *Memorie della Società degli Spettroscopisti Italiani* 1 (1872): 25–32.

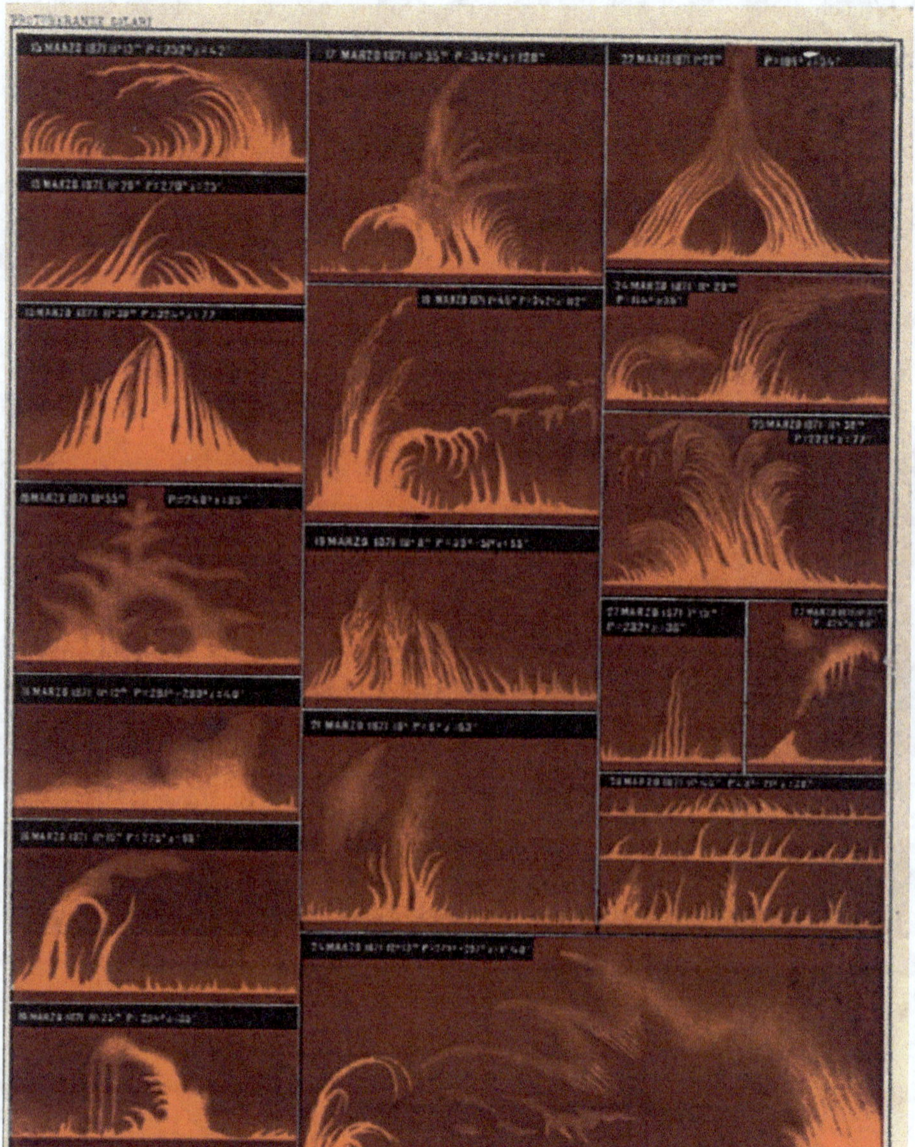

FIGURE 8.2 Solar prominences observed spectroscopically in Palermo in 1871. From Pietro Tacchini, "Osservazioni di protuberanze solari fatte alla Specola di Palermo," *Bullettino meteorologico del Reale Osservatorio di Palermo* 7 (1871): 17–27, pl. [1].

conditions, and that the prominence showed the same structure, though less detailed, when observed with smaller instruments.

2 Cooperation in Solar Research

Soon, Secchi conceived of an innovative project to further develop solar physics research. Writing to Tacchini, he proposed establishing a scientific society with other astronomers willing to be involved in a program of daily solar limb monitoring:

> A small group should be recruited that can deal with the matter of drawing the prominences, and if we formed an association of three or four [members], we could work in pairs, one to each month, in order to avoid gaps due to bad weather. [...] The work must be divided because it is not a bagatelle [...]. It is a very laborious task, and it must be done well and carefully. If the Milan equatorial is mounted there,[8] Tempel,[9] being an excellent artist, could do wonders. [...] It seems that [the astronomers] in Naples are keen to undertake [these observations] [...]. Respighi could also be invited, thus we could achieve more without wearing ourselves out.[10]

Tacchini immediately welcomed Secchi's proposal, as he recognized the potential contribution it could make to science. In this project, Secchi was the mind, Tacchini the arm—and the leader, for political convenience. He would

8 The Milan (Brera) Observatory was equipped with a Merz equatorial, which it acquired in 1862, but the instrument was not in working condition until 1874; see Enrico Miotto, Giuseppe Tagliaferri, and Pasquale Tucci, *La strumentazione nella storia dell'Osservatorio di Brera* (Milan: Unicopli Editore, 1989), 64–65.

9 Tempel was assistant at Brera Observatory from 1870 to 1875, when he was appointed astronomer at the Florence Observatory and charged with its direction until his death.

10 *Sarebbe da fare una piccola Società che si occupasse di questa faccenda di disegnare le protuberanze, e se fossimo in tre o 4 associati, si potrebbe impegnarne due alla volta, un mese per ogni paio, acciò non restassero lacune pel tempo cattivo.* [...] *Ma bisogna dividere il lavoro, perché non è una bagattella* [...] *E' cosa laboriosa assai, e bisogna farlo bene e con attenzione. Se a Milano avessero l'equatoriale montato ivi essi hanno Tempel che essendo eccellente disegnatore farebbe meraviglie.* [...] *Pare che abbiano voglia di occuparsene a Napoli* [...] *Si potrebbe anche invitare Respighi, e così senza ammazzarci faremmo di più.* Secchi to Tacchini, Rome, July 12, 1871, quoted in Chinnici and Gasperini, *Alle origini dell'astrofisica italiana*, 139; Secchi had already written to Tacchini about the idea of establishing a society (see Secchi to Tacchini, Rome, June 25, 1871, quoted in Chinnici and Gasperini, *Alle origini dell'astrofisica italiana*, 131–133), but this time he was more explicit.

mediate with political and cultural institutions to secure financial support for the life of the society, especially for its publications.[11]

To discuss the scientific program of the proposed society, Tacchini and Secchi invited other Italian astronomers possessing spectroscopic instruments to a meeting in Rome at the beginning of October.[12] Tacchini wrote to Respighi[13] and to Annibale De Gasparis (1819–92),[14] director of the Naples Observatory, while Secchi invited Arminio Nobile (1838–97),[15] astronomer at the same observatory. Secchi also proposed involving Schiaparelli,[16] one of the most important scientific advisors to the Italian government; Tacchini then sent a formal invitation to Schiaparelli, in order to secure his approval:

> After I had proposed making simultaneous observations, [Secchi] had the idea of organizing a society of Italian spectroscopists for observing the prominences: if the society does come about, we shall have the advantage both of sharing the work and collecting a continuous series of limb [drawings], because it would be agreed that in the bad season only those having good weather will work, communicating the sky conditions by telegraph: thus, within a short time we would have gathered data suitable for investigating many laws relating to the physical constitution of

11 See point 8 of the Rules of the Society, part 2 in Pietro Tacchini, "Nuova Società di Spettroscopisti Italiani," *Memorie della Società degli Spettroscopisti Italiani* 1 (1872): 3–6, here 6.

12 See Tacchini to Secchi, Palermo, August 22, 1871, quoted in Chinnici and Gasperini, *Alle origini dell'astrofisica italiana,* 148. In defining the details of the meeting, some problems arose because of Secchi's delicate political position after the annexation of Rome.

13 See Tacchini to Respighi, Palermo, August 29, 1871; AOAR, Fondo Respighi, 1264.

14 See Tacchini to Secchi, Palermo, August 29, 1871, quoted in Chinnici and Gasperini, *Alle origini dell'astrofisica italiana,* 151. This letter makes clear Tacchini's intention to extend the invitation to Donati, who was a pioneer in astronomical spectroscopy and had worked on stellar and cometary spectra since 1860. Later, Tacchini wrote to Secchi that Donati had not been invited owing to his lack of interest in studying solar prominences, as remarked by Cacciatore, director of the Palermo Observatory, who had met Donati in Florence (see Tacchini to Secchi, n.p. [Palermo], September 1871, quoted in Chinnici and Gasperini, *Alle origini dell'astrofisica italiana,* 53–54). In those years, Donati was very busy establishing the new observatory in Arcetri, and this explains his original decision not to join the society. In fact, he worked intermittently on the subject while maintaining contact with the society (see Donati to Tacchini, Florence, April 27, 1872; Archive of the Consiglio per la ricerca in agricoltura e l'analisi dell'economia agraria [Council for Agricultural Research and Economics, hereinafter ACREA] Fondo Tacchini. See also Giovan Battista Donati, "Osservazioni spettroscopiche di macchie solari," *Memorie della Società degli Spettroscopisti Italiani* 1 [1872]: 52–56). This potential collaboration was interrupted in 1873 by Donati's sudden death from cholera.

15 See De Gasparis to Tacchini, Naples, August 29, 1871; ACREA, Fondo Tacchini.

16 At that time, Schiaparelli was the director of the Milan Observatory; he became a senator in the Kingdom of Italy in 1889 in recognition of his scientific work.

> the Sun. [...] At the beginning of October, both Lorenzoni and I must go to Rome. Having made simultaneous observations with Secchi in July, the three of us together must now set them in order: on this occasion, it would be good if the others could come as well, in order to agree on procedures, on the method of observing, on the nomenclature, etc., and, finally, on the sharing of the work. It would be most fortunate if you could attend our meeting, and should your engagements not allow it, you can send Tempel. [...] At the moment, we should have Palermo, Naples, Rome, Padua, and Milan: I don't know if Donati will accept, but it would be good if Florence also took part. [...] I hope you will approve the establishment of such a society.[17]

Schiaparelli was neither in favor nor opposed to astronomical spectroscopy, as revealed in a letter to Secchi written in 1866:

> I don't know much about physics, unfortunately, and I doubt I will be ever able to carry out spectroscopic research properly: nevertheless, I beg you not to number me among those who rail against these studies and would confine astronomy to pure gravitation in perpetuity for the sole reason that the latter is the only part rigorously subject to the empire of calculus.[18]

---

17 *Dopo di avergli proposto osservazioni contemporanee, il Secchi andò all'idea di organizzare una società di spettroscopisti italiani per l'osservazione delle protuberanze: se la società riesce si avrebbe il vantaggio di dividere il lavoro e l'altro di raccogliere una serie continua di bordi, perchè si andrebbe di accordo che nella stagione cattiva lavorerà chi ha buon tempo servendoci del telegrafo per far conoscere le condizioni del cielo: in qualche maniera in poco tempo si avrebbe un materiale molto adatto alla ricerca di molte leggi sulla fisica costituzione del sole.* [...] *ai primi di ottobre io devo andare a Roma ove verrà pure Lorenzoni, avendo noi col Secchi fatte osservazioni contemporanee in Luglio che dobbiamo ordinare tutti e tre riuniti: in quella occasione sarebbe bene venissero anche gli altri per intenderci sul da fare, sul modo di osservare, nella nomenclatura, ecc* [...] *sulla divisione in fine del lavoro. Perciò noi saremmo ben fortunati se alla nostra riunione ella potesse intervenire e nel caso che le sue occupazioni non glielo permettessero potrebbe mandare il Tempel.* [...] *intanto si avrebbe Palermo, Napoli, Roma, Padova e Milano: Donati non so se accetterà, ma sarebbe bene che anche Firenze concorresse.* [...] *spero anche che lei approverà l'idea di formare una tale società.* Tacchini to Schiaparelli; Archive of the Osservatorio Astronomico di Brera (hereinafter AOAB) *Corrispondenza Scientifica*, Serie III. This undated letter can be assigned to the end of August 1871 (see Tacchini to Secchi, Rome [*sic*; Palermo], August 29, 1871, quoted in Chinnici and Gasperini, *Alle origini dell'astrofisica italiana*, 150–151).

18 *Io non so molto di fisica, pur troppo, e dubito di poter mai attendere con proposito alle ricerche spettrali: tuttavia prego V.S. di non mettermi a mazzo con quelli che bestemmiano questi studj e vorrebbero eternamente confinare l'astronomia alla pura gravitazione per la sola ragione che è la sola parte rigorosamente soggetta all'impero dell'analisi.* Schiaparelli to Secchi, Milan, August 4, 1866, in Buffoni and Manara, *G.V. Schiaparelli, A. Secchi*, 61.

Schiaparelli's reply to Tacchini was encouraging, but he excluded the contribution of the Milan astronomers, at least until their new Merz refractor[19] was ready to use.[20]

At the beginning of September, Tacchini had not yet received all the replies, and there was a risk that the meeting might not materialize. He expressed his anxiety to Secchi as well as his intention to establish the society in any case, even if it meant limiting it to Palermo, Rome, and Padua.[21] At the end of the month, Tacchini wrote to Secchi:

> There is no time left to set up an official meeting: but it matters little because the society will be established in any case, and then I will be able to present its protocol to the minister and ask for the necessary assistance, because I would like, if possible, that everything that is accomplished would be quickly published. [...] I will come in any case, and if De Gasparis joins me, the three of us will establish the rules of this society on the basis of five stations, namely Palermo, Naples, Campidoglio, Collegio Romano, Padua. The entire limb must be drawn by the observer [...]. Then we will discuss the rest together.[22]

3 The Establishment of the Società Degli Spettroscopisti

The meeting was held in Rome on October 5,[23] with the only participants being Secchi, Tacchini, and Nobile, the latter being sent by De Gasparis, probably

19 After the discovery of the minor planet Esperia in 1861, Schiaparelli obtained government funding to purchase a 218-millimeter aperture Merz refractor. The instrument was delivered in 1865, but it was only installed almost ten years later, in 1874. Schiaparelli used this instrument to carry out his famous observations of Mars in 1877.

20 See Schiaparelli to Tacchini; AOAB, *Corrispondenza Scientifica*, Serie III.

21 See Tacchini to Secchi, n.p. [Palermo], September 1871, quoted in Chinnici and Gasperini, *Alle origini dell'astrofisica italiana*, 153–154.

22 *Ormai non si è più in tempo a combinare una seduta ufficiale: ma ciò poco importa per la ragione che la società si farà lo stesso e dopo potrò io presentare al Ministro le basi di detta società domandando quei sussidi che crederemo necessarii, perchè io vorrei, se sarà possibile, che tutto ciò che si farà venisse presto pubblicato.* [...] *io verrò egualmente e se De Gasparis vorrà accompagnarmi allora noi tre combineremo lo statuto di detta società in base a 5 specole, cioè Palermo Napoli Campidoglio, Collegio Romano, Padova. I bordi debbono esser fatti per intiero dall'osservatore* [...]. *Il resto poi discuteremo assieme.* Tacchini to Secchi, Palermo, September 23, 1871, quoted in Chinnici and Gasperini, *Alle origini dell'astrofisica italiana*, 155–156.

23 It was probably held at the Collegio Romano Observatory: Secchi preferred to avoid entering government buildings, perhaps because he feared that he would not obtain the

at Tacchini's insistence. The three men drafted a program for the society and a work plan, with the intention of publishing the results of their studies as soon as possible.

Unfortunately, the society's program as proposed by Secchi and Tacchini was immediately subject to criticism. De Gasparis (a senator in the Italian government) was the first to question the work plan; he was probably irritated by the leading role of Secchi and Tacchini in taking the initiative and drafting the program. In any event, the participation of the astronomers in Naples never materialized. The way in which Secchi and Tacchini had monopolized the decision-making when formulating the society's program may have played a role in this outcome. It was the same with Respighi, who was engaged in personal disputes with Secchi (see Chapter 9) that had not yet been completely resolved by that time and would drag on over the years that followed.[24] Tacchini endeavored to mediate the uneasy relationship between the two astronomers in Rome, albeit with meagre results.[25] The reasons for Respighi's discontent were made explicit in a letter to Tacchini:

> When you first invited me to take part in the Society of Italian Spectroscopists, I confess that I was quite reluctant to accept, fearing that I should not be generally welcomed by its membership, or that my participation would be of no value. [...] Despite these difficulties, I have decided to join this association, being convinced of its usefulness and persuaded that its true and only appointed purpose was to promote the advancement of science and to maintain the honor of our country. Unfortunately, [...] my suspicions have been confirmed [...]. I received no letter telling me what had been decided, yet only one month later, the first issue appeared, carrying the program.

relevant permission to do so from his superiors: he was also still embarrassed as a result of the La Sapienza affair (see Chapter 5). See Secchi to Respighi, Rome, September 28, 1871; AOAR, Fondo Respighi, 1171. See also to Tacchini, Rome, September 18, 1871, quoted in Chinnici and Gasperini, *Alle origini dell'astrofisica italiana*, 154–155.

24 Respighi was initially willing to cooperate, but he then withdrew (see Chapter 9). His contributions were minimal, as he published only three articles in the *Memorie* (see *Indice generale delle Memorie della Società degli Spettroscopisti Italiani, vol. I–XL* [Catania: Stabilimento Tipografico S. Di Mattei e C., 1912], 23–24). The first, extracted from another journal and inserted by Tacchini, provoked a complaint from Respighi, whereas the other two, significantly, date from immediately after Secchi's death.

25 See Tacchini to Secchi, Palermo, February 28, 1872, quoted in Chinnici and Gasperini, *Alle origini dell'astrofisica italiana*, 180–183; see also Lorenzoni to Tacchini, Padua, June 2, 1872 (ACREA, Fondo Tacchini), and Tacchini to Secchi, Palermo, July 10, 1872, quoted in Chinnici and Gasperini, *Alle origini dell'astrofisica italiana*, 203–205.

FIGURE 8.3 Lorenzo Respighi's diploma of the Italian Spectroscopic Society, dated 1888. Courtesy of INAF-OAR.

> Forgive me if I speak to you so frankly, but you too, in my place, would have thought that in the society I was considered a mere cipher.[26]

Tacchini, president of the nascent society—an office he held until his death—promptly sought financial support from the government, and in January

---

26 *Sino da quando tu mi invitasti a prender parte alla Società degli Spettroscopisti Italiani, ti confesso che io provava una qualche ripugnanza ad acconsentirvi, nel timore che la mia persona nella medesima società non potesse essere gradita a tutti quelli che vi avrebbero preso parte, o che almeno fosse considerata come inutile la mia cooperazione.* [...] *Malgrado queste difficoltà io mi decideva a prendere parte a questa associazione nella convinzione della sua utilità, e nella persuasione che il vero ed unico scopo prefisso alla medesima fosse quello di promuovere il progresso della scienza e di sostenere il decoro del nostro paese. Ma purtroppo* [...] *ho dovuto maggiormente confermarmi nel mio sospetto* [...] *non ho trovato nessuna lettera su quello che era stato stabilito, e soltanto un mese dopo mi vedo comparire il primo fascicolo in cui era inserito il programma. Perdonami se ti parlo franco, ma anche tu al posto mio avresti pensato, che io era considerato nella Società come zero.* Respighi to Tacchini: Rome, June 11, 1871 [1872]; ACREA, Fondo Tacchini.

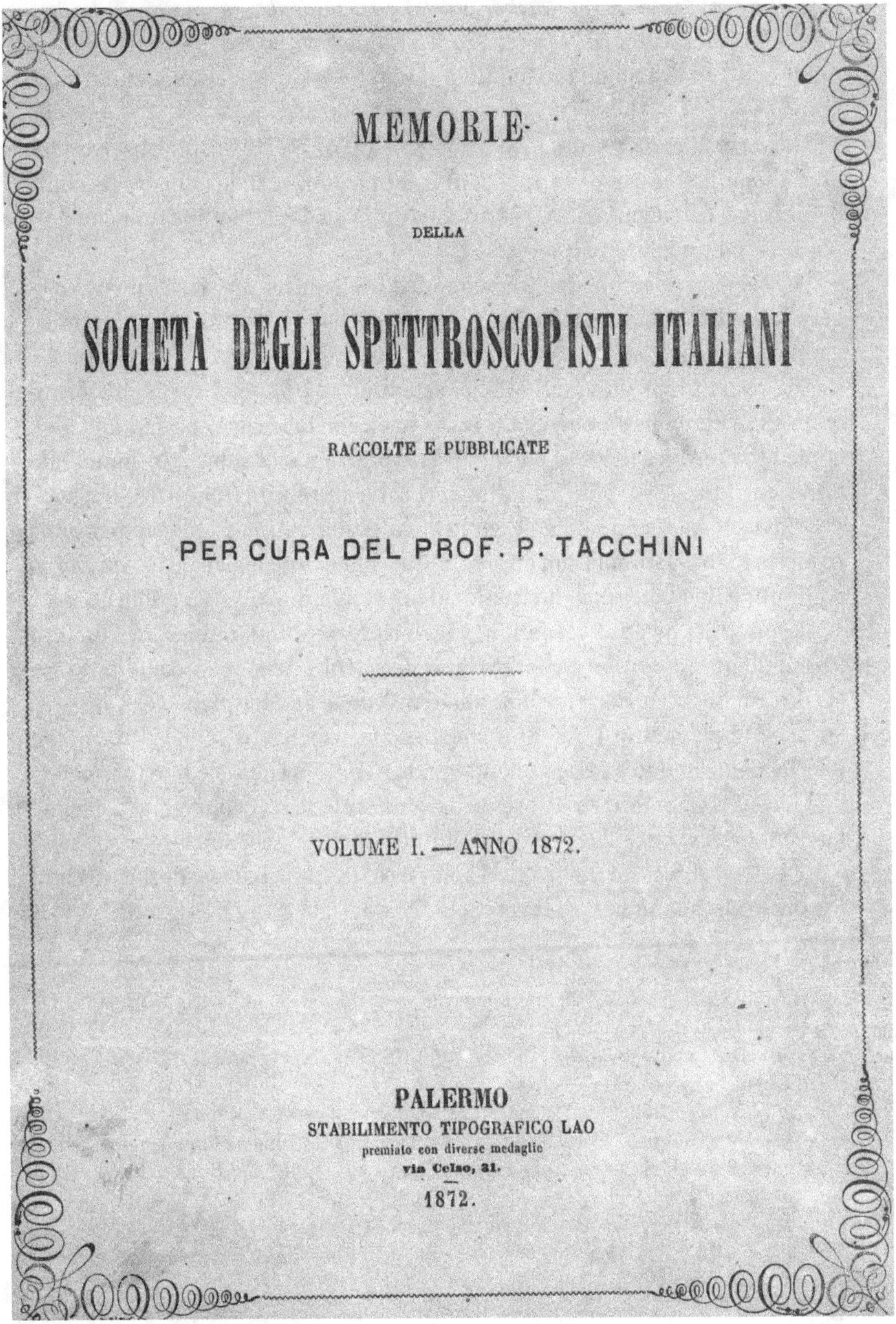

MEMORIE

DELLA

SOCIETÀ DEGLI SPETTROSCOPISTI ITALIANI

RACCOLTE E PUBBLICATE

PER CURA DEL PROF. P. TACCHINI

VOLUME I. — ANNO 1872.

PALERMO

STABILIMENTO TIPOGRAFICO LAO

premiato con diverse medaglie

via Celso, 31.

1872.

FIGURE 8.4 Cover of the first issue of the *Memorie della Società degli Spettroscopisti Italiani*. Courtesy of INAF-OAPa.

1872, the first issue of the *Memorie della Società degli Spettroscopisti Italiani* (Memoirs of the Italian Spectroscopical Society), edited by Tacchini, was published and distributed to about one hundred correspondents in Italy and abroad.[27]

The journal mainly contained a comparison of spectroscopic observations of the solar limb carried out in 1871 in Palermo, Rome, and Padua by Tacchini, Secchi, and Lorenzoni, as well as articles on related matters, including one by physicist Pietro Blaserna (1836–1918).

The society's membership was extended to chemists, physicists, meteorologists and other scientists in order to stress its collaborative intent and interdisciplinary nature.[28]

The response of the international scientific community was highly favorable: all of the most renowned foreign scientific journals reported and welcomed the birth of the new society.[29] The program was subject to some mild criticisms from Faye, but this ultimately served to further publicize the society.[30] The *Memorie* immediately gained widespread success: "I keep receiving requests for our journal from abroad,"[31] Tacchini wrote to Secchi in April 1872, confirming the increasing international interest in the society's publications.

Despite this positive reception, the society was undermined by some of its members' opposition to Secchi, and its scientific program could not proceed regularly because contributions from Nobile and Respighi were missing. Secchi complained in 1872: "Our spectroscopic society is reduced to myself and Tacchini, and to a lesser extent, to Lorenzoni."[32] The welcome the society had received from foreign colleagues was nevertheless encouraging: "Schellen has sent me a very kind letter about our new society," Tacchini wrote, "he said, among other things 'In the spectral analysis of the Sun, Italy leads the way and the other nations let her overtake them.'"[33]

---

27 See Tacchini to Secchi, Palermo, January 27, 1872, quoted in Chinnici and Gasperini, *Alle origini dell'astrofisica italiana*, 169.

28 See article 10 of the Rules of the Society, in Tacchini, "Nuova Società di Spettroscopisti Italiani," 6.

29 See Lorenzoni to Tacchini, Padua, May 5, 1872; ACREA, Fondo Tacchini.

30 See Lorenzoni to Tacchini, Padua, May 5, 1872; ACREA, Fondo Tacchini. See also Tacchini to Secchi, Palermo, April 22, 1872, quoted in Chinnici and Gasperini, *Alle origini dell'astrofisica italiana*, 194–195.

31 *Mi vengono sempre domande dall'estero per il nostro giornale* [...]. Tacchini to Secchi, Palermo, April 22, 1872, quoted in Chinnici and Gasperini, *Alle origini dell'astrofisica italiana*, 195.

32 *La nostra società spettroscopica è ridotta a me e Tacchini e a Lorenzoni, in qualche piccola cosa.* Secchi to Schiaparelli, Rome, April 13, 1872, quoted in *G.V. Schiaparelli, A. Secchi*, 196.

33 *Schellen mi ha scritto una lettera gentilissima riguardo la nostra nuova società: egli dice fra le altre cose "Dans l'analyse spectroscopique du soleil l'Italie marche à la tête et les autres*

FIGURE 8.5 Comparison of solar prominences observed at Palermo, Rome, and Padua Observatories. From Pietro Tacchini, "Protuberanze solari osservate contemporaneamente a Palermo, Roma e Padova nel luglio ed agosto 1871," *Memorie della Società degli Spettroscopisti Italiani* 1 (1872): 25–32, pl. 3 (detail).

Indeed, Secchi and Tacchini's decision to build a network of collaborative scientists was groundbreaking: they understood that, in a context of scarce human and technical resources, as was the case in Italy, only cooperation could ensure the quality of their research. Yet the Italian astronomical community was still wedded to a model of individual scientific research, which led astronomers to cultivate personal scientific interests rather than to engage in broader, collaborative projects.[34]

Moreover, while the *Memorie* became increasingly successful, many in the academic community, both in Italy and the rest of Europe, were still opposed[35] to astronomical spectroscopy. The division of the community into two

nations se laissent devancer". Tacchini to Secchi, Palermo, April 6, 1872, quoted in Chinnici and Gasperini, *Alle origini dell'astrofisica italiana*, 190.

34 See Ileana Chinnici, "Per una storia istituzionale degli Osservatori Astronomici in Italia," *Giornale di astronomia* 41 (2015): 11–21.

35 For example, in a letter to Tacchini, Lockyer claimed that "the Astronomer Royal [Airy], who is very jealous of the spectroscope, has determined not to use that instrument [for observing the transit of Venus]." Tacchini to Secchi, Palermo, March 4, 1874, quoted in Chinnici and Gasperini, *Alle origini dell'astrofisica italiana*, 338.

FIGURE 8.6 Giuseppe Lorenzoni. From *Memorie della Società degli Spettroscopisti Italiani*, n.s. 3 (1914): 132.

parties—those favorable to spectroscopy and those who opposed it—became more and more evident. As elsewhere in Europe, astronomers who worked on celestial mechanics did not welcome the emergence of physical astronomy. Tacchini complained about the situation:

> It seems impossible, but so it is. In all men accustomed to formulae there exists an excessive aversion to physical astronomy: while in the sciences it seems to me of equal interest to know the composition of a prominence as well as the orbit of a planet or a star: moreover, the spectroscope has

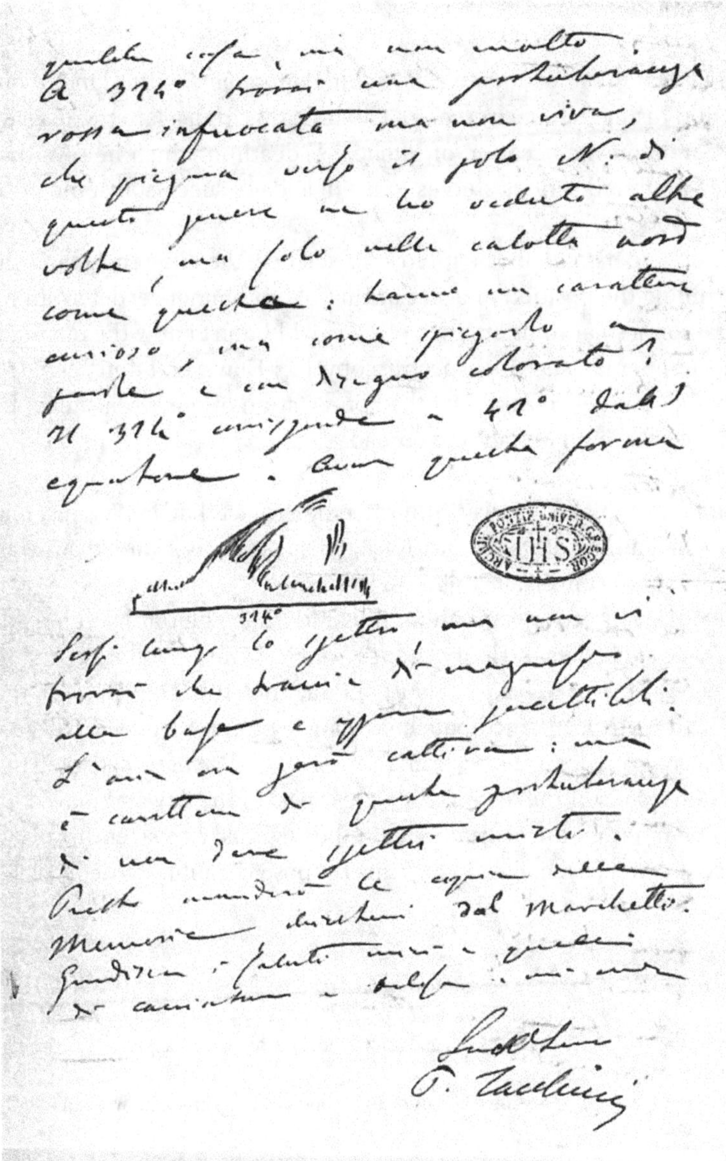

FIGURE 8.7 Page of a letter from Tacchini to Secchi with drawings of solar prominences. Courtesy of APUG.

given results on star motions [...] as good as will ever be obtained through the observations of star positions.[36]

36 *Pare impossibile, ma pure è così, in tutti gli uomini avvezzi alle formole si trova una ripugnanza eccessiva per l'astronomia fisica: mentre per la scienza a me pare che sia di eguale interesse l'arrivare a conoscere la composizione di una protuberanza come l'orbita di un*

And, a month later:

> It seems that in our country we have some opponents to this kind of observation [using the spectroscope]. I assume this from a letter written by Schiaparelli on the occasion of Michez's[37] death in which he says that Michez preferred serious studies to a single day's success at some fashionable work. [...]
>
> I don't understand, in the interest of science, the difference between determining the position of a star or that of a prominence: determining a radiant or a solar rotation time, etc. etc. [...] I don't know the meaning of all these remarks against spectroscopy. [...] However, I don't care too much, and I am encouraged by the appreciation of our work shown by reputable men both in Italy and abroad.[38]

The international appreciation was confirmed by the fact that, in 1873, just one year after the beginning of its publications, the society was awarded a medal and a diploma at the Universal Exhibition in Vienna.

The *Memorie* soon became an international journal. The third volume, for instance, contained papers by Henry Draper from New York, Fyodor Bredikhin (1831–1904) from Moscow, Lockyer from London, and Wolf from Geneva; the fourth and fifth issues contained papers by Langley from Allegheny, Huggins from London, and Charles A. Young from Princeton. The *Memorie* collected the main scientific results of the leading astrophysicists of the day and played an important role in promoting astrophysics and its "status" as a scientific discipline.[39] Tacchini and Secchi clearly had this purpose in mind: "Little by little,

---

pianeta o di una stella: mentre poi lo spettroscopio ha dato sui movimenti delle stelle risultati così felici, che l'osservazione della posizione delle stelle [...] *non arriverà forse ad ottenere.* Tacchini to Secchi, February 8, 1873, quoted in Chinnici and Gasperini, *Alle origini dell'astrofisica italiana*, 243.

37 Jacopo Michez (1839–73) was the director of the Bologna Observatory; he worked mainly on celestial mechanics.

38 *Pare che nel nostro paese noi avessimo dei nemici a questo genere di osservazioni. Ciò lo desumo da una lettera di Schiapparelli scritta in occasione della morte di Michez ove si dice che il Michez preferiva gli studi serii ai trionfi di un giorno ai lavori di moda.* [...] *Io non capisco che differenza ci passa per l'interesse della scienza dal determinare la posizione di una stella o quella di* [una] *protuberanza: dal determinare un radiante o un periodo di rotazione solare, ecc. ecc.:* [...] *non so cosa siano tutte queste espressioni contro la spettroscopia.* [...] *Del resto poco m'importa e mi conforta il vedere che i nostri lavori sono apprezzati e in Italia e fuori e da uomini serii.* Tacchini to Secchi, Palermo, March 31, 1873, quoted in Chinnici and Gasperini, *Alle origini dell'astrofisica italiana*, 262.

39 See Chinnici, "Società degli Spettroscopisti Italiani."

FIGURE 8.8 Illustration of the Medal for the Progress (*Dem Fortschritte*) at the Universal Exhibition of Vienna in 1873. From *L'Esposizione Universale di Vienna del 1873 illustrata* (Milan: Edoardo Sonzogno, 1873), 1, 7.

we will win the battle. At Greenwich also they have begun to do something in spectroscopy,"[40] wrote Tacchini to Secchi with satisfaction in 1876.[41]

4 The Evolution of the Society and Its Achievements

From the beginning of the society's activity, Secchi and Tacchini looked for an opportunity to have a station for spectroscopic observations in the

40 *A poco a poco finiremo per guadagnare la battaglia. Anche a Greenwich hanno incominciato a fare qualche cosa di spettroscopia* [...]. Tacchini to Secchi, Palermo, March 28, 1876, quoted in Chinnici and Gasperini, *Alle origini dell'astrofisica italiana*, 413.

41 At Greenwich Observatory, spectroscopic observations, mainly undertaken by Edward Walter Maunder (1851–1928), were connected with astrometric studies, in particular to determine the radial component of stellar proper motions by means of the Doppler shift of the spectral lines. Although the famous Paris Observatory had a "physical astronomy" section, its only important contribution to astrophysics was the identification of a new class of stars (Wolf-Rayet) in 1867.

winter months in order to undertake a continuous program of solar monitoring.[42] Although the first attempt by Secchi to establish an affiliated observatory in Malta[43] was unsuccessful, another attempt to establish a spectroscopic station in India seemed to have met with success, at least initially. The opportunity to establish a station in India came with a scientific expedition to Bengal that was organized by Tacchini to observe the transit of Venus in 1874 (see Chapter 7). During the expedition, Tacchini met Eugène Lafont, S.J. (1837–1908), rector and professor of physics at St. Xavier's College in Calcutta, who had been invited to join the expedition. Lafont took part in the transit observations and showed a deep interest in spectroscopic research:

> We have with us Fr. Lafont who has observed the transit in the ordinary way, and after seeing and observing with our spectroscopes, he is tempted to establish a spectroscopic station in Calcutta in accordance with our program: we hope he will succeed, because much, indeed much work, could be done in winter here.[44]

The observatory was built at St. Xavier's College in 1875.[45] In 1877, it was fully equipped, and in 1878 Lafont sent his first spectroscopic observations of the solar limb.

However, unfortunately, due to health problems, Lafont was forced to interrupt his research, and thus the collaboration with Calcutta faded away.[46]

In 1876, an important milestone in the collaboration between Secchi and Tacchini took place with the erection of an astronomical station on Mount

42 See Tacchini to Secchi, Palermo, November 25, 1872, quoted in Chinnici and Gasperini, *Alle origini dell'astrofisica italiana*, 224–225, here 225.

43 See Tacchini to Secchi, Palermo, January 18, 1873, quoted in Chinnici and Gasperini, *Alle origini dell'astrofisica italiana*, 234–236, here 234.

44 *Con noi abbiamo il P. Lafont il quale osservò il passaggio al modo ordinario, e vedendo ed osservando nei nostri spettroscopii si è invogliato ad impiantare in Calcutta una stazione spettroscopica di accordo col nostro programma: speriamo che vi riesca, perchè qui in inverno si può far molto, ma molto.* Tacchini to Secchi, Muddapur [India], December 25, 1874, quoted in Chinnici and Gasperini, *Alle origini dell'astrofisica italiana*, 377–378, here 378.

45 The role Secchi played in the establishment of Calcutta Observatory is still to be explored. Lafont was a physicist, highly renowned in Calcutta, who contributed significantly to the development of scientific education in India; see Ileana Chinnici, "An 'Italian' Observatory in India: The History of the Calcutta Observatory," *Studies in the History of Medicine and Science* 14 (1996): 91–115.

46 In the same period, Lockyer was also trying, unsuccessfully, to establish a solar observatory in India, on the pretext of studying the influence of solar activity on terrestrial meteorology in order to forecast famine, a recurrent scourge in that country. See Meadows, *Early Solar Physics*, 107ff.

FIGURE 8.9 Spectroscopic observatory at St. Xavier's College in Calcutta. Courtesy of INAF-OAR.

Etna, in Sicily, for solar observations at high altitudes. It constituted the original nucleus of the first Italian astrophysical observatory, established in Catania in 1880. In this case, Tacchini leaned on Secchi's scientific authority and, in order to obtain financial support from the local municipality, asked the Jesuit astronomer to write a letter[47] explaining the advantages of a mountain station for solar research. Secchi's support was decisive.

The initial idea of building an astronomical station on Mount Etna was proposed by Tacchini in 1876.[48] During this period, many attempts had been made to observe the solar corona when the Sun was uneclipsed. The physical nature of the glowing halo surrounding the Sun, visible only during the total eclipses, was one of the main mysteries of early solar physics. In order to ascertain whether it was due to diffraction effects by the Earth's atmosphere or if it was a phenomenon pertaining to the Sun, some attempts to detect it were made on high mountains in order to observe it through a cleaner and thinner air layer. One of these attempts was made by Young on Mount Sherman in 1872 and

47 See Secchi to Tacchini, October 14, 1876, quoted in Chinnici and Gasperini, *Alle origini dell'astrofisica italiana*, 433–435.

48 See Pietro Tacchini, "Della convenienza ed utilità di erigere sull'Etna una stazione astronomico-meteorologica," *Atti Accademia Gioenia* 3 (1878): 7–26.

confirmed the usefulness of these kinds of observations, so that Tacchini, after some tests, proposed installing an astronomical and meteorological station on the slopes of Mount Etna.[49]

However, it was difficult to persuade the minister of public education to establish an additional observatory when it was already thought that there were too many Italian observatories in existence and a drastic reduction of their number was also being requested. For this reason, Tacchini chose to address the proposal to the local political authorities—the province and municipality. In so doing, he was seeking to capitalize on the celebrations surrounding the transfer of the ashes of the famous musician Vincenzo Bellini (1801–35) from Paris to his native town, Catania, with Tacchini suggesting that the proposed observatory could be named in Bellini's honor. Nonetheless, he also asked Secchi to write a supporting letter, as he was sure that, given the Jesuit's undisputed scientific fame, it would have had "great effect" in cultural and political circles of Catania.[50] Secchi's letter reads as follows:

> I hear with great pleasure that you are cultivating the idea of an observatory on Mount Etna. You cannot do anything better than this; nature has given us this location that, I believe, is the best of all for physical observations of every kind, about heaven and Earth, and we must not leave it dormant. You know that, since 1870 when I was in Augusta [to observe the total solar eclipse, see Chapter 5], I was surprised at the clarity of the Sicilian sky there as in Palermo, and I wished to set up a station, at least temporarily, on Mount Etna. I believe that we could not have a more appropriate location for [studying] the Sun, because the air is not only clear but lacks the whitish [vapor] that is typical of many very high stations. [...] I am convinced that you would see more than Young on Mount Sherman [...]. The dry climate, and the outline of desert vegetation, would give a perfect transparency. [...] The good people from Catania [...] would proudly point out their solar station on Etna, like people of Clermont-Ferrand indicate the meteorological station on the Puy de Dôme, which is barely half of what theirs would be. Since special high altitude stations are being proposed, they could hold up their record against those [from Clermont-Ferrand] who would be truly pygmies in comparison [...]. The

49 See Ileana Chinnici and Carlo Blanco, "L'Etna e le stelle: La fondazione dell'Osservatorio Astrofisico di Catania," in *Atti del XXXIII Convegno annuale della Società Italiana degli Storici della Fisica e dell'Astronomia*, ed. Lucio Fregonese and Ivana Gambaro (Pavia: Pavia University Press, 2016), 67–83.

50 See Tacchini to Secchi, October 12, 1876, quoted in Chinnici and Gasperini, *Alle origini dell'astrofisica italiana*, 431.

station would be sited in the middle of the Mediterranean, and thanks to its height, winds would be completely free of the influence of low chains of hills, which so disturb [the air] on continents [...]. We may in this way learn about the circulation of the upper winds in the Mediterranean, which is so little known, and compare it with [that of the] lower [winds]: we would know at which heights the seasonal squalls dominate, [...] and we may learn a lot about clouds, and hail formation, etc. etc. Thus the station would be useful not only for astronomy but also for meteorology [...]. How many beautiful problems await their solution from this observatory: the relationship between the decrease in rainfall and temperature with height; the diurnal currents of winds or land and sea breezes; African desert dust falls: how many agronomic and botanical scientific questions will get their solution! [...]. And volcano physics, how many advantages would there be from placing seismometers and apparatus there for the study of ground motions?[51]

---

51 *Sento con grande piacere che voi coltivate il progetto di una stazione di osservazione sull'Etna. Voi non potete fare cosa migliore di questa, la natura ha dato a noi italiani questa località che la* [sic] *credo la migliore di tutte per le osservazioni fisiche di ogni specie di cielo e di terra e non dobbiamo lasciarla inoperosa. Voi saprete che fino da quando nel 1870 fui ad Augusta io restai sorpreso dalla chiarezza del cielo siciliano tanto colà che a Palermo, onde anch'io vagheggiai l'erezione di una stazione temporanea almeno sull'Etna. Credo che non si possa avere stazione più opportuna pel Sole perché l'aria non solo è chiara ma manca di quel biancastro che è proprio di molte stazioni assai elevate* [...] *Sono persuaso che voi vedreste più di quanto ha visto Young sul monte Sherman* [...] *Il clima asciutto, e il contorno deserto di vegetazione darebbe una perfetta trasparenza* [...] *I bravi Catanesi* [...] *sono generosi assai e suppliranno lautamente ai comodi indispensabili del coraggioso soggetto che andrà lassù a passare alcuni mesi. Essi potranno additare con orgoglio la stazione solare Etnea come quelli di Clermont-Ferrand indicano la loro stazione meteorica sul Puy de Dôme che pure sarebbe appena la metà della loro. Giacché siamo in cercare stazioni alte ed opportune, non si lascino togliere la palma da questi che sono veramente pigmei in confronto.* [...] *La stazione resterebbe nel mezzo del Mediterraneo, e per la sua altezza le correnti sarebbero affatto libere dall'influsso delle basse catene che tanto le sturbano nei continenti* [...]. *Potremo conoscere di là quale è nel Mediterraneo il giro de' venti superiori che è tanto poco conosciuto e confrontarlo cogli inferiori: sapremmo fino a che altezza dominino le burrasche secondo le stagioni,* [...], *e potremo conoscere tante cose sulle nubi e la formazione della grandine ecc. ecc. Sicché la stazione sarebbe utilissima non solo alla astronomia, ma anche alla meteorologia* [...] *Quanti bei problemi aspettano la loro soluzione da questo vostro osservatorio: la legge del decremento delle piogge, e della temperatura colle altezze; le correnti diurne di venti o brezze di terra e di mare; le cadute di polveri del deserto africano: quanti problemi di botanica agronomica e scientifica avranno la loro soluzione!* [...] *La fisica del Vulcano quanto non ne vantaggerebbe mettendovi sismometri ed apparati per lo studio de' moti del suolo?* Secchi to Tacchini, October 14, 1876, quoted in Chinnici and Gasperini, *Alle origini dell'astrofisica italiana*, 433–435.

FIGURE 8.10 The astronomical station on Mount Etna, known as "Bellini Observatory." Courtesy of INAF-OAR.

The letter had the effect that Tacchini desired, and the project was approved. The Etna station, named Bellini Observatory, was ready for use in 1880.

An additional branch located downtown was also created, which became the Catania Astrophysical Observatory. The adjective "astrophysical" (instead of "astronomical") indicated a special orientation toward spectroscopic research, as in the case of the other solar observatories that were established in those years throughout Europe (see introduction). Secchi, however, was not able to see the completion of the Bellini Observatory, because he died about two years after writing the letter that would pave the way for its establishment.

5 A Long-Lasting Friendship

The friendship between Secchi and Tacchini was fed by a rich exchange of opinions and views, reciprocal confidences, and concrete gestures of affection: frequent visits by Tacchini to Rome, invitations to Modena, words of encouragement and solace in delicate moments, a search for political support to promote shared scientific projects and to preserve Secchi's observatory from confiscation. This is particularly evident in the fascinating correspondence[52] between the two astronomers, which reveals the strength of their friendship.

It is worth looking deeper into Tacchini's role in the events surrounding the Collegio Romano Observatory. When the confiscation of church properties was approved by law in 1873 (see Chapter 10), Secchi hastened to inform him:

> The Chambers have decided to free themselves from friars, and I am a friar, therefore, out! What is now my property will be that of others! I will be told to hit the road and, after forty years of hard work, in my old age I will have to go wandering, looking for a loaf of bread, perhaps in Australia or America.[53]

Tacchini replied:

> The idea of losing you saddens me. [...] I think [...] that a position for you in Italy will always be found [...]. I believe that your observatory would be left to you, it would not be touched but maintained and, perhaps, even improved, so that you might devote the rest of your life to your precious research. For my part, you know that I would use all of the little that I can, in order to better arrange things for your benefit. [...] The attention and esteem I have for you is so much that I would like to see you in your observatory in Rome forever [...].[54]

---

52 See Chinnici and Gasperini, *Alle origini dell'astrofisica italiana.*

53 *Alle camere è deciso di disfarsi de' frati, e io sono frate, dunque via! quello che ora è mio diverrà proprietà altrui! Io sarò in mezzo a una strada e dopo 40 anni di fatiche nella mia vecchiaia andrò ramingo a cercare un tozzo di pane, forse nell'Australia o nell'America.* Secchi to Tacchini, Rome, April 10, 1873, quoted in Chinnici and Gasperini, *Alle origini dell'astrofisica italiana*, 270.

54 *L'idea di perder lei mi rattrista.* [...] *Io penso* [...] *che per lei una posizione in Italia potrà sempre trovarsi* [...] *io credo, che il suo osservatorio resterebbe ancora per lei e non sarebbe molestato, ma conservato e forse anche migliorato, e lei potrebbe così consacrare il resto della sua vita alle sue preziose ricerche. Per parte mia, s'immagini che userei di tutto quel poco che io posso per meglio combinare le cose a di lei vantaggio.* [...] *è tanta la stima ed attenzione che ho per lei, che vorrei vederla sempre nel suo osservatorio di Roma* [...]; Tacchini to Secchi, Palermo, April 14, 1873, quoted in Chinnici and Gasperini, *Alle origini dell'astrofisica italiana*, 273.

Tacchini was ready to offer his services and his political support. In a confidential letter to Secchi, he wrote:

> The Senate too will approve the law [...]; it seems to me that the time has come to discuss your position and your observatory: as I wrote you once, I have friends in high places who share with me a full interest of maintaining you and your observatory, and they are prepared, through me, to negotiate the matter [...]. And I hope you will accept my sincere advice, which is dictated only by a love of science and a friendship [...] that connects us in our scientific lives [...] so that your observatory might have good funding, you would certainly have an important position in astronomy with the government in Italy; in short, I consider such a combination to be a great service that we, you and me, could make to Italian science and astronomy in general.[55]

Thanks to the intervention of the minister, Sella, the observatory remained under the direction of Secchi, its ultimate destiny hanging on a forthcoming reform of the Italian observatories—a reform that was proposed by Tacchini a year later and decreed in 1876 (see Chapter 10).

Secchi also confided to Tacchini the reasons that had led him to withdraw from the expedition to observe the transit of Venus in India in 1874 (see Chapter 5). The Jesuit had decided to turn down Tacchini's invitation because of the still uncertain situation surrounding his observatory and his difficult personal position:

> You know that I am ready to come, and I confirm it again, but my support does not depend just on me. [...] Some days ago, I made another attempt to obtain permission, and I was told that it would be better for me to leave the observatory than to keep insisting [...]. Card[inal] Antonelli [...] even refused to maintain any subsidy for my person and the employees

---

55 *Anche il Senato approverà la legge* [...] *mi pare dunque che sia arrivato il tempo di trattare della di lei posizione e del suo osservatorio: come le ho scritto altra volta io tengo amici altolocati, che dividono pienamente con me l'interesse di conservar lei e il suo osservatorio, e sono disposti col mio mezzo a trattare la faccenda* [...] *Ed io spero che lei vorrà accettare un mio consiglio sincero dettato soltanto dall'amore alla scienza e dai rapporti di amicizia che ci legano nella nostra vita scientifica* [...] *il suo osservatorio potrà avere un buon assegno, lei avrà sicuramente importanti incarichi dal governo per riguardo all'astronomia in Italia, insomma, una tale composizione io la considero un servizio ben grande che lei e noi possiamo rendere alla scienza italiana e in generale all'astronomia.* Tacchini to Secchi, Palermo, May 22, 1873, quoted in Chinnici and Gasperini, *Alle origini dell'astrofisica italiana*, 282.

> because the government did not want to recognize the observatory as pontifical, and he told me clearly that I had to leave […]. I am far from wanting to exhaust myself, to toil and suffer to prepare a bed to be enjoyed by another! Nothing of the sort! If the observatory would be declared either mine or pontifical, I would be ready for anything, to face travels, costs, fatigues: but the prospect of this being for someone else's benefit brings me down. Plus there is the overt hostility that has been recently developed against everything I do. You are the only exception who appreciates what I am doing, and Schiaparelli a bit as well. All the others are racing to put down what I am doing. It is a clique established a long time ago, which has now developed into a hostile force: I have been expelled from the Academy [of Lincei; see Chapter 10], deprived of the right to my pension, deprived of honors from Philosophical College, everything is conspiring to sink me! This is a formal declaration of war! Nobody accepts anything I do: even if I wear myself out, if I produce a prodigious amount of observation and results, everything would be criticized, neglected, or at least buried in a conspiracy of silence! Why should I exhaust myself with such extraordinary efforts, now that I am already quite overcome, and my health suffers so much? […] Now, being closer to sixty than fifty years old, I am a bit wiser, and I don't want to endure such pains and troubles, to seethe with rage and let my enemies take pleasure in this. […]. Here is my final answer: I am ready to come if it is allowed by my superiors, but if they deny it, I am not willing at all to antagonize them or do anything that would displease them.[56]

---

56 *Voi sapete che io sono pronto a venire, e ve lo confermo nuovamente, ma io non dipendo da me solo.* […] *Giorni sono feci fare un tentativo di nuovo per aver il permesso, e mi fu fatto rispondere che farei meglio a lasciare l'osservatorio che a stare a insistere* […] *il Card*[*inal*] *Antonelli* […] *perfino rifiutò di continuare a darmi nulla pel mantenimento della mia persona e degli addetti, perché il governo non voleva dichiarare l'osservatorio pontificio, e mi disse tondo che dovea andarmene.* […] *io sono ben lungi dal volermi anche mazzare, faticare e penare per preparar ad altri un letto da godere! Manco per idea! Se l'osservatorio sarà dichiarato o mio o pontificio io sono pronto a tutto, a cacciarmi in viaggi, in spese, in fatiche: con questo orizzonte di giovare a chi sa chi, io non me la sento. A questo aggiungete la dichiarata ostilità, sviluppatasi non ha molto, verso tutto quello che io faccio. Voi siete l'unica eccezione che stimate le mie cose, e un poco ancora Schiaparelli. Tutti gli altri fanno a gara a deprimere ciò che da me si fà. È una cricca ordinata da gran tempo, ora si sviluppa con schifoso vigore: espulso dall'accademia, privato dei diritti che avea alla mia pensione, dei compensi pel Collegio filosofico, tutto in somma tende a deprimermi! La dichiarazione di guerra è formale! Non si vogliono le cose mie: se anche mi facessi a pezzi, se anche riuscissi a fare un prodigio di osservazioni e di risultati, tutto sarebbe criticato, negletto, o almeno sepolto nella congiura del silenzio! A che dunque ammazzarmi, e ciò ora che sono già abbastanza affranto, e che sensibilmente la mia salute se ne risente di ogni fatica straordinaria?* […]*Ormai sto più*

Tacchini was ready to search out possible mediators who could facilitate getting permission from Secchi's superiors,[57] but he had to accept the Jesuit's final decision to turn down the appointment, which was probably also influenced by Secchi's anxiety surrounding his own personal situation.

Tacchini again took the initiative to invite Secchi to participate in the Congress of Italian Scientists held in Palermo in 1875 (see Chapter 10). A meeting to discuss the reforms proposed by Tacchini[58] was also to have been held, and Tacchini wrote him: "Remember well that, I want you in Palermo at any cost for the congress of scientists: there we shall be at home and could take advantage of that opportunity to our benefit."[59]

The relationship between Secchi and Tacchini was not restricted solely to science but extended to the private sphere. During his stay in India, for instance, Tacchini contracted recurrent tropical fevers, and Secchi worried about his friend's health: "If you are not yet well recovered, don't come here, because your fevers might outbreak again in Rome and be more malignant than the Indian ones."[60] The Jesuit also sent him some natural remedies. Tacchini wrote him:

> Yesterday, I received a small box containing two bottles of eucalyptus elixir from Rome: I suppose that the shipment was made by you and thank you distinctly [...]. I am pleased to say to you that since I have taken the eucalyptus I have no longer had a fever: and if things stay like this, indeed I could say that I have been healed by you.[61]

---

vicino ai 60 che ai 50 anni e ho messo un po' di giudizio, e non mi voglio procurare pene e triboli per schiattare di rabbia e far godere i miei nemici. [...] *eccovi la mia risposta finale = Io son pronto a venire qualora ciò mi sia concesso dai miei superiori, ma se questi me lo negano io non sono niente disposto a nemicarmi loro, e far dei passi che li contristino.* Secchi to Tacchini, Rome, May 16, 1873, quoted in Chinnici and Gasperini, *Alle origini dell'astrofisica italiana*, 353–354.

57 See Tacchini to Secchi, Palermo, May 21, 1874, quoted in Chinnici and Gasperini, *Alle origini dell'astrofisica italiana*, 354–355.

58 See Poppi, Bònoli, and Chinnici, "Il progetto Tacchini e la riforma degli Osservatori italiani," 123–171.

59 *Si ricordi bene che pel Congresso degli scienziati la voglio in Palermo a qualunque costo: là siamo in casa nostra e potremo approfittare di quell'occasione a nostro vantaggio* [...]. Tacchini to Secchi, Modena, June 22, 1873, quoted in Chinnici and Gasperini, *Alle origini dell'astrofisica italiana*, 383.

60 *Se non è ben rimesso non torni qua perché a Roma le febbri potrebbero ritornare e forse più maligne che le indiane* [...]. Secchi to Tacchini, Rome, May 30, 1875, quoted in Chinnici and Gasperini, *Alle origini dell'astrofisica italiana*, 381.

61 *Ieri ho ricevuto un cassettino contenente due bottiglie d'elisir d'eucaliptus, proveniente da Roma: m'imagino che la spedizione è stata fatta da lei e la ringrazio distintamente* [...]. *Piacemi intanto dirle che dopo che prendo l'eucaliptus non ho più avuto febbre: e se le cose*

A few days later, he added: "Regarding eucalyptus, I haven't yet finished the first bottle: I take one teaspoon every morning, dissolved in water: but I ask you, who now are my doctor, should I continue? Is it dangerous, could it hurt me in the long run?"[62] Secchi reassured him,[63] and having learned that Tacchini also took concoctions of rhubarb and cinchona[64] without much benefit, he advised and encouraged him in almost paternal tones:

> The use of rhubarb is not indicated because it causes relaxation: against fevers, you have to take tonics, and good wine, but not fortified like yours. Ferrous wines, such as Médoc and Bordeaux, would be preferable. Hold on, and the fever will go away: watch out for sweating and feeling worn out.[65]

For his part, Tacchini, during his frequent family visits to Modena (he never married, but was constantly in contact with his numerous relatives),[66] was always sure to invite Secchi to stop at his family home during his travels: "I am abducting you now, namely I want you to stop at least one day at Modena in my house so that we will eat the *zampone* there that we had in Rome [...]."[67]

---

continueranno così, potrò davvero dire che mi ha guarito lei. Tacchini to Secchi, Palermo, February 15, 1876, quoted in Chinnici and Gasperini, *Alle origini dell'astrofisica italiana*, 400.

62 *Riguardo all'eucaliptus non ho ancora finito la prima bottiglia: ne prendo un cucchiaino ogni mattina sciolto nell'acqua: ma domando a lei che ora è il mio medico, si deve continuar sempre? Non vi è pericolo che faccia male a lungo andare?* Tacchini to Secchi, Palermo, February 26, 1876, quoted in Chinnici and Gasperini, *Alle origini dell'astrofisica italiana*, 403.

63 See Secchi to Tacchini, Palermo, March 2, 1876, quoted in Chinnici and Gasperini, *Alle origini dell'astrofisica italiana*, 405. The success of this therapy led to a trade in bottles of eucalyptus elixir among Tacchini's friends (see correspondence exchanged between October 1876 and June 1877, in Chinnici and Gasperini, *Alle origini dell'astrofisica italiana*, 433ff).

64 See Tacchini to Secchi, Palermo, March 6, 1876, quoted in Chinnici and Gasperini, *Alle origini dell'astrofisica italiana*, 409.

65 *L'uso del rabarbaro non è tanto indicato perché provoca rilassatezza: colle febbri ci vogliono tonici, e buon vino ma non de' vostri spiritosi. Vino ferroso, come Medoc e Bordò sarebbe preferibile. Fatevi coraggio, e le febbri se ne andranno: guardatevi dal sudare, e affaticarvi.* Secchi to Tacchini, Rome, March 9, 1876, quoted in Chinnici and Gasperini, *Alle origini dell'astrofisica italiana*, 410.

66 See Mario Umberto Lugli, *Pietro Tacchini* (Modena: Edizioni Il Fiorino, 2001), 9.

67 *La sequestro fin d'ora, cioè voglio che al ritorno si fermi almeno un giorno a Modena in casa mia, così mangeremo là il zampone che dovevamo mangiare a Roma* [...]; Tacchini to Secchi, November 2, 1875, quoted in Chinnici and Gasperini, *Alle origini dell'astrofisica italiana*, 396–397.

When the symptoms of the illness that would lead Secchi to his grave began to appear, Tacchini, unaware of their seriousness, suggested in turn: "I think you would do well to use eucalyptus, because I believe that there is a remnant in you of that bad fever that sometimes occurs to me: I too will take this harmless drug."[68]

This is the last known letter between the two men. Secchi's deteriorating health led to an interruption of their exchanges, although Tacchini was constantly kept informed about Secchi's health by Fr. Ferrari, who wrote him in the last days of Secchi's life:

> Unfortunately, you are right to be anxious about our Father Secchi, because after several days of relative calm, the day before yesterday the illness worsened and this morning alarming symptoms appeared [...]. Fr. Secchi is grateful to you and your director for the care to ask for news about him by telegraph, and I am obliged to let you know how he is feeling [...].[69]

Secchi died two days later (see Chapter 10), and the feverish Tacchini was unable to attend the funeral. His friend Lorenzoni, who was able to attend, wrote him a few days later: "I went to Rome hoping to see Secchi again and there I arrived in such a time that his corpse was still lukewarm so that I could just do the sad office of attending his funeral. Oh what a serious loss for Italian astronomy!"[70]

Tacchini must have had similar feelings: he had lost his main scientific mentor, the man with whom he had worked for many years and with whom he had

68 *Credo che farebbe bene ad usare dell'Eucaliptus, perché io ritengo che in lei vi sia un resto di quel brutto elemento di febbre che qualche volta si ripresenta a me: io pure riprenderò questo innocuo medicamento.* Tacchini to Secchi, Naples, December 10, 1877, quoted in Chinnici and Gasperini, *Alle origini dell'astrofisica italiana*, 468.

69 *Purtroppo ha ragione di non esser tranquillo sul conto del nostro P. Secchi poiché se vi furono alcuni giorni di relativa calma, da jeri l'altro però il male si è esacerbato e questa mattina sono comparsi dei sintomi allarmanti* [...] *Il P. Secchi è gratissimo verso di lei e del Sig. Direttore per le premure che si son dati in richiedere telegraficamente sue notizie e mi commette di manifestar loro questi suoi sentimenti* [...]. Ferrari to Tacchini, Rome, February 24, 1878; ACREA, Fondo Tacchini.

70 *Andai a Roma colla speranza di vedere ancora una volta il Secchi e vi giunsi in tempo che il suo cadavere era ancora tepido, perciò non potei se non che compiere il mesto ufficio di assistere al suo funerale. Oh quale grave perdita è stata questa per l'astronomia italiana!* Lorenzoni to Tacchini, Padua, March 10, 1878; ACREA, Fondo Tacchini.

shared hopes for astronomy. It may be not too far from the mark to assume that Tacchini considered himself Secchi's scientific heir.

However, despite the close relationship between the two men, Tacchini's participation in the expulsion of the Jesuits from the Collegio Romano Observatory (see Chapter 10) and its confiscation would lead to accusations of opportunism from some Jesuit circles.[71] In fact, Fr. Ferrari, who had expected to become Secchi's successor, had asked Tacchini to act as an intermediary with the government, so that he could be left as the head of the observatory: "In his will, Father Secchi leaves me as his friend and companion heir specifically to that which devolves to him personally; having already gotten that vote of confidence from such a man; could you add your vote to Secchi's? I won't say more and hope in your goodness."[72] Even though Schiaparelli also recommended this measure, the minister came to a different decision. The issue, which is beyond the scope of this work, deserves further study.

If Tacchini really regarded himself as Secchi's heir, he probably also felt morally obliged to prevent Secchi's observatory being shut down. "In May 1879, the government moved me from Palermo to Rome," Tacchini wrote, "charging me to establish a Central Bureau of Meteorology in the premises of Collegio Romano […]. At the same time, the minister of education invited me to take over the astronomical observatory of the Collegio Romano, entrusting to my care its preservation and improvement."[73] In fact, the site of the Collegio Romano was probably chosen by Tacchini precisely to preserve the observatory, which had become superfluous in the context of the reform of the Italian observatories. Deprived of funds and staff, the observatory was annexed by Tacchini to the Central Bureau of Meteorology. But this was merely a symbolic attempt to prolong the life of an institution that had in fact died with Secchi, as Tacchini and his successors tried to maintain the astronomical tradition of the Collegio Romano Observatory, but with only poor results: astrophysical research, in crisis

---

71 See Felice Ciampi, "La confisca della Specola del Collegio Romano (1879)," *Societas* 42 (1993): 113–116.

72 *Il P. Secchi nel suo testamento mi lascia erede specialmente per ciò che gli spetta personalmente come a suo amico e compagno, ciò è già un voto di fiducia e di un tal uomo, potrebbe ancora V.S. unirsi al voto del Secchi? Non dico altro e spero nella sua bontà.* Ferrari to Tacchini, Rome, March 5, 1878; ACREA, Fondo Tacchini.

73 *Nel maggio del 1879 il Governo mi traslocava da Palermo a Roma, affidandomi l'incarico di impiantare in locali del Collegio Romano un ufficio centrale di meteorologia* […] *Alla stessa epoca il Ministro della P. istruzione mi invitò a prendere in consegna l'Osservatorio astronomico del Collegio Romano affidandolo alle mie cure per la conservazione e miglioramento.* Lugli, *Pietro Tacchini*, 128.

in Italy at the end of the century, would revive only in 1925 with the completion of the solar tower at Arcetri[74] by Giorgio Abetti and the revival of those solar studies in which Secchi and Tacchini had been pioneers. Not by chance, the Arcetri Observatory today holds the precious correspondence between Secchi and Tacchini, donated by the heirs of the latter to the prestigious Florentine institution in recognition of the legacies of the two astronomers.[75]

74 See Antonella Gasperini, Massimo Mazzoni, and Alberto Righini, "La costruzione della Torre Solare di Arcetri nel carteggio Hale-Abetti," *Giornale di astronomia* 3 (2004): 23–30.

75 See Chinnici and Gasperini, *Alle origini dell'astrofisica italiana*, 21–25.

CHAPTER 9

A Controversial Personality: Secchi and the Scientific Debates of His Time

Secchi's success was due in no small part to his personality. He was an intriguing figure, unusual for being both a Jesuit and a scientist. Secchi himself, in recounting a meeting with some professors of Florence, reported: "There was the usual squawking that I failed to have a long beard and old face [...]."[1] His teacher of classics, Angelini, said about him: "He was a man of strong feelings and elevated thoughts and could not suffer narrow-mindedness and meanness of heart [...]. He was of ready, sharp, penetrating intelligence, and grasped at first sight the point of the matter."[2]

Some of Secchi's confrères noted that he did not hesitate to rebuke blasphemers and was inflexible with those who left the Society, reporting the case of an ex-Jesuit[3] who had met Secchi, but when he greeted him, Secchi claimed he did not know him, despite recognizing his former companion.

Sensitive to praise as well as to criticism, Secchi liked to mingle with ordinary people without seeking special privileges; however, when it was offered, he nevertheless appreciated such attention. He had considerable self-regard but, in his later years when he had lost his status, he was close to depression and often accused his opponents of victimizing him. However, he never lost his sense of humor. Being forthright and outspoken, given sometimes to a lively polemical style, he easily attracted criticisms and disparagement: once he entered into a controversy, he was scathing and ready to ridicule his opponent. Not surprisingly, then, he had disputes of various kinds. Here, I will briefly mention only the liveliest ones.

1 Caught in the Crossfire: Between Anti-Clericalism and Ultra-Conservatism

The historical period in which Secchi lived was certainly difficult for the Jesuits. Even after the exile of 1849 and subsequent return to Rome, the Jesuits

1 *Vi fu la solita diatriba del non aver io la barba lunga e faccia vecchia* [...]. APUG, FS 23.I.

2 *Egli era di un forte sentire, e di alti pensamenti, e non poteva patire certe picciolezze d'idee, e meschinità di cuore* [...] *Era d'ingegno pronto, acuto, penetrante, afferrava di primo tratto il punto della questione.* APUG, FS 23.IV.

3 The ex-Jesuit was Giuliano Giordano, professor of physics in Naples. See APUG, FS 23.III.

 | DOI:10.1163/9789004387331_011

were only barely tolerated in Italy: the Risorgimento spirit that animated the Italian patriots aimed to make Italy "one and independent," a plan whose achievement included the end of the temporal power of the popes. As the "elite troops" of the pope, the Jesuits were continuously exposed to anti-clerical attacks.[4] The hostile climate that raged against the Jesuits is well summarized in a threatening letter received by Secchi at Palermo Observatory when he was preparing the expedition for the solar eclipse of 1870 (see Chapter 5):

> Oremus [Let us pray] that the Rev[erend] Angelo Secchi lets us remain at peace. We respect the famous Italian astronomer in him—we hate the Jesuit. Your tremendous cult, with its professed principles, is well known to us. Italy was always a slave because of your horrible cult. You act neither for God nor for humanity, but for the pleasure of domination. If the blood you have spilled under pretext of religion should come back upon you, humanity would be freed from a thousand other misfortunes, persecutions, and deaths. May you be damned forever!!! The best thing would be if you would go away.[5]

Being a well-known figure, Secchi was often the target of the anti-clerical press, which repeatedly circulated the rumor that he was going to leave the Jesuit order over alleged disagreements with his superiors (see Chapter 7). On the contrary, Secchi had a strong sense of attachment to the Society of Jesus and often intervened, publicly and passionately, to defend it in moments of political persecution.[6] In 1877, Secchi's name was even mentioned by Giovanni Bovio (1837–1903), member of parliament and exponent of Masonry, in a parliamentary speech that had wide repercussions; the passage quoted here gives

---

4 In those years, the figure of Galileo was taken as an example to demonstrate the obscurantism of the church, which was considered the enemy of science and progress, with the Jesuits as its paladins. This view, primarily arising for political or personal reasons, deliberately ignored the important contribution that the church had made to the development of sciences, especially astronomy; see Ileana Chinnici, "Personaggi e istituzioni della Chiesa Cattolica nella fondazione degli osservatori astronomici in Italia," Documentazione interdisciplinare di scienza e fede (2009); http://www.disf.org/altriTesti/Chinnici.asp (accessed April 5, 2018).

5 *Oremus il Rev[erendo] Angelo Secchi a lasciarci tranquilli. Rispettiamo nella di lui persona il celebre astronomo Italiano—odiamo il Gesuita; la sua setta tremenda, coi principi che professa, ci è nota. L'Italia fu sempre schiava per la vostra orribile setta. Non per Dio, né per l'umanità, ma per il piacere di dominare voi agite—se il sangue che avete fatto versare sotto pretesto della religione ricadesse su voi, l'umanità si libererebbe da mille altre sventure, persecuzioni e morti. Che siate per sempre maledetti!!! Per il vostro meglio andate via.* Anonymous to Secchi, [Palermo], January 10, 1871; AOAPA, Fondo G. Cacciatore 85, 38.

6 See APUG, FS 9.V.

an idea of how Secchi was lumped together with atheists and materialists, and used as an argument against the Society of Jesus:

> Catholicism, quickly paganized, was forced to move from the fury of Saint Dominic to the cunning of Saint Ignatius. And this saint devised, in order to implement the dream of a universal monarchy, a tremendous black oligarchy that for 333 years has invaded most of the Earth; behind the white pope, hid the black pope. [...] Father Secchi, celebrating the unity of physical forces, turns the key in the lock of the order and throws the key into the Tiber. Behind Father Secchi, there is Pomponazzi;[7] perhaps having been absorbed by the photosphere, he doesn't realize it very well [...]. Father Secchi is, in the history of [the Society of] Jesus, the first Jesuit who dares to think that he is just as worthy of infallibility as the pope; that is the absurdity of it, by proclaiming himself infallible, he voids [the claim].[8]

2 Polemics with neo-Thomists

The anti-clerical front, however, was not the only battle that Secchi had to engage in. Another front, perhaps even more insidious, had formed within the church and the same Society of Jesus. During this period, there were many apologetic writings aimed at counteracting the risk of atheism inherent in the new scientific theories. Secchi often remarked on the errors in physics that he found in them, in the belief that truth cannot contradict itself.

Around the middle of the nineteenth century, a circle of intellectuals including some Jesuits strongly advocated a return to St. Thomas's philosophy. With the founding of the journal *La civiltà cattolica*, they had an effective organ

7 Philosopher Pietro Pomponazzi (1462–1525) supported the irreconcilability between reason and faith, in disagreement with Thomist theology.

8 *Il cattolicismo, rapidamente impaganito, fu costretto passare dai furori di San Domenico all'astuzia di Sant'Ignazio. E questo santo escogitò, per attuare il sogno della monarchia universale, una tremenda oligarchia nera che per trecento trentatrè anni invase gran parte della terra e dietro al papa bianco nascose il papa nero.* [...] *Il padre Secchi, celebrando l'unità delle forze fisiche, volta gira la chiave nella toppa dell'ordine e butta la chiave nel Tevere. Sotto il padre Secchi v'è Pomponazzi; ei, forse assorbito dalla fotosfera, non se ne accorge bene* [...] *Il padre Secchi è, nella storia di Gesù, il primo gesuita che osa pensare, ed è però degno contemporaneo della infallibilità del papa, cioè dell'assurdo, che, proclamandosi infallibile, si esaurisce.* Giovanni Bovio, "Discorso," in *Atti del Parlamento Italiano, Camera dei Deputati, Sessione del 1876–77: Discorsi 1* (Rome: Tipografia Eredi Botta, 1877), 690–697, here 691.

of propaganda for the neo-Thomist movement.[9] However, within the Society of Jesus there was a trend headed by the Jesuits of the Gregorian University, which, although not being strictly "anti-Scholastic," was opposed to neo-Thomist fundamentalism.[10] This camp was represented by Salvatore Tongiorgi, Domenico Palmieri (1829–1909), and Secchi himself.[11] In 1861, Tongiorgi published a well-regarded course of philosophy, *Institutiones philosophicae*, used as textbook for colleges and seminaries. In this work, the Jesuit theologian departed from Scholastic tradition, rejected the peripatetic theory of matter and form as well as the Leibnitzian monads, and adopted chemical atomism:[12] essentially, he denied the real distinction between accidents and substance, claiming that natural phenomena were simply the result of mechanical and chemical forces—a view that was shared by Secchi (see Chapter 6). Among the most active non-Jesuit supporters of Thomistic hylomorphism was the Dominican Vincenzo Nardini (1830–1913).[13] Argumentative and combative, Nardini incessantly promoted scientific study as a way to achieve his ideological program of providing adequate scientific support for Thomistic philosophy. In 1862, in a note on a paper against the neo-Thomist Matteo Liberatore, S.J. (1810–92), whom Nardini accused of being unfaithful to true Thomistic philosophy, the Dominican criticized Tongiorgi's views about mechanicism and atomism. He thus provoked a reaction from Secchi, who wrote a short essay in the form of a letter to Nardini.[14] Secchi stated that the Aristotelian–Thomist concepts of substance, matter, and form were entirely inadequate to explain natural phenomena in the light of experimental data, and that all physical

9 See Heinrich Schmidinger, "Centros tomistas en Roma, Nápoles, Perugia, etc.: S. Sordi, D. Sordi, L. Taparelli D'Azeglio, M. Liberatore, C.M. Curci, G.M. Cornoldi y otros," in *Filosofía cristiana en el pensamiento católico de los siglos XIX y XX*, ed. Emerich Coreth, S.J., Walter M. Neidl, and Georg Pfligersdorffer (Madrid: Ediciones Encuentro, 1994), 2:103–122, here 110.

10 See Schmidinger, "Centros tomistas en Roma, Napoles, Perugia, etc.," 110–111.

11 Palmieri, unlike Tongiorgi and Secchi, was a supporter of dynamism; see Luciano Malusa, "Palmieri, Domenico," in *Dizionario biografico degli Italiani* 80 (Rome: Treccani, 2014); http://www.treccani.it/enciclopedia/domenico-palmieri_(Dizionario-Biografico)/ (accessed June 3, 2018).

12 See Giuseppe Filograssi, "Teologia e filosofia nel Collegio Romano dal 1824 ad oggi," *Gregorianum* 35, no. 1954 (1965): 512–540, here 516.

13 On Nardini, see Rita Fioravanti, "Nardini, Vincenzo," in *Dizionario biografico degli Italiani* 77 (Rome: Treccani, 2012); http://www.treccani.it/enciclopedia/vincenzo-nardini_(Dizionario-Biografico)/ (accessed June 3, 2018).

14 See Secchi, *Intorno alla soluzione di un problema fisico-cosmologico*. On the debate between Secchi and Nardini, see Rita Fioravanti, "*Intorno alla soluzione di un problema fisico-cosmologico*: Il domenicano Vincenzo Nardini e il gesuita Angelo Secchi," in *Angelo Secchi, astronomo e fisico: Attualità scientifica e luoghi storici a Roma*, ed. Tullio Aebischer and Rita Fioravanti (Rome: Edizioni Nuova Prhomos, 2014), 23–48.

and chemical phenomena were attributable to matter in motion. Nardini was surprised by Secchi's intervention; he replied with a pamphlet[15] where he rejoined that his own work was mostly about metaphysics rather than physics, and reaffirmed that physical forces, which were considered efficient causes in Scholastic philosophy, are the real principles of each effect, including motion. Moreover, he accused Secchi of supporting theories that were Cartesian and materialistic, from a philosophical point of view, and scientifically absurd.

Secchi reacted in the pages of the *Bullettino*[16] and, referring to Nardini's booklet—without openly mentioning the name of the author—denounced the confusion between physics and metaphysics that he found in it. He thus specified that he "never intended to speak about the atoms of the metaphysicists, but only about those as understood by physicists,"[17] and indirectly accused the Dominican of being ignorant of recent advances in physics.

Despite these disputes, however, Secchi appreciated the initiative of Nardini to establish a small astronomical observatory in 1864 at the convent of St. Mary above Minerva—the third observatory in Rome at that time—and mentioned it in one of his lectures.[18]

Two years later, the publication of the treatise *Sull'unità delle forze fisiche* drew fierce criticism from some of the neo-Thomist fundamentalists.

Liberatore declared the book's content "false and dangerous": Secchi remarked that his confrère had made so many errors, especially in the application of the concepts of matter and form to modern physics, that he was "worthy of compassion" for his "intolerable confusion": "The concept of matter and form, as intended by Scholasticism, is beautiful and wide; but as intended by Liberatore, it becomes limited, erroneous, and foolish [...]."[19]

Secchi's main opponent, however, was Giovanni Maria Cornoldi, S.J. (1822–92),[20] one of the main proponents of neo-Thomism, who published a work[21]

15 See Vincenzo Nardini, *Risposta* [...] *alla lettera del p. Angelo Secchi* (Rome: Tipografia di Giovanni Cesaretti, 1862).

16 See Angelo Secchi, "Elettricità atmosferica," *Bullettino meteorologico dell'Osservatorio del Collegio Romano* 1 (1862): 158–160, here 159n.

17 *Non mai noi intendemmo parlare degli atomi de' metafisici, ma di quelli come si comprendono dai fisici* [...]. Secchi, "Elettricità atmosferica," 159n.

18 See Angelo Secchi, *Sulla struttura delle macchie solari* (Rome: Tipografia delle Belle Arti, 1866), 3.

19 *Il concetto di materia e forma inteso al modo scolastico è bello e vasto, ma inteso al modo di Liberatore diviene gretto, inesatto e sciocco* [...]. APUG, FS 5.VIII.3.

20 On Cornoldi, see Mario Casella, "Cornoldi, Giovanni Maria," in *Dizionario biografico* 29 (Rome: Treccani, 1983); http://www.treccani.it/enciclopedia/giovanni-maria-cornoldi_(Dizionario-Biografico)/ (accessed June 3, 2018).

21 See Giovanni Maria Cornoldi, *I sistemi maccanico e dinamico circa la costituzione delle sostanze corporee considerati rispetto alle scienze fisiche* (Verona: Tipografia Vicentini e Franchini, 1864).

L'UNITÀ

DELLE

FORZE FISICHE

SAGGIO

DI FILOSOFIA NATURALE

DEL

P. ANGELO SECCHI D. C. D. G.

PROFESSORE DI ASTRONOMIA,
E DIRETTORE DELL'OSSERVATORIO DEL COLLEGIO ROMANO;
MEMBRO DEL COLLEGIO FILOSOFICO DELL'UNIVERSITA' ROMANA,
DELLA SOC. ITAL. DE' XL, DELL'ISTITUTO DI FRANCIA,
DELLA SOC. R. DI LONDRA EC. EC.

ROMA
TIPOGRAFIA FORENSE
1864

FIGURE 9.1 Title page of the first edition of Secchi's treatise on the unity of physical forces. Courtesy of Specola Vaticana.

where he vehemently attacked Secchi, Tongiorgi, and Palmieri. Cornoldi accused Secchi of rejecting Aristotelian physics and adopting the Epicurean atomic theory in physics and Cartesian vortices in astronomy. It was, in essence, an accusation of atheism, implying the denial of Transubstantiation. Behind these allegations, there was the need of the neo-Thomists to defend Catholic doctrine in respect of the new scientific theories, and their fear that the rapid changes brought about by progress in physics could undermine its foundations. Cornoldi sent a copy of his booklet to Secchi: "You will find your name mentioned therein, but I could not abstain from so doing, due to your authority [in the field]. I am highly sorry that we do not agree in some important questions,"[22] he wrote in the accompanying letter.

Secchi asked his superiors for permission to reply to Cornoldi, considering that he had distorted the ideas contained in his book, which were neither Epicurean nor Cartesian;[23] and he prepared a long and detailed paper[24] where he replied to the criticisms point by point.[25]

However, that was just the beginning of an anti-modernism campaign, with Cornoldi, who was on the editorial board of *La civiltà cattolica*, as a leading figure. Some fundamentalists even published a petition to the First Vatican Council,[26] requesting that it close the Collegio Romano Observatory and asking it to tell Father Secchi to desist from continuing to use his telescopes, being "evil machines" that needed to be "powerfully exorcized," so as to be freed "from any Satanic influence." In their view, astronomy was ultimately a science that drew one away from God, and consequently, it should be neither taught nor practiced within the Catholic Church.[27]

---

22 *Troverà citato il suo nome, ma non mi sono potuto astenere di ciò fare attesa l'autorità di cui gode* [*in questo campo*]. *Mi dispiace altamente che non possiamo andar d'accordo in certe questioni di gran rilievo* [...]. Cornoldi to Secchi, Verona, December 16, 1864; AOAR, Fondo Secchi, I.E.2.1, 35.

23 See APUG, FS 5.VIII.6.

24 See APUG, FS 26.VII.C.

25 In order to prevent the spread of conflicts inside the Society, Superior General Beckx was forced to intervene in the controversy (see Agustín Udías, *Jesuit Contribution to Science: A History* [New York: Springer, 2015], 216); Cornoldi was also removed from his philosophy professorship in northern Italy (see Luciano Malusa, *Neotomismo e intransigentismo cattolico: Il contributo di Giovanni Maria Cornoldi per la rinascita del tomismo* [Milan: IPL, 1986], 110).

26 The council was convened in June 1868 and was held in four sessions, between December 1869 and July 1870; it aimed at confirming the condemnation of rationalism and materialism and resulted in defining the doctrine of the Catholic faith and papal infallibility *ex cathedra*.

27 See Leone Paladini, *Petizione ai molto reverendi Padri del Concilio per la soppressione dell'Osservatorio Romano* (Turin: Tipografia del giornale "Il Conte di Cavour," 1869).

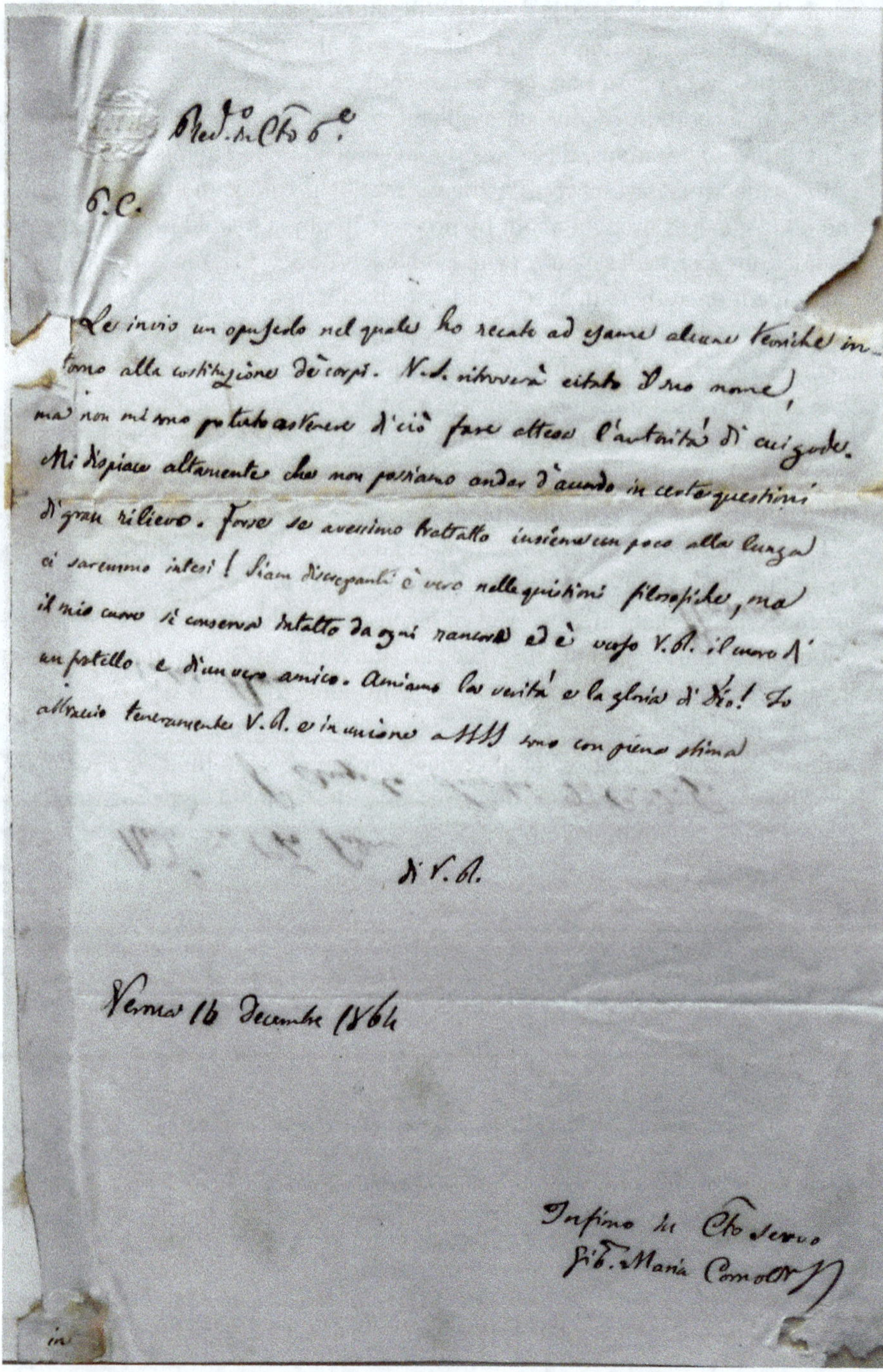

Rev.do in Cto P.e

P.C.

Le invio un opuscolo nel quale ho recato ad esame alcune teoriche intorno alla costituzione de' corpi. V.R. ritroverà citato il suo nome, ma non mi sono potuto astenere di ciò fare atteso l'autorità di cui gode. Mi dispiace altamente che non possiamo andar d'accordo in certe questioni di gran rilievo. Forse se avessimo trattato insieme un poco alla lunga ci saremmo intesi! Siam discrepanti è vero nelle quistioni filosofiche, ma il mio cuore si conserva intatto da ogni rancore ed è verso V.R. il cuore d'un fratello e d'un vero amico. Amiamo la verità e la gloria di Dio! Io abbraccio teneramente V.R. e in unione a SS. sono con pieno stima

di V.R.

Verona 16 decembre 1864

Infimo in Cto servo
Gio. Maria Cornoldi

FIGURE 9.2 Letter from Cornoldi to Secchi, which accompanied a copy of Cornoldi's booklet of 1864. Courtesy of INAF-OAR.

In 1874, Cornoldi obtained (according to some sources, by deception) the approval of Pius IX for the foundation of the Accademia Filosofico–Medica di San Tommaso d'Aquino (Academy of Medicine and Philosophy of St. Thomas Aquinas) in Bologna, which was intended "to spread true science," consistent with the principles of Saint Thomas, and to preserve youth "from false doctrine predominant in schools and universities."[28] In the first page of a printed pamphlet containing a list of members of the academy,[29] Secchi wrote the following annotation: "It is time to stop this imposture by which you swindle the world and seduce the people, and even dare to falsify the writings of the pope, much less me, in this iniquity."[30] In 1876, Cornoldi also established a periodical, *La scienza italiana* (Italian science), to support his side of the battle. Nothing could irritate Secchi more than this title, having the pretension of representing Italian science. He expressed his severe judgment to his friend Schiaparelli:

> Lately [...], a kind of cult, or something like that, has been formed; they want to resurrect the most ridiculous peripatetic [philosophy] [...]; certain recent articles, published by them, were really disgusting to me. What is certain is that their doctrines are neither Jesuit nor Catholic at all, but a fruit of their brain [...]; they consider me as anathema, and my poor conversation on the *Unity of Physical Forces* was represented by them as a work not only full of errors, but also dangerous for faith [...].[31]

Secchi was well aware that the neo-Thomists' attitude was dangerous to Catholicism. Among Secchi's papers, there is a draft of a letter to Cornoldi[32] in which Secchi accused his confrère of spreading ideas that were deleterious to the church:

28 Casella, "Cornoldi, Giovanni Maria."

29 See *Elenco dei membri dell'Accademia Filosofico–Medica di San Tommaso d'Aquino* (Bologna: Istituto Tipografico, 1876). Together with Cornoldi and Liberatore on the board, there was a long list of archbishops and bishops, professors of theology, philosophy, natural science, physicians, and surgeons.

30 *È ora di finirla con questa impostura con cui gabbate il mondo e seducete la gente, osate perfino falsare i rescritti del papa, lungi da me tale iniquità.* APUG, FS 5.VIII.7.

31 *Ultimamente* [...] *si è formata una specie di setta o quid simile, che vuol far risorgere la peripatetica più ridicola* [...] *certi ultimi articoli da loro pubblicati mi hanno schifato realmente. Quel che è certo si è che le loro dottrine non sono punto né le dottrine gesuitiche, né le necessariamente cattoliche, ma un frutto del loro cervello* [...] *essi mi stimano anatema da Loro, e la mia povera chiacchierata sull'Unità delle Forze Fisiche era per essere rappresentata come una opera non solo piena di errori, ma anche pericolosa per la fede* [...]. Secchi to Schiaparelli, November 1, 1876, quoted in Buffoni, Manara, and Tucci, *G.V. Schiaparelli, A. Secchi*, 238.

32 It is unclear if this is a reply to a letter by Cornoldi, expressing solidarity with Secchi, on the occasion of Bovio's speech (see Cornoldi to Secchi, Bologna, February 1, 1877; APUG, FS 22.III).

> I have long hesitated deciding if I should answer you. I am doing it because I believe that the time has come to abolish an abomination slithering through the church [...]; you tend to believe that nobody can be Catholic if he does not retain the *materia prima* and other twaddle of peripatetic [philosophy]. You have become a champion of this sect that does immense damage and that is the only one that justifies the great cry of the unbelievers—that the church is opposed to science. Because of such folly, we are insulted, mocked, and trampled [...]. It is time to put an end to this sect that pretends to exhibit papal infallibility to support its own personal truth, and hides its ignorance. [...] I myself am a victim of your foolishness! I tell you, it is time for it to end.[33]

It is unclear whether the letter was ever sent. However, it clearly shows the split between the scientific world and some ultra-conservative clerical circles, the consequences of which many religious scientists had to bear.[34]

Secchi described his paradoxical position, caught in a crossfire coming from outside and inside the church:

> I believed that, in order to stay untroubled in the world, it sufficed not to bother anyone and deal with things of distant worlds. But I see that, although I have done thus, I have not succeeded. One person tortures me with superlatives and unmerited praise, making me ridiculous; someone else wracks me by stretching scripture to make me appear blasphemous or a dullard. I mean, one person slams a censer on my nose to his advantage, while another drags me through the mud! [...] While some see disbelief and atheism in my writings, others see an exalted theology that falsifies physics to support the Bible [...]. One person complains about not finding therein the discoveries he expected, another does not find

33 *Ho esitato molto se dovessi risponderle. Lo fo perché credo che sia venuto il tempo di abolire una infamia serpeggiante nella Chiesa!* [...] *essi tendono a far credere che nessuno possa esser cattolico se non tenga la materia prima e le altre scempiaggini del peripateticismo. Ella si è fatto campione di questa setta che fà danni immensi ed è la sola che giustifica il gran detto degl'increduli = che la chiesa osteggia la scienza = Per queste loro follie noi tutti siamo insultati, derisi e calpestati* [...] *E' ora di finirla con questa setta che pretende stendere l'infallibilità pontificia a sostenere le loro verità personali, e coprire la loro ignoranza.* [...] *Io stesso sono vittima delle vostre insensataggini! Vi dico, è ora di finirla.* Secchi to Cornoldi, [Rome], s. d. [1877]; APUG, FS 9.VII.

34 The accusations of heresy against Secchi and other opponents of neo-Thomist fundamentalism were finally eased by Pius IX, who, in response to the controversy over hylomorphism inside and outside the Society of Jesus, issued an official document in 1877 on the freedom to assume philosophical attitudes different from Scholasticism (see Malusa, "Palmieri, Domenico").

> therein the physics of St. Thomas. To the latter, I will only say that physics has progressed a little since St. Thomas, and if St. Thomas had been in our times, he would not have adopted the physics that he did adopt but the physics that is in use in schools nowadays, just as he adopted then the one in use in those days. These gentlemen want me to pass for a metaphysician, which I neither am nor claim to be […]. Despite its progress, however, science has not been able to get rid of God, and those who hoped that science would, as well as their successors, will never obtain this.[35]

Secchi's views are surprisingly modern: a theology that is neither drawing on nor feeding from the spirit of the times, but instead remains distant and disconnected from its historical context is a "disembodied" and sterile theology.

However, that was not the position adopted by the church. Tongiorgi's and Secchi's early deaths spared them from seeing the triumph of neo-Thomism with the accession to the papacy of Leo XIII. The new pope had the Jesuits align the teaching in their schools to Scholastic philosophy, which led to the purge of non-Thomist professors like Palmieri[36] at the Collegio Romano. Moreover, with the encyclical *Aeterni Patris* (1879), Leo XIII indicated that the teaching of philosophy in Catholic schools was to be conducted following the metaphysics and cosmology of St. Thomas.[37]

35 *Io mi credeva che per star quieto al mondo bastasse non dar fastidio a nessuno e occuparsi delle cose de' mondi lontani, ma vedo che anche così non vi riesco. Chi ti strazia con elogi superlativi e immeritati che ti rendono ridicolo, chi ti dilania stiracchiando le scritture per farti comparire un empio o un babbeo. Insomma, chi ti sbatte il turibolo sul naso per trarne suo vantaggio, chi ti strascina nel fango!* […] *Mentre alcuni vedono l'incredulità e l'ateismo nei miei scritti, altri vi vede invece un'esaltata teologia che falsifica la fisica per appoggiare la bibbia* […] *Chi si lamenta di non trovarvi le scoperte che aspettava, chi non vi trova la fisica di S. Tommaso. A questi dirò solo che la fisica dopo San Tommaso ha camminato un poco, e che se S. Tommaso fosse stato a' tempi nostri, non avrebbe adottato la fisica che adottò ma avrebbe preso quella adesso in uso nelle scuole ai tempi nostri, come allora prese quella in uso a tempo suo. Questi signori mi vogliono far passare per un metafisico, mentre io né lo sono né pretendo di esserlo* […]. *Coi suoi progressi però la scienza non è arrivata a fare a meno di Dio, né quelli che speravano che la scienza vi arrivi avranno mai, né essi né i loro successori, questo.* APUG, FS 23. I. 5.

36 See Malusa, *Neotomismo e intransigentismo cattolico*, 258–261, 278–280.

37 It is important to note, however, that this philosophy cannot really *be accused of being opposed to the advance and development of natural science.* The Scholastics *well understood that nothing was of greater use to the philosopher than diligently to search into the mysteries of nature*: St. Thomas, Blessed Albertus Magnus, and other prominent Scholastics *give large attention to the knowledge of natural things; and, indeed, the number of their sayings and writings on these subjects* […] *is by no means small* (from Louis Caruana, S.J., private communication, March 19, 2018).

3 Philosophical Controversies

In 1863, Renan published *La vie de Jésus*; following the method of historical rationalism, he proposed a reconstruction of the life of Christ from a purely historical perspective. The publication of this book ignited a debate on the divinity of Jesus Christ, denied by Renan, who defended the historical existence of Christ but only acknowledged the exemplary value of his teaching. Catholics replied with a triduum of prayers against blasphemy, and Secchi took a stand against Renan's conclusions in a long unpublished manuscript. According to Secchi, one "unmistakable bit of evidence" of the divine origin of Christianity is its duration. It was born from humble beginnings but has lasted for ages, enduring persecutions, falls of empires, changing cultures, and so on: "If Jesus of Nazareth [were] but a talkative agitator of people, if he [were] nothing but an impostor, a Cagliostro," then "his disciples would have been scattered like sheep once they lost their shepherd. [...] If this were a congregation of fools who follow a misleading magician," they would scarcely have endured prolonged and cruel persecutions. Secchi considered the conversion of St. Paul a significant example:

> The change of mind in a man with the temperament of St. Paul is a greater prodigy than a dead man's resurrection. There could be some kind of illusion in this latter case, not in the former. A top-rank man in culture and society who fiercely hounds a cult and then becomes its apostle, not gradually but in an instant [...], this is a marvel [...]; a man amenable to reason in such a case does not give in on a whim, but only upon meeting a superior force.[38]

After recalling how Christianity has defied history and its changes of civilizations and cultures, inspiring great thinkers and laying the groundwork for justice and coexistence of civil society, Secchi delivered an attack on the French philosopher without mentioning him by name:

> The miracle of the preservation of Christianity has not yet ceased, [and] for a pathetic man to clutch at straws in distorting and misinterpreting

38 *La mutazione di maniera di pensare in un soggetto della tempra di Paolo è prodigio maggiore che la resurrezione di un morto. In questa vi può essere illusione, ma non in quella. Un uomo di alta sfera nella scienza e nella società che dal perseguitare accanitamente una setta, passa ad esserne l'apostolo, e ciò non a gradi a gradi ma in un istante per un colpo che anche fisicamente l'atterra l'annienta, questo è un prodigio* [...] *l'uomo suscettibile di ragione a un tal colpo non cede per capriccio, ma solo per forza superiore.* APUG, FS 24.VI.

> the scriptures, a crowd of sublime intellectuals glories in humiliating the homage of his labors and the fruit of his studies. [...] All those who abandon the faith do so only for base emotional reasons [...] or for greed and speculation.[39]

The Jesuit therefore concluded that, despite "the handful of miserable men trying to destroy the work of centuries founded by God [...], *portae inferi non prevalebunt* [the gates of hell will not prevail]."[40]

This text was considered by Secchi's assistants as a proof of his convictions about faith and a denial of the materialistic views of which he was shortly to be charged. Renan was highly popular and visited many scientists of his time; in 1875, for example, he was invited to attend the Congress of Italian Scientists. During the congress, Secchi kept his distance from the French philosopher so as not to reignite the accusations of heresy that were still circulating against himself (see Chapter 10).

It is interesting to note that, later, another of Secchi's books, *Die Sonne* (1872), which was the German translation of *Le soleil*, was studied by materialist philosopher Friedrich Engels (1820–95), who mentioned and quoted Secchi in many fragments of his unfinished work *Dialectics of Nature*, which he started to write in 1873. Engels included Secchi among the believer scientists who did not find an appropriate place for God in nature:

> God is nowhere treated more poorly than by the natural scientists who believe in him. [...] What God has had to suffer at the hands of his defenders! [...] Father Secchi bows him out of the solar system altogether, with all canonical honors it is true, but nonetheless categorically for all that, and he only allows him a creative act as regards the primordial nebula.[41]

---

39 *Il prodigio della conservazione del cristianesimo non è ancor cessato, per un infelice che si arrabbatta a falsare scritture e a stravolgerne il senso, una detta di più sublimi intelletti si gloria di umiliare l'omaggio delle sue fatiche e il frutto de' suoi studi.* [...] *tutti quelli che abbandonano questa fede non per altro lo fanno che o per vile passione* [...] *o per avarizia e speculazione.* APUG, FS 24.VI.

40 APUG, FS 24.VI.

41 *Gott wird nirgends schlechter behandelt als bei den Naturforschern, die an ihn glauben.* [...] *Aber was hat Gott von seinen Verteidigern erdulden muessen!* [...] *P*[*ater*] *Secchi komplimentiert ihn, zwar mit allen kanonischen Honneurs, aber darum nicht weinger kategorisch, aus dem Sonnensystem ganz heraus und erlaubt ihm nur noch in Beziehung auf den Urnebel einen Schoepfungsakt.* Friedrich Engels, *Dialektit der Natur* (Berlin: Hofenberg, 2017), 175–176.

Engels cited many quotes about nebulae, stars, and the solar system from Secchi's book, leaving some notes incomplete, and making use of them in the second part of the introduction. This raises a question about the impact of Secchi's scientific texts in the cultural context of that age—an aspect of Secchi's life that would clearly benefit from further research.

4 Scientific Quarrels

Secchi entered into scientific controversies on several fronts, often to defend himself from the charge of taking credit for other people's discoveries: such controversies thus encroached on a personal level. This notoriety accompanied him from the beginning of his career and quickly spread across the Alps, especially in the United Kingdom, where Secchi had scientific rivals and his reputation was severely damaged by those who sought to prevent his work being published in English (see below). In France, on the contrary, the controversies remained limited to the scientific context, and while the disagreements he had (especially with Faye and Janssen) were marked by antipathy or resentment, they stimulated a scientific debate on some unsettled points of solar physics. In some cases, an anti-Jesuit and/or anti-Catholic spirit that permeated some scientific contexts of the time may have influenced some of this opposition.

4.1 *In France: Faye, Radau, Janssen*

In 1852, Secchi's early studies on solar radiation provoked a reaction from some academics in Paris. When Faye communicated the results of these studies, Arago claimed that Secchi was taking credit for his own contributions in this field. Secchi reported:

> From the words of Mr. Arago some newspapers took the occasion to label this research as an usurpation of the scientific rights of others, which certainly would have been at the very least a discourtesy from anyone working at this observatory, which is much obliged to Mr. Arago, in particular for the generous and warm welcome he gave to the past director, Fr. de Vico.[42]

---

42 *Dalle parole del sig. Arago qualche giornale prese occasione di riguardare queste ricerche come una occupazione degli altrui scientifici diritti, che certo sarebbe stata almeno scortesia in una persona addetta a questo osservatorio, tanto obbligato al signor Arago, specialmente per le generose e cordiali accoglienze prestate da esso al passato direttore P.F. de Vico.* Secchi, "III: Sull'intensità del calore nelle varie parti del disco solare," lxiin.

Secchi then defended his good faith, claiming to have been unaware of those earlier studies.[43] However, this episode put the scientific community on guard and darkened the beginning of Secchi's career, with long-lasting results.

A truly vitriolic attack against Secchi came from France in 1867, at the time of the Exposition Universelle. The Prussian physicist Rudolph Radau (1835–1911) was charged by the Parisian journal *Le moniteur scientifique* (The scientific monitor) with writing a review of the instruments on display at the exposition. He seized this opportunity to attack Secchi vigorously on two points: the originality of the meteorograph and the validity of the physical–mathematical theory on which its operation was based, particularly that of the barometrograph. The tone used by Radau was offensive enough: on the first point, indeed, he accused Secchi of bad faith in taking credit for discoveries that he did not deserve, since other kinds of recording instruments had already been designed and built by other scientists. Even worse, on the second point he accused Secchi of serious miscalculations, implying that the work of the Jesuit was of little scientific value: "What value would it be possible to grant now to the work of Fr. Secchi, following these samples of his knowledge, his logic, and his good faith?"[44] The controversy carried into the Académie des Sciences and resulted in a long back-and-forth in which Secchi refuted the accusations, point by point, explaining that he did not intend to steal anything, defending the originality of his work, and proving the empirical rather than theoretical nature of his results. Thanks to the mediation of Abbé Moigno, the polemical tone was muted, but each of the two scientists held their ground, and Secchi did not hesitate to bring up the dispute with Radau in a later booklet describing the meteorograph.[45]

This diatribe essentially derived from two different approaches: See Brenni, "Il meteorografo di Padre Angelo Secchi," 229. while Radau gave privilege to theoretical and mathematical analysis with detailed calculations, Secchi preferred an empirical approach, looking in particular at the final result—namely easing the tasks of recording the data and comparing various series of meteorological parameters. Indeed, Secchi's pragmatic approach can be seen in all of his works: he mistrusted purely theoretical analyses and always put objective facts and observations first, continually insisting on the reality of the phenomenon as a starting point for its explanation and interpretation.

---

43 See Angelo Secchi, "Extrait d'une lettre à M. Arago," *Comptes rendus de l'Académie des Sciences* 34 (1852): 949–950.

44 *Quelle valeur il sera possible d'accorder désormais aux travaux du P. Secchi, après ces spécimens de son savoir, de sa logique et de sa bonne foi?* Rudolph Radau, "Académie des Sciences," *Le moniteur scientifique* 9 (1867): 888–891, here 890, quoted in Brenni, "Il meteorografo di Padre Angelo Secchi," 228.

45 See Angelo Secchi, *Descrizione del meteorografo dell'Osservatorio del Collegio Romano* (Rome: Tipografia delle Belle Arti, 1870), 88–96.

Such a vicious dispute can only be explained if it was informed not only by scientific but also by personal animosities, perhaps a veiled anti-clericalism in Radau.[46] This was not the only such instance, as we shall see below. Secchi's scornful judgment about Radau was well described in a letter that he sent to Tacchini in 1868, who had read a "ridiculous"[47] article by Radau against Italian astronomers:

> Radau is a bad guy. He is a poor, hungry German, who came to Paris to eat and peddle the science that earns him no bread in Germany. He is paid as many other scribblers to flatter one and discredit another, and does this job willingly and with success, because he is superb and uncommonly malignant. He is puffed up with a lot of pride because some academy member has him do some calculations that he then passes off as his own, and some professor has tried to correct his blunders of physics. Hence he believed himself to be important, and superior to the academicians and professors. He started in on me, but I did show three or four big blunders that he had made, and the academy has not given much space to his communications. He is very petulant and dirties scientific journals with his scribbles, and since he works a lot and with scarce remuneration, and he barely gets breakfast from one and lunch from another, then everybody calls him, and therefore he has credit [...] at the dupes! [...] I will not take the trouble to read his insults. [Italian astronomy] will not turn red in the face because of a pedantic German.[48]

Much more academic in nature was the dispute with Faye on the cyclonic nature of sunspots, which has already been mentioned (see Chapter 6). The

46 See Brenni "Il meteorografo di Padre Angelo Secchi," 229.

47 See Tacchini to Secchi, Palermo, 2 March 1868, quoted in Chinnici and Gasperini, *Alle origini dell'astrofisica italiana*, 85.

48 *In quanto al Radau esso è un cattivo soggetto. Esso è un povero tedesco affamato venuto a Parigi per mangiare e spacciare la scienza che non dà pane in Germania. Esso è pagato come tanti altri fogliettantisti per adulare l'uno e screditare l'altro, e fa questo mestiere molto volentieri e con successo, perché è superbo e maligno in modo non comune. Esso è salito in molta superbia perché qualche membro dell'Accademia lo adopera in far calcoli che poi spaccia per suoi, e qualche professore lo ha adoperato a correggere i suoi sfondoni di fisica. Quindi esso si crede un gran chè, e superiore agli accademici e ai professori. Avea cominciato a pigliarsela con me, ma io ho fatto vedere tre o quattro grossi sfondoni che ha fatto, e l'Accademia non ha dato seguito alle sue comunicazioni. È petulante assai e sporca tutti i giornali scientifici colle sue chiacchiere, e siccome lavora molto e per poco, e gli basta avere da chi la colazione da chi il pranzo, quindi tutti se ne servono, e perciò esso fa l'opinione* [...] *presso i gonzi!* [...] *io non mi prenderò la briga di leggere le sue insolenze.* [*L'astronomia italiana*] *non arrossirà in faccia a un pedante tedesco.* Secchi to Tacchini, Rome, March 4, 1868, quoted in Chinnici and Gasperini, *Alle origini dell'astrofisica italiana*, 86–87.

ensemble of publications related to this debate was published by Tacchini in an appendix to the second volume of the *Memorie della Società degli Spettroscopisti* (1873), which contained the entire collection of related articles, and was later reviewed by Young in his book on the Sun.[49] Though they strongly disagreed, Faye and Secchi restricted their reciprocal criticisms to the science, keeping on a plane of mutual respect; in 1878, in a report on Secchi's work, Faye wrote: "Each time that physics has created a new instrument or a new method, Secchi was among the first ones to seize it, often he was the first one."[50]

Whereas the disputes with Faye focused on theories about the solar surface and interior, those with Janssen centered on spectroscopic observations. Janssen entered the same field of solar physics and used the same spectroscopic and photographic techniques as Secchi. After the incident of 1862–63, when Secchi had welcomed Janssen "as a colleague"[51] before banning him from the Collegio Romano Observatory (see Chapter 6), there was residual resentment between the two astronomers. Janssen judged that Secchi had acted improperly toward him and joined his friend Lockyer (see below) in discrediting the Jesuit astronomer.[52] They seldom clashed openly, however.[53] The Jesuit considered his French colleague a beginner in solar observations, without the necessary expertise.[54] However, he appreciated Janssen's efforts in the field of solar photography and hoped the French astronomer could carry out regular photographic monitoring of the Sun. On this point, he wrote to Tacchini: "I wish him [i.e., Janssen] heartily [to continue his photographic work] because [his photographs] are really important and could be very useful, provided they are taken regularly and well."[55]

4.2 *In Italy: Zantedeschi, Volpicelli, Respighi*

In Italy, outside of Rome, Secchi's main dispute was with Abbot Francesco Zantedeschi (1797–1873), a physicist and meteorologist in Padua. Starting in 1862, Secchi published several of Zantedeschi's works in the *Bullettino*, especially concerning the origin of atmospheric electricity. Nevertheless, in 1866

49 See Young, *Sun*, 181–187.

50 Aubin, "Orchestrating Observatory, Laboratory, and Field," 630.

51 *Comme un confrère*, Janssen, quoted in Launay, *Un globe-trotter de la physique céleste*, 38.

52 See, for instance, Launay, *Un globe-trotter de la physique céleste*, 40–41, 48–49.

53 See Secchi to Tacchini, Rome, April 13, 1872, quoted in Chinnici and Gasperini, *Alle origini dell'astrofisica italiana*, 192.

54 See, for instance, Secchi to Tacchini, Rome, May 28, 1877, quoted in Chinnici and Gasperini, *Alle origini dell'astrofisica italiana*, 458.

55 *Gliel'auguro di cuore perché* [*le sue fotografie solari*] *sono veramente importanti, e possono rendere grandi servizi, ma alla condizione che siano costanti, e ben dirette*. Secchi to Tacchini, Rome, May 3, 1877, quoted in Chinnici and Gasperini, *Alle origini dell'astrofisica italiana*, 451.

the Jesuit much resented the publication of an anonymous booklet[56]—almost certainly written by Zantedeschi—that accused him of bad faith and "scientific theft": "The purpose of this paper is to prove me ignorant of things done in Italy about the prediction of squalls, and moreover, calls me unfair, freebooter, and pirate because, it says, I usurped Zantedeschi's findings."[57]

Secchi and Zantedeschi were studying the same phenomena, both carried out research on electricity and spectroscopy, and the two men often had differences of opinion; but up to then, their conflict had been limited to refuting the arguments of the other in scientific publications. This booklet now portrayed Secchi in a bad light because he was accused of having omitted, knowingly or otherwise, mention of the work of Zantedeschi on the matter. Ignorance of the work of others and disloyalty were both unbearable shames to Secchi.

Secchi recognized that Zantedeschi was the anonymous author and replied with an article in the *Bullettino* where he disputed Zantedeschi's theories and accused him of hiding the authorship of the booklet. Zantedeschi replied via a long essay accusing Secchi of making "free and malignant assertions [...] distilled from his brain to make the public believe that they are my inaccuracies."[58] He also recalled that Secchi himself, in 1862, had appealed to his authority in the field, inviting him to publish some works in the *Bullettino*. The controversy faded away in the following years, probably because of the increasing blindness of the scientist in Padua.

In 1868, further hostilities erupted with physicist Paolo Volpicelli (1804–79) over a matter of a purely historical nature, relating to the assessment of the time of the complete blindness of Galileo. In July 1867, the French mathematician Michel Chasles announced the discovery of some unpublished letters by Blaise Pascal (1623–62) containing the universal gravitation theory, which consequently would mean that it should not be attributed to Newton. The finding resulted in an international controversy, and other letters by Newton and Galileo were made known by Chasles. Secchi wrote a letter of protest to the Académie des Science, suggesting that the letters had been taken out of

---

56 *Delle dottrine del R.P. A. Secchi intorno ai presagi delle meteore e delle burrasche e di quelle del prof. Zantedeschi con documenti storici* (Padua: Tipografia di A. Bianchi, 1866).

57 *Lo scopo di questo scritto è quello di dimostrare me ignorante delle cose fatte in Italia sulla previsione delle burrasche, e di più sleale, predone e pirata per essermi, dice esso, usurpato le scoperte del Zantedeschi.* APUG, FS 5.VIII.8.

58 *Asserzioni gratuite e maligne* [...] *distillate dal suo cervello per far credere al pubblico che sono inesattezze mie* [...]. Francesco Zantedeschi, *Risposta documentata del professore Francesco Zantedeschi all'articolo del p. A. Secchi inserito nel Bullettino meteorologico dell'Osservatorio del Collegio Romano: Vol. . pag. 107, nº di Ottobre 1866 intorno ai presagi delle meteore e delle burrasche con documenti storici* (Padua: Tipografia di A. Bianchi, 1866), 14.

context.[59] The Jesuit questioned the authenticity of the letters because they referred to writings by Galileo at a time when he would no longer have been able to write as he had gone completely blind.[60] Secchi's letter was considered offensive to Chasles by many French newspapers; even Moigno judged that Secchi had been unwise to enter into the question and to have used undiplomatic language,[61] and the director of the Collegio Romano Observatory was persuaded to give explanations to the academy. Secchi had entered the dispute as the result of his own research: he had examined the letters in question (acquired at great expense by the French mathematician), and compared the signatures of Newton and Galileo with the originals, finding many differences.[62] Meanwhile, Volpicelli stated that Secchi was wrong and that, at the time of the contested letters, Galileo had not yet completely lost his sight. Secchi replied to Volpicelli in some presentations at the Accademia Tiberina, using such provocative tones toward the Roman physicist that "a dispute [...], having becoming too lively, as usual, from [Secchi], was interrupted by Mr. President of the academy."[63] Volpicelli replied in kind, summing up the whole story in a three-hundred page monograph:

> We know by experience how Fr. Secchi is soft-spoken, how he is moderate in discussions especially about science; we also know, by experience, how kind and noble is his spirit with everybody, and especially with those who always praise his productions; therefore, we cannot doubt at all, that [his letters to Moigno, who disagreed with Secchi's conduct] could contain insults [...].[64]

---

59 See Angelo Secchi, "Observations sur les documents relatifs à Galilée qui ont été publiés par M. Chasles," *Comptes Rendus de l'Académie des Sciences* 65 (1867): 1018–1019.

60 See Angelo Secchi, "Sull'epoca vera e la durata della cecità di Galileo," *Giornale Arcadico di Scienze, Lettere ed Arti* 54 (1867): 197–248.

61 See Paolo Volpicelli, *Sull'epoca della completa cecità di Galileo: Risposta al Ch. e Rev. P. Angelo Secchi* (Rome: Tipografia delle Belle Arti, 1868).

62 In a communication to the Accademia Tiberina, Secchi explained that he had considered his duty to defend the verity of the historical facts, to be considered *no less sacred than religion* (*ho creduto dovere sostenere la storica verità de' fatti, la quale* [...] *deve essere non meno sacra che la religione*), also adding that scientists should care to preserve history of science from any "corruption" (*E' dovere di chi rappresenta la scienza* [...] *conservare intemerata e pura da ogni macchia la storia di quella ereditata dai nostri maggiori*). See Secchi, "Le recenti scoperte astronomiche," 195.

63 *Nacque un dibattimento, che per essere divenuto troppo vivo da parte, secondo al solito, del nominato astronomo e fisico, fu interrotto dal sig. presidente dell' accademia.* Volpicelli, *Sull'epoca della completa cecità di Galileo*, 5.

64 *Conosciamo tutti per prova quanto il p. Secchi è pacato, quanto è moderato nelle discussioni specialmente se scientifiche; conosciamo altresì per prova, quanto sia gentile e nobile l'animo suo con ciascuno, e specialmente con quelli che sempre hanno lodato le sue produzioni;*

This time, Secchi chose not to reply directly to his opponent, but instead addressed a letter[65] to the *Giornale arcadico*, whose pages had hosted the controversy, in order to clarify his position. Within a few months, however, the alleged letters from Pascal and others were recognized as fakes,[66] and the dispute died out, leaving a strained relationship between the two scientists.

The conflict continued inside the Pontificia Accademia dei Nuovi Lincei, the prestigious institution founded in the seventeenth century under the name of Accademia dei Lincei, which, after a period of inactivity, had been re-founded by Pius IX in 1847. Volpicelli was secretary of the academy for about thirty years from its re-foundation, and Secchi, a member of the same academy from 1850, accused him of mismanaging its academic activities and even of abuses in carrying out his role. In 1868, Secchi wrote a letter with a list of these abuses to the president of the academy, Benedetto Viale Prelà (1796–1874), formulating proposals "to see the academy flourish" and to restore it to its previous prestige. Secchi protested that the publication of the proceedings was constantly delayed because the secretary used them to publish his lengthy papers without taking into account the page limit allocated to each member; there was no system for available pages, and "members who wished to publish [some works]" were forced "to beg right to use their [free] pages from members who were not using them";[67] the minutes were written negligently, on random sheets, without making the corrections required, and the names of substitutes were included or excluded at the whim of the secretary, just to form a group of three: "This is a real insult," Secchi claimed, "especially if you go around by saying that this guy's [name] was put in just to fill the blanks, being sure that he will not be chosen."[68] Moreover, the Carpi Award[69] had not yet been assigned and the

---

quindi è che non possiamo punto dubitare, che [*le sue lettere*] *contenga*[*no*] *cose che offender possano* [...]. Volpicelli, *Sull'epoca della completa cecità di Galileo*, 51.

65 See Angelo Secchi, "Sull'ultima pubblicazione del prof. Volpicelli intorno alla cecità del Galileo: Lettera del p. Secchi al ch.mo prof. cav. Betti uno dei compilatori del Giornale Arcadico," *Giornale arcadico di scienze, lettere ed arti* 57 (1868): 233–246.

66 The deception was revealed when a letter supposedly written by Galileo was compared with original manuscripts in Florence; Secchi also remarked that the fake letters were written in different handwriting and hypothesized the existence of accomplices of the falsifier, Denis Vrain-Lucas (1818–83); see Angelo Secchi, "Lettera a Salvatore Betti intorno alla condanna del falsario degli autografi di Galileo Galilei ec.," *Giornale arcadico di scienze, lettere ed arti* 207, no. 62 [1869 (*sic*; 1870)]: 170–171.

67 *Non obbligare i socii che stampano a mendicare dai socii che non stampano il diritto di servirsi de' loro fogli* [...]. APUG, FS 5.VIII.1.

68 *Questo un vero insulto, tanto più se si va dicendo che quel tale si mette per riempire, sicuro che non verrà scelto.* APUG, FS 5.VIII.1.

69 An annual grant of one hundred ecus, established on the legacy of the mineralogist Pietro Carpi (1792–1861), to award selected scientific works.

call not yet announced, perhaps because the funds for it had been improperly used by the secretary for other purposes.

Secchi's complaints bore little fruit, however, and two years later, Secchi finally asked to be excused from the task of reviewing the academy's budget, "having determined to abstain [himself] from taking part in any act of academic committees," and considering that the review was useless because the secretary was doing what he wanted, contrary to the decisions of the academy itself. However, Secchi's protests were often misinterpreted in high places as unnecessarily disruptive. In fact, however, because of his sense of propriety, righteousness, and morality, he did not resign and refused to compromise.

As another example, when at last the Carpi Award was published, Secchi resigned from the competition commission in order to avoid being responsible for infringements by the commission. The confidentiality of the names of those in contention for the prize had been violated, and Secchi recognized one candidate as incompetent: the competitor was the grandson of the award testator,[70] and the essay that he presented, Secchi said, was certainly not

> his flour sack. This young man doesn't know how to calculate with logarithms, how to deal with spherical triangles, much less how to operate with least squares. He does not know German, and German authors are cited! The work is corrected by another hand,[71] whose handwriting I think I recognize.[72]

Secchi decided to minimize his interactions with Volpicelli. Meanwhile, other controversies were beginning to emerge.

It is clearly understandable that no scientist in Rome, especially another astronomer, could avoid confrontation with Secchi's brilliant personality. This was the case with Lorenzo Respighi, former director of the Observatory of Bologna, who was called in to direct the University Observatory of the Campidoglio in Rome in 1865, on the advice of Calandrelli, his predecessor and teacher.

---

70 He probably was the young Serra-Carpi, whom Secchi knew well, since he had collaborated with Secchi in making weather charts for the meteorological network (see Chapter 4, note 27).

71 Secchi recognized the handwriting of German geophysicist Filippo Keller, assistant of Volpicelli.

72 *Certamente farina del suo sacco. Questo giovane non sa calcolare coi logaritmi, non sciogliere triangoli sferici e molto meno lavorare ai minimi quadrati. Esso non conosce il tedesco e si citano autori tedeschi! Il lavoro è corretto da mano estranea di cui parmi conoscere il carattere* [...]. APUG, FS 5.VIII.9.

FIGURE 9.3 Portrait of Lorenzo Respighi. Courtesy of INAF-OAR.

Relations between Secchi and Respighi were often rather confrontational, according to the correspondence of other astronomers[73] as well as their publications, which often contained reciprocal criticisms of each other's work.

Secchi undoubtedly had a strong personality, and his outspokenness and self-regard could often irritate his colleagues. He often claimed to have arrived at other people's discoveries before they did; sometimes, these assertions were founded, sometimes they were a fiction. For this reason, Secchi was occasionally accused of wanting to highlight his own work to the detriment of others. His excessive tendency to stand out probably drew animosity toward him.

73 See, especially, Secchi's letters to Schiaparelli (Buffoni, Manara, and Tucci, *G.V. Schiaparelli, A. Secchi*) and Tacchini (Chinnici and Gasperini, *Alle origini dell'astrofisica italiana*) and Respighi's letters to Tacchini (ACREA, Fondo Tacchini).

Respighi was one of Secchi's main accusers. Nevertheless, it appears that he did not have the personality to address the Jesuit openly and, after a few clashes in which he took a beating, decided to spread accusations underground (see below).

At the beginning, the relationship between the two directors was good, one of mutual courtesy.[74] The instruments that Respighi purchased for the Campidoglio Observatory were delivered at the address of the Collegio Romano Observatory along with those of Secchi,[75] the two astronomers exchanged frequent visits, and Respighi was informed about the movements of his Jesuit colleague.[76] Secchi, for his part, is said to have facilitated the appointment of Respighi as director of the Campidoglio Observatory[77] and to have introduced him to the practice of spectroscopic observations,[78] but he later complained at the excessive secrecy of Respighi on the studies that he subsequently conducted.[79]

The first serious incident between the two astronomers occurred in January 1870, on the occasion of the establishment of Pontifical Geodetic Commission (see Chapter 4). The minister of commerce appointed Secchi to the presidency of the commission, arousing resentment from Respighi. The memory of that unfortunate episode was still alive in Secchi a couple of years later, when he wrote:

> I know what I have suffered for the Geodetic Commission from Respighi, who, resentful of not being the head, made a thousand impertinences toward me and spoke ill of me behind my back with my superiors, trying to denigrate and disrepute me. I generously gave up, considering him as good as me I wanted to install him as the head, but my superiors did not agree, and he, having apparently quieted down, later burst out into continuously contradicting all my work.[80]

---

74 It can be easily deduced from the correspondence between Secchi and Schiaparelli; for example, in 1866 Secchi informed Schiaparelli about the "painful news" of the sudden death of Respighi's wife (see Buffoni, Manara, and Tucci, *G.V. Schiaparelli, A. Secchi*, 40).

75 See, Respighi to Ertel, Rome, April 9, 1868; AOAR, Fondo Respighi, II.E.3.1.

76 See Respighi to Ertel, Rome, April 24, 1869; AOAR, Fondo Respighi, II.E.3.1.

77 See Secchi to Tacchini, Rome, April 13, 1872, quoted in Chinnici and Gasperini, *Alle origini dell'astrofisica italiana*, 192.

78 See Secchi to Tacchini, Rome, February 19, 1872, quoted in Chinnici and Gasperini, *Alle origini dell'astrofisica italiana*, 179; also Secchi to Schiaparelli, Rome, August 22, 1866, quoted in Buffoni, Manara, and Tucci, *G.V. Schiaparelli, A. Secchi*, 68.

79 See Secchi to Tacchini, Rome, March 10, 1870, quoted in Chinnici and Gasperini, *Alle origini dell'astrofisica italiana*, 114.

80 *So quanto ho patito da Respighi per la commissione geodesica, il quale piccato di non essere egli il capo, mi fece mille impertinenze e disse ogni male di me presso i superiori, cercando di*

Respighi was also probably deceived about being excluded by the Italian commission for the observations of the eclipse of 1870 (see Chapter 5) and probably expected to be supported by Secchi. Nevertheless, the director of the Collegio Romano Observatory often came out in support of Respighi's scientific credentials.[81] He proposed that Tacchini invite him to be a member of the Spectroscopic Society[82] (see Chapter 8); he also suggested that Respighi host the first meeting of the society on the premises of his observatory and invited him to send a delegate if he was unable to participate directly.[83] Behind this insistence there was probably also Secchi's belief that it was better to have Respighi as an ally rather than as an enemy.

Unfortunately for Secchi, Respighi was not exempt from that "invincible repugnance, which many people have, to play the role of the satellite, be it wrongly or rightly."[84] He was reluctant to accept the invitation "in fear that [his] person [might] not be welcome to all those who would [have] take[n] part in the same society, or even that [his] cooperation would be considered useless." Once the society's program had been discussed without his contribution and he had not been informed about it, he expressed his discomfort to Tacchini (see Chapter 8).

The conflict culminated in 1872. The bone of contention this time was the early use of the objective-prism for studying stellar spectra (see Chapter 6). Respighi was certainly the first to take up the original idea of von Fraunhofer; as early as 1868, he commissioned the construction of a glass prism to be placed in front of the objective lens of his equatorial telescope.[85]

Secchi, however, was the first to use this device in a systematic way, but—either on purpose or accidentally—he disseminated his results without

---

mettermi in mala vista e discredito. Io generosamente cedetti, lo pareggiai in tutto a me stesso, voleva metter lui alla testa, ma i superiori non vollero, ed egli, abbonacciato in apparenza ruppe poi una continua contraddizione a tutte le cose mie. Secchi to Schiaparelli, Rome, April 20, 1872 quoted in Buffoni, Manara, and Tucci, *G.V. Schiaparelli, A. Secchi*, 200–201.

81 See Angelo Secchi, "Ricerche nelle protuberanze solari," *Nuovo cimento* 3 (1870): 217–220, here 217.

82 See Secchi to Tacchini, Rome, July 12, 1871, quoted in Chinnici and Gasperini, *Alle origini dell'astrofisica italiana*, 139.

83 See Secchi to Respighi, Rome, September 28, 1871; AOAR, Fondo Respighi. Indeed, Respighi's participation was unnecessary given that the Collegio Romano Observatory would provide the data from the Rome station, but it was still desirable, because he possessed a different instrument and had already carried out similar studies. Secchi's insistence that Respighi should be a member of the society appears to have been an act of courtesy.

84 *Invincibile ripugnanza, che hanno molti a fare la parte del satellite, dove a torto, od a ragione.* Lorenzoni to Tacchini, Follina [Treviso], September 19, 1871; ACREA, Fondo Tacchini.

85 See Respighi to Merz, Rome, February 3, 1868; also Respighi to Ertel, Rome, June 21, 1868; AOAR, Fondo Respighi, II.E.3.1.

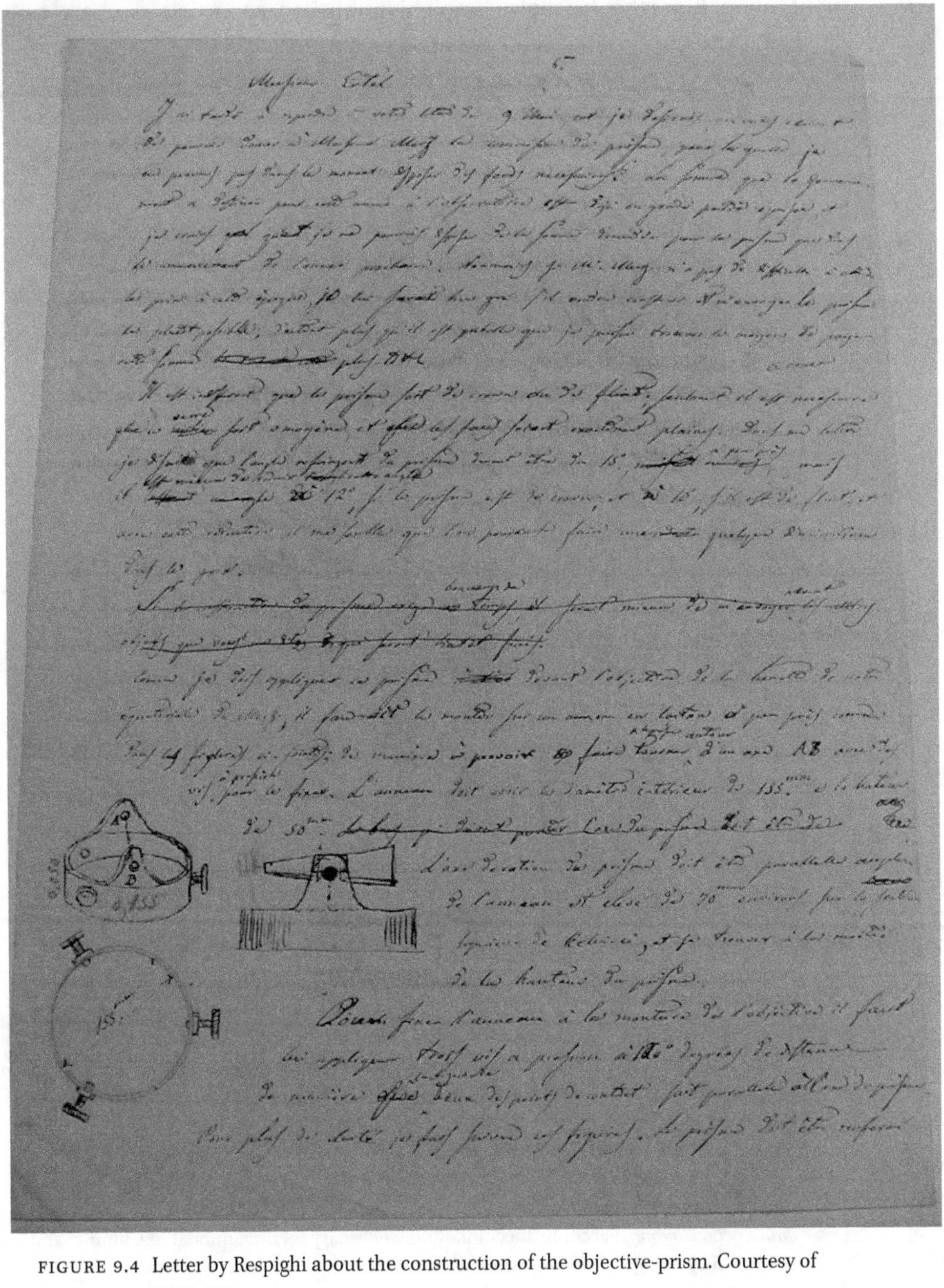

FIGURE 9.4 Letter by Respighi about the construction of the objective-prism. Courtesy of INAF-OAR.

mentioning Respighi. In reply, Respighi published an article[86] where he challenged Secchi's results and his claims to have made some discoveries for the first time. The Jesuit unloaded on Respighi in a letter to his friend Tacchini: "He [i.e., Respighi] wrote a big long memoir […] to prove that <u>all</u> that I said was bullshit. Hooray! I think he is a man disoriented."[87] Tacchini was not surprised: he had known Respighi for many years and had already remarked in him "an excessive spirit of opposition."[88] Secchi was genuinely wounded by the attitude of Respighi, whom he judged "rude and full of jealousy" against him; he even suspected that Respighi may have engineered the failure of the expedition to the Fréjus Tunnel (see Chapter 5)[89] and accused him of bad faith in claiming

> the curious method of using the prism in front of the slit, which I discovered, he was saying that he had discovered as well, but being a trifle he did not care! Even if he used it, he clearly did not understand it and therefore he kept silent. But I think this is a blatant lie.[90]

Tacchini had to use all his diplomacy to persuade Respighi to remain in the Spectroscopic Society and, at the same time, to mediate the conflict with Secchi, who did not want "to lick his paws," fearing that he could use the society to spread further invective: "I declare, frankly speaking, as customary for Reggio's people, that if he comes back, I cannot remain. If you want to publish his stuff, you are the boss, but then I will send my miserable [works] elsewhere."[91]

---

86 See Lorenzo Respighi, "Sulle osservazioni spettroscopiche del bordo e delle protuberanze solari, fatte all'Osservatorio dell'Università Romana sul Campidoglio," *Atti della Reale accademia dei Lincei* 25 (1872): 247–316.

87 *[Respighi] Ha scritto un lungo memorione* […] *per dimostrare che <u>tutto</u> quello che ho detto io è minchioneria. Evviva! Mi pare un uomo disorientato.* Secchi to Tacchini, Rome, January 20, 1872, quoted in Chinnici and Gasperini, *Alle origini dell'astrofisica italiana*, 168.

88 *Uno spirito eccessivo di opposizione.* Tacchini to Secchi, Palermo, January 27, 1872, quoted in Chinnici and Gasperini, *Alle origini dell'astrofisica italiana*, 171.

89 See Secchi to Tacchini, Rome, April 13, 1872, quoted in Chinnici and Gasperini, *Alle origini dell'astrofisica italiana*, 192.

90 *Il modo così curioso di usare il prisma avanti alla fessura, che io ho scoperto, esso và dicendo che pure l'avea scoperto, ma essendo una inezia non se n'era curato! Anche ammesso che l'abbia usato, si vede che non l'ha capito e perciò ha taciuto. Ma io credo che questa sia una bugia bella e buona.* Secchi to Tacchini, Rome, February 19, 1872, quoted in Chinnici and Gasperini, *Alle origini dell'astrofisica italiana*, 179.

91 *Io non approvo di andargli a leccare le zampe* […] *Vi dichiaro pertanto collo schietto parlare Reggiano che se lui rientra, io non vi posso restare. Se voi volete metter le cose sue voi siete padrone, ma io allora mi volterò altrove per le mie miserie.* Secchi to Tacchini, Rome, July 6, 1872, quoted in Chinnici and Gasperini, *Alle origini dell'astrofisica italiana*, 202.

Tacchini had to send him a long and patient explanation by letter,[92] which succeeded in preventing his defection: "Your considerations make me take the following resolution: I will submit my works until Respighi sends you criticism or invective. As soon as something of this kind appears in the memoirs, I immediately retire—I want to be left quiet."[93]

Needless to say, Respighi indeed attacked Secchi again, in a very contentious article[94] about the distribution of solar prominences, where he claimed to be the first to have carried out this kind of research. Now, while monitoring the solar limb in 1871 and preparing the work plan of the Società degli Spettroscopisti, Tacchini proposed a classification of prominences,[95] which was then adopted by Secchi. Respighi wrote:

> If [Fr. Secchi] had bothered to read, with some attention, the papers on solar prominences that I have published before Fr. Secchi and Mr. Tacchini had started their regular observations [...], he would be convinced that [...] a certain class of solar prominences had already been noticed and clearly described by me [...]. Without even taking the trouble to read these memoirs, Fr. Secchi could easily assure himself, by the simple inspection of my 140 drawings of the solar chromosphere already published at this time, that the filamentous structure of prominences and upper parts of the photosphere are distinctly reproduced therein, despite the smallness of the scale. [...] Indeed, Fr. Secchi must know that my observations, established long before his own, were regularly continued [...] from October 26th, 1869 to July 23rd, 1871 [...].[96]

---

92 See Tacchini to Secchi, Palermo July 10, 1872, quoted in Chinnici and Gasperini, *Alle origini dell'a strofisica italiana*, 203–205.

93 *Le vostre riflessioni mi fanno prendere la risoluzione seguente = manderò i lavori finché Respighi non manderà da stampare critiche o invettive. Alla prima produzione di questa specie nelle memorie, mi ritirerò subito = Io voglio stare in pace.* Secchi to Tacchini, Rome, July 12, 1872, quoted in Chinnici and Gasperini, *Alle origini dell'astrofisica italiana*, 205.

94 See Lorenzo Respighi, "Réponse à une note précédente du P. Secchi sur quelques particularité de la constitution du soleil," *Comptes rendus de l'Académie des Sciences* 74 (1872): 1387–1390.

95 See Tacchini, "Fisica solare," 93–95.

96 *Si [le P. Secchi] s'était donné la peine de lire, avec quelques attention, les Mémoires que j'ai publiés sur les protubérances solaires avant que le P. Secchi et M. Tacchini eussent commencé leurs observations régulières* [...] *il se serait convaincu que* [...] *une certaine classe de protubérances solaires avait déjà été remarquée et clairement décrite par moi* [...] *Sans même se donner la peine de lire ces Mémoires, le P. Secchi se serait facilement assuré, par la simple inspection de mes 140 profils solaires, déjà publiés à cette époque, que la structure filamenteuse des protubérances et des sommités de la photosphère s'y trouvent distinctement reproduite, malgré la petitesse de l'échelle adoptée.* [...] *Le P. Secchi doit bien savoir, en effet, que mes*

The Jesuit astronomer wrote to Tacchini:

> Have you seen the pride of Respighi, who treats me as a thief and [says] that I stole his studies, which he made one year before I presented them as new discoveries in the C[omptes] R[endus]? He is a real stinker. He made a vague mention of this distribution, which came out of a casual look at the profiles [of the solar limb], arranged one under the other. He did not discuss them quantitatively; but after I told him of my result, and a month after me, he published his calculations. Now he claims to have discovered all of it. But I gave him a rebuke that he will remember for a while.[97]

The "rebuke" was nothing but the publication of an article in which Secchi replied to Respighi's criticisms:

> This is not the first time that Mr. Respighi has tried to show that what I have done is inaccurate or untrue, or is others' work that I have appropriated. He declared himself the victim of my robberies before the Academy of Paris [...]. First, he began attacking me with much criticism, which I did not care to reply to; then he made allegations that my facts were inaccurate; now he comes to slander me of plagiarism and appropriation of others' work. It would be better to give no answer but instead disregard anything written in such a spirit, and I did that for a long time; but since my silence is thought to derive from my guilt, and I appear to be the usurper of others' [work], I shall briefly make some points. [...] About the use of a large prism placed in front of the slit, we have never pretended that this combination was ours, even though we first dared to make a prism of six inches by expending a lot of money from our purse. He speaks about our application in such a manner as if it too had

---

observations, instituées longtemps avant les siennes, ont été continuées régulièrement [...] *du 26 octobre 1869 au 23 juillet 1871* [...]. Respighi, "Réponse à une note précédente du P. Secchi," 1388.

97 *Avrete veduto l'orgoglio di Respighi che mi tratta da ladro e che io gli ho rubato le sue scoperte avendo esso dato un anno prima ciò che io ho pubblicato come scoperta nuova nei C*[*omptes*] *R*[*endus*]. *E' un vero birichino. Egli non diede che un cenno vago di questa distribuzione, quale risulta da una occhiata gettata a casaccio nei profili posti uno sotto l'altro. Non venne alla discussione numerica se non dopo che io gli dissi del mio risultato, e un mese dopo di me stampò le sue cifre. Ora pretende di aver fatto lui la scoperta di tutto. Ma gli ho dato una piccola lavatina di testa che se ne ricorderà per un pezzo.* Secchi to Tacchini, Rome, June 9, 1872, quoted in Chinnici and Gasperini, *Alle origini dell'astrofisica italiana*, 200–201.

been robbed from him [...]. We offer him a formal denial. While we have always been very liberal with him, showing him all our work without any reservation, we have usurped nothing, nor asked of any favors from him, even when in some cases his experience could have saved us much labor. Even knowing that he was carrying out some special studies, we were careful to reference them, except when necessity forced us not to pass on scientific inaccuracies. And these adjustments were done mostly without calling him out, nor giving him any reason for regret. [...] We will conclude by saying that we greatly regret having to dispute with Mr. Prof[essor] whose merits we have always respected and honored, and tried to give him every favor as best we could; but we are now obliged to do this to protect our reputation. [...] I do not think I deserve to be treated in such manner by Mr. Respighi, and I hope he will recognize his wrong, and if he continues in the same way, let him know that, from now on, I shall put him among those with whom I cut off every discussion.[98]

---

98 *Non è questa la prima volta che il Sig. Respighi ha cercato di far vedere che quello che io faccio o è inesatto, o non è vero, o è roba d'altri che io mi vado appropriando. Egli stesso si è dichiarato vittima de' miei latrocinii avanti all'Accademia di Parigi* [...] *Cominciò dapprima ad attaccarmi con molte critiche, che passai senza curarmene: venne poscia alle accuse di fatti inesatti; adesso viene alle calunnie di plagio e di essermi appropriato cose altrui. A scritture da tale spirito dettate non si dovrebbe dare altra risposta che del dispregio, e così ho fatto per lungo tempo, ma vedendo che il mio silenzio vien creduto derivare dal torto mio, e trattandosi del non comparire usurpatore dell'altrui, farò brevemente alcune riflessioni.* [...] *Quanto* poi *all'uso di un gran prisma posto avanti all'obiettivo, noi non abbiamo mai supposta cosa* nostra *questa combinazione, se non perché* per i primi *abbiamo osato far fare un prisma di* sei pollici *con molta spesa di* nostra *borsa. Egli parla però in modo di questa nostra applicazione, come se anch'essa fosse stata a lui derubata da noi, perché ne fissa con diligenza la priorità* [...] *Noi gli opponiamo una formale smentita. Mentre noi siamo sempre stati liberalissimi con lui facendogli vedere tutti i nostri studi, senza nessuna riserva, noi ci siamo ben guardati dall'usurpargli cosa alcuna, o dal sollecitare presso di lui simili favori, anche in alcuni casi in cui la sua pratica ci poteva risparmiare grandi fatiche. Anzi sapendo che Egli si occupava di certi studi speciali, ci siamo guardati di fargli concorrenza e di entrare in essi, salvo quando la necessità vi ci ha costretto, onde non passassero inesattezze nella scienza. E le rettifiche stesse sono state fatte per lo più senza nominarlo, né dargli ragione di dispiacere.* [...] *Concluderemo col dire che a noi dispiace grandemente di dover venire a discussione col sig. Prof*[essore] *cui sempre abbiamo stimato e onorato secondo il suo merito, e cercato di prestargli ogni servigio che per noi si potesse migliore; ma siamo obbligati a farlo per tutelare la nostra riputazione.* [...]*Io non credo di aver meritato di esser così trattato dal sig. Respighi, e spero che riconoscerà il suo torto, che se continuerà lo stesso, sappia che d'ora innanzi metterò anche lui tra quelli coi quali cesso ogni discussione.* Angelo Secchi, "Questioni spettroscopiche," *Bullettino meteorologico dell'Osservatorio del Collegio Romano* 11 (1872): 53–57.

Respighi wrote to Secchi very resentfully:

> I will ignore your article [...] to avoid the danger of making a response that would force me against my will to show that the author of this article, in order to support his views, rather than making use of solid arguments instead resorts to mean and vile tactics, not excluding slander and falsehood.[99]

However, the effect was achieved: hostilities were suspended, although at the price of ceasing communication for a number of years. Despite this silence, Secchi kept an open attitude toward Respighi; but the latter misinterpreted some of his attempts at reconciliation, at the risk of reigniting the controversy.[100] Secchi remained fixed in his resolution not to reply to Respighi, but expressed his bitterness when, in the face of a divergence of scientific opinions with Tacchini, he wrote to him:

> Our debate can and should be amicable, unlike some other people who do it with poison on the tip of the pen and, worse, abusing the ignorant about what little they know. The other day, someone was telling me that R[espighi] is so furious with me that he has cut off all connections, not even in speaking. [...] I'm sorry, because I have done nothing but good for him, I did him great good.[101]

The long-standing rivalry, though never really healed, was eased by time; a favorable, though sad, circumstance for reconciliation occurred about one year later:

---

99 *Le respingo il suo articolo* [...] *per togliermi dal pericolo di fare una risposta, colla quale sarei costretto, mio malgrado, a mostrare che l'autore di questo articolo, per sostenere le sue opinioni, anziché valersi di solidi argomenti, ricorre invece alle armi più basse e vili non esclusa la calunnia e la menzogna.* Respighi to Secchi, Rome, July 22, 1872; AOAR, Fondo Respighi, II.E.3.1.

100 See Ileana Chinnici and Sabino Maffeo, "Angelo Secchi e gli astronomi del suo tempo," in Altamore and Maffeo, *Angelo Secchi*, 65–88, here 78. Moreover, Respighi even addressed his polemical tone toward Tacchini, whom he considered an "ally" of Secchi: he probably resented not having been invited by Tacchini to take part in the expedition to India for the transit of Venus, and contested the scientific value of the results obtained there (see Tacchini to Secchi, Modena, June 22, 1875, quoted in Chinnici and Gasperini, *Alle origini dell'astrofisica italiana*, 383).

101 *La discussione può e deve farsi amichevolmente non come fanno certi altri col veleno sulla punta della penna e peggio, abusando tra gli ignoranti di quel poco che sanno.// L'altro giorno uno mi diceva che R*[*espighi*] *è inviperito talmente verso di me che non connetteva più né anche nel parlare.* [...] *A me dispiace perché a costui non ho fatto che benefizi, e benefizi grossi.* Secchi to Tacchini, Grottaferrata, August 15, 1873, quoted in Chinnici and Gasperini, *Alle origini dell'astrofisica italiana*, 317.

> Respighi took the opportunity of Fr. Rosa, who was dying, to make us a visit, and as it was my desire, I welcomed him with every effusion of friendship, and the past is forgotten. I have been invited by him to go to Monte Mario where he is occupied by geodetic matters, and there I shall go, to show that I have nothing against him.[102]

The volleys between Respighi and Secchi were tempered in tone, but they smoldered beneath the ashes. In the minutes of the meeting for the reform of the Italian observatories,[103] held in Palermo in 1875 (see Chapter 10), Respighi's uncooperative attitude clearly emerges (he voted against all the resolutions and disagreed with everything that was claimed by Secchi or Tacchini), leading to the suspicion that he was attempting to boycott the entire reform project.[104]

Considered in retrospect, the relationship between Secchi and Respighi looks like a missed opportunity. From a scientific point of view, there was certainly great potential for collaboration, but the personalities of the two astronomers got in the way, leading to a rivalry that every now and then would flair up, to the detriment, ultimately, of the peace of mind and scientific productivity of both scientists.

Unfortunately for Secchi, this rivalry had serious consequences. Respighi had adopted the practice of disseminating, often furtively, his accusations against Secchi among the scientific community. In Italy, Secchi could count upon the diplomatic *savoir faire* of Schiaparelli, who replied to Respighi:

> It really surprises me that Fr. Secchi wants to *leverrieryze*;[105] because even if he is jealous of the paternity of his own inventions, I had never seen in him a tendency to steal the work of others. And perhaps it may be that any such occurrence is just an oversight on his part. When you are so busy with your own work, you don't have the time to read and rightly appreciate the work of others.[106]

---

102 *Respighi all'occasione del P. Rosa che stava moribondo venne a farci una visita, e come era mio desiderio lo accolsi con ogni espansione di amicizia, e il passato è per me dimenticato. Mi invitò ad andare a Monte Mario dove ora si occupa di cose geodesiche, e vi sono stato, per mostrare che io non ho nulla con lui.* Secchi to Tacchini, Rome, July 26, 1874, quoted in Chinnici and Gasperini, *Alle origini dell'astrofisica italiana*, 367.

103 See Poppi, Bònoli, and Chinnici, "Il progetto Tacchini e la riforma degli Osservatori italiani," 165.

104 See Secchi to Tacchini, Rome, October 29, 1875, quoted in Chinnici and Gasperini, *Alle origini dell'astrofisica italiana*, 396.

105 To *leverrieryze*: to behave in a similar way to Leverrier; the French astronomer was well known for his irascible behavior.

106 *Mi sorprende davvero che il P. Secchi voglia* leverrierizzare*; perché se da una parte egli è geloso della paternità delle sue invenzioni, non mi era mai sembrato di scorgere in lui una*

However, Respighi found allies across the Channel, who helped spread and reinforce accusations of plagiarism against Secchi, grounded or not, which in part compromised his reputation.

4.3 *In Great Britain: Lockyer and Huggins*

Respighi's main ally against Secchi was Lockyer, the influential editor of *Nature*.

The controversies between Secchi and the English astronomer deserve to be examined in depth; here, they are solely mentioned in brief, deferring more extended examination of the subject to later work.

Secchi and Lockyer frequently clashed over scientific questions, especially those related to studies of solar physics. In 1869, the English astronomer found a redshift or blueshift in the F line[107] of the chromospheric spectrum when it was observed at the solar limb and interpreted these distortions as "extreme rates of movements in the chromosphere."[108] After questioning some of Secchi's observations, which strongly differed from his own, he added that, however, "Father Secchi does not combat this theory; indeed it is not to be gathered from any of his communications that he has seen any of the papers communicated by myself to the Royal Society."[109]

Secchi, who had observed similar displacements due to the Doppler effect, gave another explanation: they were due to solar rotation.[110] Unfortunately, he only gave a qualitative estimate, thus incurring much criticism.[111] In 1871, however, Vogel confirmed Secchi's conclusions, showing that the Doppler shift

---

tendenza ad appropriarsi le cose altrui. E forse può essere che qui abbia avuto luogo una dimenticanza da sua parte. Quando si è molto occupati ne' proprii lavori, non si ha il tempo di leggere e di giustamente apprezzare i lavori altrui. Schiaparelli to Respighi, [Milan], January 5, 1870, quoted in Alfonso Di Legge, Francesco Giacomelli, and Antonio Prosperi, "Lorenzo Respighi: Una pagina di storia dell'astronomia romana," *Rivista di astronomia e scienze affini* 6, no. 9 (1912): 625–640, here 629.

107 In the mapping of the spectrum made by Fraunhofer, F was one of the hydrogen lines, which was later denominated H-beta. When observed at the solar limb, it appeared to shift toward the red or the blue ends of the spectrum, and this was explained by the Doppler effect (change in frequency or wavelength of a wave for an observer who is moving relative to the wave source), giving evidence of approaching (blueshift) and drifting (redshift) motions in the chromosphere in the direction joining the observer and the solar limb.

108 See Hearnshaw, *Analysis of Starlight*, 149.

109 Norman J. Lockyer, "Observations of the Sun: No. V," *Proceedings of the Royal Society* 18 (1869): 74–79, here 77.

110 See Secchi, "Ricerche nelle protuberanze solari."

111 See Meadows, *Science and Controversy*, 63.

FIGURE 9.5 Norman Lockyer, leader of the English anti-Secchi party; https://it.wikipedia.org/wiki/Norman_Lockyer#/media/File:Lockyer-Norman.jpg (accessed June 8, 2018).

detected solar rotation and giving an empirical confirmation that the Doppler effect worked for light as for sound.[112]

Lockyer also contested the existence of a "reversing layer," which had been anticipated by Kirchhoff and supported by Secchi:

> Father Secchi states that the chromosphere is often separated from the photosphere, and that between the chromosphere and the photosphere there exists a stratum giving a continuum spectrum, which he considers to be the base of the solar atmosphere, and in which he thinks that the inversion of the spectrum takes place.[113]

112 See Hearnshaw, *Analysis of Starlight*, 148.
113 Lockyer, "Observations of the Sun: No.V," 77.

The inversion of the spectrum was observed for the first time during the eclipse of 1870 by Young, and this discovery confirmed the validity of Kirchhoff's solar model, largely adopted by Secchi.

Respighi was not the only one who claimed to have been the first to study solar prominences. In fact, in 1869 Lockyer had also carried out an extensive survey of the shape of prominences and distinguished them into two basic types, eruptive and nebulous. When Secchi presented his classification system without referring to Lockyer's work, the English astronomer immediately reacted with one of his "anti-Secchi papers,"[114] which was so polemic as to be rejected for publication in the *Proceedings of the Royal Society*.

The two astronomers had a different approach to scientific research. While Secchi gave priority to the observations and contested the mixing of theory and observations, Lockyer adopted the same un-Baconian approach of Faye, considering that "a good theory is as necessary as a good telescope."[115] In Lockyer's view, without a working hypothesis, the risk was "to observe blindly or haphazardly."[116] Lockyer was another person with a fiery temper, ready to ally with other scientists as well as to enter into conflicts; he was a friend of Janssen, to whom Secchi had showed the door at the Collegio Romano Observatory (see Chapter 6) and sympathized with Respighi, whom he considered a "victim" of Secchi (indeed, this was how Respighi represented himself). Those two astronomers kept on good terms: on many occasions, Lockyer came out in support of Respighi and invited him to join the English expedition to India to observe the total eclipse of the Sun of 1871, probably to compensate for the "forgetfulness" of the Italian government in 1870.

In 1872, Respighi wrote to Lockyer,[117] discrediting Secchi and raising suspicions about his claims to have made certain discoveries for the first time. He also added that the program of the Spectroscopic Society had been drafted by Secchi and Tacchini and published without being communicated to the members, and that he could not find "that harmony of feeling and mutual confidence which are indispensable in a society of persons formed for the pursuit of a common object."[118] Lockyer then openly attacked Secchi in the pages of the journal *The Academy*.[119] He wrote a review of the book *Le soleil* and criticized

114 Meadows, *Science and Controversy*, 63.
115 Meadows, *Science and Controversy*, 63.
116 Meadows, *Science and Controversy*, 63.
117 See Respighi to Lockyer, Rome, June 13, 1872; University of Exeter, Historical Archives (hereinafter UEHA), Lockyer Papers.
118 Meadows, *Science and Controversy*, 64.
119 See Norman J. Lockyer, "Physical Sciences, etc.," *The Academy* 3, no. 53 (1872): 288–291; see also Hearnshaw, *Analysis of Starlight*, 63–66.

Secchi by stating that his conclusions were so contradictory that, in the case of sunspots, for example, his "objections [were] those of one ignorant of the matter":[120]

> The other portions of the work are chiefly remarkable for the skillful manner in which the author gives the idea that in all the recent progress he alone has borne the heat and the burden of the day […]. In the account of the use of the pile, we find no mention of Henry,[121] while the types of stars are not referred to Rutherford [*sic*],[122] as they should have been, and so on. Surely there should be nothing more disastrous for the reputation of a scientific man than such unfair dealing as this.[123]

In the face of this attack, Secchi defended the originality of his work:

> Neither by the method of the division nor by the principle governing the classification do my types have anything in common with Mr. Rutherfurd's groups. This scientist has relied on general color, and this is in agreement with my first classification published […] in 1863 […]. The differentiation between the types that I have undertaken does not belong to the American astronomer, who, in fact, has never claimed it. In order to firmly establish these types, it is necessary to examine all the first magnitude stars and a considerable number of second magnitude, which Mr. Rutherfurd has never done, but which I accomplished […].[124]

Moreover, Lockyer, having noticed some additions in the German edition of *Le soleil*, commented:

> If we are not dealing with the work of the German editor, I am certain many scientific men in many lands will hail it as an indication that Father Secchi will probably do them justice in future; that he has not done justice in the past is an opinion I share with every man of science

120 Lockyer, "Physical Sciences, etc.," 289.

121 Joseph Henry had dealt with this topic; see Joseph Henry, "On the Application of the Thermo-Galvanometer to Meteorology, etc.," in *Scientific Writings of Joseph Henry*, ed. Smithsonian Institution (Washington, DC: Smithsonian Institution Press, 1886), 1:215.

122 In 1863, Rutherfurd had also proposed a spectral classification of stars (see Chapter 6).

123 Lockyer, "Physical Sciences, etc.," 289.

124 Hearnshaw, *Analysis of Starlight*, 66.

with whom I have spoken on the subject, including many of his own countrymen.[125]

Once he read Lockyer's letter, Secchi privately wrote him that he knew the individual who might have inspired these accusations of larceny and thievery (he indirectly referred to Respighi); he called on Lockyer to exhibit the precise motivation for the accusations and threatened to respond publicly, accusing him of slander. Finally, he noted that, except for a few cases, the additions in the German edition of *Le soleil* were made by his own hand, not by the German editor (i.e., Schellen), as Lockyer had suggested.[126] In his private reply, the English astronomer seized the opportunity to reproach Secchi:[127]

> How have you treated [Respighi]? As an impartial witness I am bound to say that I think no treatment could have been worse. It is because I believe with you that the shameful things to which you refer are intolerable among men of science, that I have attempted to prevent them in the future.[128]

Secchi replied with a letter defending himself, inviting Lockyer to see both sides. He explained that he had sent Janssen away from the Collegio Romano Observatory because of his ungrateful attitude, that Respighi was jealous of him and his accusations were ungrounded, and he justified the missing mention of Lockyer's work in his book as the result of the oft-complained two-years' delay in the delivery of the English journals to his observatory.[129] He then concluded by asking for a timely correction from Lockyer.[130]

In the meantime, Respighi was continuing to discredit Secchi in his correspondence with the English astronomer, feeding Lockyer's prejudices by stating that the Jesuit would often reply with "sophisms and insolences."[131]

---

125 Norman Lockyer, "Father Secchi on Solar Discoveries," *The Academy* 3, no. 61 (1872): 451–452, here 451.

126 See Secchi to Lockyer, Rome, June 14, 1873; UEHA, Lockyer Papers.

127 Secchi also mentioned Lockyer's letter defending Respighi in a letter to Tacchini; see Secchi to Tacchini, Rome, February 13, 1873, quoted in Chinnici and Gasperini, *Alle origini dell'astrofisica italiana*, 249.

128 Meadows, *Science and Controversy*, 64.

129 This was indirectly confirmed by Respighi, who complained about the two-year delay in the delivery of the *Monthly Notices* and asked Lockyer to send him copies; see Respighi to Lockyer, Rome, June 13, 1872; UEHA, Lockyer Papers.

130 See Secchi to Lockyer, Rome, January 27, 1873; UEHA, Lockyer Papers.

131 See Respighi to Lockyer, Rome, January 17, 1873; UEHA, Lockyer Papers.

As might be expected, the controversies between Lockyer and Secchi had a serious effect on the circulation of Secchi's works in Britain. Being written in Italian and/or French, it was necessary for them to be translated into English in order to reach the wider British public beyond the academic context. Secchi, who was very attentive to the dissemination of science, began to work on the English edition of *Le soleil* in July 1872[132] after the publication of the German edition and during the preparation of the second French edition. However, while the editing of the manuscript was in progress, Lockyer somehow vetoed its publication. This was probably done as a form of punishment for the alleged plagiarism, but also as a sort of "protectionist" act, to prevent competition with Lockyer's treatise *Contributions to Solar Physics*, published in 1874—a text in which, furthermore, Respighi's works were mentioned with great emphasis.[133] Lockyer used the pages of his journal, *Nature*, to give the following information to the readers: "Father Secchi is preparing, at Gauthier Villars [Jean-Albert (1828–1898), publisher], a second edition of his work on the Sun, on an enlarged scale. He has quoted so largely from Mr. Lockyer's *Solar Physics* that an intended translation of this work is abandoned for the present."[134]

Secchi read the note "with great surprise" and wrote to the journal:

> I have the honor to inform you that the complete original of my second edition has been in the hands of M. Gauthier for more than a month, so far as that part that may have something in common with Mr. Lockyer's work is concerned, and that I had not seen Mr. Lockyer's work until a fortnight ago, when I bought it from M. Loescher [Ermanno (1831–1892), publisher] here in Rome. Mr. Lockyer of course is quoted, but only from his original memoirs, and not from his new publication, nor in such a manner that his publication will render my work useless.[135]

The letter was published with a note by Lockyer, the journal's editor:

> The following explanation has been sent us by the Paris correspondent who furnished us with the note referred to by Father Secchi: I was told by his [Secchi's] editor himself, when I spoke to him about publishing a French edition of Mr. Lockyer's "Solar Physics," the substance of what I have written to you. I think that the note I have written is a

132 See Secchi to Tacchini, Rome, July 6 and 14, 1872, quoted in Chinnici and Gasperini, *Alle origini dell'astrofisica italiana*, 202 and 206.

133 See Norman J. Lockyer, *Contributions to Solar Physics* (London: MacMillan, 1874), ix–x.

134 Norman J. Lockyer, "Notes," *Nature* 9 (1874): 389–391, here 390.

135 Norman J. Lockyer, "Father Secchi's Work on the Sun," *Nature* 10 (1874): 3–4.

> recommendation of Father Secchi's work; but not so his statement that he did not possess "Solar Physics" until it was too late to use it. There is nothing dishonorable in quotation.[136]

Lockyer, being the editor of a widely read journal, was in a favorable position to circulate criticisms and potentially demolish Secchi's reputation; the Jesuit astronomer regretfully wrote to Tacchini: "Lockyer has filled his journal, Nature, with invectives against my work *Le soleil* and hastened to collect all the stupid things that have been written against me."[137]

The anti-Secchi campaign was successful, and other English astronomers joined the list. More discreetly, another scientist who "was generally disposed to distrust Secchi's claims"[138] and, though not publicly, accused Secchi of theft, was William Huggins. In 1878, the American astronomer Simon Newcomb had published the book *Popular Astronomy*, where he stated that, in the winter of 1864–65, the spectrum of the Orion nebula "was examined independently by Secchi and Huggins, who found that it consisted of three bright lines, and hence concluded that the nebula was composed, not of stars, but of glowing gas."[139] Huggins wrote to his American colleague, Edward S. Holden (1846–1914), claiming that he was the first to have made the discovery and affirming that "Secchi had no inkling of the existence of such lines until he heard about them from Otto Wilhelm von Struve," who visited Secchi and told him about Huggins's discovery, as von Struve himself reported to the English astronomer. Huggins's wife, Margaret (1848–1915), who closely collaborated in her husband's works, added at the bottom of the letter that "in the interests of morality" she denounced Secchi's conduct "as shameful thievery."[140] Secchi, however, was dead in 1878, and, moreover, the error in the attribution was made by the author of the book, Newcomb. Consequently, Huggins's judgment appears excessively harsh and, as in Lockyer's case, probably contained some degree of prejudice. It is difficult to say how much anti-Catholicism may have influenced the opinion of the British astronomers who insisted on discrediting Secchi's reputation.

Consequently, in some of the obituaries that were published in Britain, Secchi's figure is often depicted in a negative light. In an obituary of Secchi

136 Lockyer, "Father Secchi's Work on the Sun," 4.

137 *Lockyer, ha empito il suo giornale, il Nature, di invettive contro il mio lavoro del Soleil e si dà premura di raccogliere tutte le sciocchezze che si scrivono contro di me.* Secchi to Tacchini, Rome, October 29, 1875, quoted in Chinnici and Gasperini, *Alle origini dell'astrofisica italiana*, 395–396.

138 Becker, *Unravelling Starlight*, 91.

139 Newcomb, *Popular Astronomy*, 447.

140 Becker, *Unravelling Starlight*, 194.

by mathematician Rev. Philip M.A. Kelland (1808–79) for the Royal Society of Edinburgh, we read:

> Secchi, though an excellent observer and a man of great power, was of discursive turn of mind. He had little power of concentration, and appears to have tired of the monotony of astronomical observations, and to have turned his attention to the more popular studies of terrestrial magnetism and solar physics. His attention to the latter subject had probably been aroused by his having assisted Professor Henry, when in America, in making the first experiments on the heat radiated by different portions of the Sun's disc, by means of the thermo-electric pile. [...] Secchi's reputation was undoubtedly very great and wide-spread. [...] His great merit consisted in industry and activity—his error, in want of definiteness of aim, in over-production. The Royal Society's Catalogue of Scientific Papers contains a list, carried down to 1863 only, of no less than 230 contributions to scientific journals. [...] This list must have been greatly extended. Their value is probably not in proportion to their extent; but it cannot be doubted that they contain much that will help on the future progress of science.[141]

Similarly, in the obituary of Secchi (anonymous, but attributable to the editor Lockyer) that was published in *Nature*,[142] it was stressed that, under the Papal States,

> no funds were spared to make the observatory as complete as possible. Secchi had instruments and assistants in abundance [...]; his position gave him great facilities for giving the widest publicity to his work.[143] What he lacked in originality he made up in assiduity, and hence, although he has left no great life work on any one subject behind him, there is, we think, hardly any question which has turned up touching observations in

141 Kelland, "Father Secchi."

142 A version "purged" of this obituary appeared (in French) in the celebratory booklet on Secchi by Moigno; see Moigno, *Le révérend Père Secchi*, 206–208.

143 Actually, with regard to the resources of the observatory, Secchi always stated that he had invested his money in the purchase of the instruments; nor did the number of assistants appear oversized compared with the needs of the research activities carried out therein. The author of the obituary seems to want to assert that Secchi profited from his position as a "pontifical" astronomer in publishing his works; it is unclear whether this refers to the publication of the *Bullettino meteorologico*, funded in fact by Prince Boncompagni, or attributing his ease in finding approval from various scientific institutions, out of respect for papal authority.

> astronomy, magnetism or meteorology on which a multitude of papers have not been written by his busy pen. Many of these papers are very admirable and show great penetration and power of generalization as well as a wide grasp of many subjects.[144]

Then, switching to a discussion of Secchi's personal affairs upon settlement of the Italian government in Rome, the offer of the chair at La Sapienza University (see Chapter 7) is presented as a compromise, suggested by none other than the king of Italy, Victor Emmanuel II (r.1849–61): indeed, Secchi is presented as his "foster brother," and the subsequent denial is described as an act required by the superiors of his order. Such details were so confidential that it seems unlikely that Lockyer could have had them directly from Secchi; it is more likely that the source was Respighi, whose impartiality, however, for the reasons outlined above, one may quite well doubt. The text ended with the phrase: "It must not been forgotten that if there may have been traits of Secchi's character open to criticism, the exigencies of his post, rather than the inclinations of the man, may have been to blame."[145]

The most serious consequence of the British campaign against Secchi was that no work by the Jesuit astronomer was ever translated into English. Among the English-speaking audience, his ideas could circulate only thanks to quotations in American texts. In fact, in the United States, in a more open context where Jesuits were generally respected and admired, Secchi's reputation remained unfiltered,[146] and his works were much appreciated. For example, in the successful book *The Sun* by Young, Secchi's name is mentioned many times,[147] as in other already-mentioned American popular books (see introduction); not by chance, it was mainly US astronomers who would inherit Secchi's scientific legacy (see conclusions).

5 Science and Faith

It is useful in this context to briefly examine the connection between faith and science in Secchi's work in order to highlight some elements that could be better developed in future studies. The basic statement by Secchi that reveals his attitude in this field was that "science is a gift from God."[148]

144 Norman J. Lockyer, "Father Secchi," *Nature* 17 (1878): 370.

145 Lockyer, "Father Secchi," 370.

146 See William F. Rigge, "Father Angelo Secchi," *Popular Astronomy* 26 (1918): 589–598.

147 Secchi is mentioned in the foreword of Young's book as one of the scientists to whom the author was "specially indebted"; see Young, *Sun*, vii.

148 *La scienza è dono di Dio*. APUG, FS 2.VII.a, 17.

In 1856, Secchi concluded the volume of the *Memorie del Collegio Romano* with some general considerations that well expressed his conviction about the connection between science and faith:

> True faith is not opposed to science, [...] both are rays of the same Sun, aimed at illuminating the way to Truth to our blind and weak minds. Deprived of this noble purpose, these studies are mere curiosities and often just result in pain or, at least, unrewarding struggles. It is stimulating to consider how it is wonderful to show the Creator's work, even when any other excitement fades away; this raises the mind over the materiality of the calculations and makes these struggles a sublime and divine work.[149]

Secchi was aware that, in the face of new scientific discoveries that were sometimes challenging for the faith, there was the risk of locking the door to progress and of shielding it with an ultra-conservative attitude, by giving up rational and scientific investigations. He strongly opposed this, often declaring that this kind of investigation was very important to understand creation and that every addition to knowledge was a further revelation of God's glory:

> Yes, God wants the homage of our intellect in the belief of things that exceed our intelligence, but he does not want that we renounce reason in the examination of those motivations that form the reason for our belief.[150]
>
> The understanding of the sciences is therefore also a gift from God, and this is one of the gifts of the Holy Spirit. We could then, courageously but also rightfully, ask for this gift from God, when in our studies we share the same purpose of every Christian: namely not the vanity to compete with other scientists, nor the pride and arrogance that comes after surpassing them, but only to ask the light of intellect for understanding the Lord's works, for knowing his greatness and our obligations.[151]

---

149 *La vera fede non è ostile alla scienza,* [...] *ambedue sono raggi di uno stesso Sole diretti ad illuminare le nostre cieche e deboli menti alla via della Verità. Senza quest'alto scopo, tali studi sono una mera curiosità, e spesso solamente fruttiferi di pene o almeno di non rimunerate fatiche. Il pensare quanto sia magnifico il manifestare le opere del Creatore è uno stimolo che sprona anche quando vien meno ogni altro eccitamento; questo solleva la mente sopra la materialità delle cifre, e forma di queste fatiche un'opera sublime e divina.* Angelo Secchi, "Conclusione," *Memorie dell'Osservatorio del Collegio Romano 1852–55* (1856): 157–158, here 157.

150 *Dio vuole bensì l'omaggio del nostro intelletto nella credenza delle cose che superano la nostra intelligenza, ma non vuole che rinunziamo alla ragione nell'esame di que' motivi che formano la ragione di nostra credenza.* APUG, FS 24.VI.

151 *E' dunque dono di Dio anche l'intelligenza nelle scienze ed uno dei doni del Santo Spirito. Allora, pertanto con coraggio, e dirò quasi con un certo diritto, potremo chiedere questo dono a*

> Profane studies, by opening the mind, facilitate the intelligence of the divine things. [...] The heavens show the glory of the Lord and the wildflower and the small flying insect show his endless wisdom. Now, intelligence itself is a gift from God [thus] every potential bit of knowledge comes from God.[152]

For this reason, in contrast with many of his opponents, Secchi trusted deeply in the progress of technology and in the extension of knowledge that comes from new theories: "Ten years ago, who could have imagined the wonders to be revealed by the spectroscope? Every new advance in technique leads to another in science [...]":[153]

> Although it is impossible for us to comprehend completely the mystery of the constitution of the world, nevertheless following the work of the astronomers up to now we have obtained so many things to illuminate our ignorance. In this field, we must always keep in view the wise warning of Herschel to avoid two extremes; the first, to fabricate worlds in our imagination, because in so doing we cannot reach a knowledge of nature [...]; the other, to hesitate too much in making conjectures, because by doing so our observations became fruitless, since conjectures are necessary to arrive at a knowledge of the composition and structure of the universe [...].[154]

Dio quando nei nostri studi noi ci proporremo lo scopo che si deve prefiggere ogni cristiano. Cioè non la vanità di superare gli emoli, non la boria o la superbia dopo averli superati ma solo col chiedere il lume dell'intelletto a comprendere le opere del Signore, a conoscere le sue grandezze e i nostri doveri. APUG, FS 2.VII.a, 13.

152 *Gli studi profani aprendo la mente le facilitano l'intelligenza delle cose divine.* [...] *i cieli mostrano la gloria del Signore, e il fiore del campo e l'insettuccio volante mostrano la sua infinita sapienza. Ma questa stessa intelligenza è dono di Dio* [...] *ogni potenza nostra è da Dio.* APUG, FS 2.VII.a, 13.

153 *Chi avrebbe immaginato dieci anni or sono le meraviglie che stava per rivelarci lo spettroscopio? Ogni nuovo perfezionamento dell'arte ne porta uno alla scienza* [...]. Angelo Secchi, "La grandezza del Creato. Discorso secondo," in *Lezioni elementari di fisica terrestre*, 205–218, here 218.

154 *Malgrado che ci sia impossibile penetrare completamente il mistero della costituzione del Mondo, tuttavia dai lavori eseguiti finora dagli astronomi, abbiamo già molti materiali per illuminare un poco la nostra ignoranza. In questa materia dobbiamo sempre avere avanti agli occhi il bello avviso di Herschel che si devono evitare i due estremi; il primo di fabbricar mondi a nostra fantasia, perché così non arriveremmo a conoscere la natura* [...] *l'altro è la troppa timidità di congetturare, perchè così si perde il frutto delle osservazioni, le quali appunto si devono fare, affine d'arrivare a conoscere la composizione e la struttura dell'Universo* [...].Secchi, *Le stelle: Saggio di astronomia siderale*, 299.

Secchi viewed the universe as a unity, animated by motions originated in forces; these ideas were synthesized in the last book he published, *Le stelle*:

> The matter composing [the celestial objects] is, however, always the same. The elements that are studied by the chemist in his laboratory are the same ones, existing in far nebulae and in stellar atmospheres, that the spectroscope reveals to us.[155]

The laws ruling matter are the same here for us and in the far deep universe.[156]

> Heat is the primary force animating the universe; and heat is no more than motion. Energy is transmitted from a body to another through a continuous medium that we call the ether, and we are connected to the far regions of the space through this mysterious medium, whose vibrations form heat, light, and chemical activity, whose density fluctuations provoke attraction and all electric and magnetic phenomena [...]. Gravity is the force holding together all creation, from the pebble falling down to the ground, to the nebula undergoing condensation in faraway space. It is the primary reason for the incandescence of the stars [...]. But it is not the only force dominating the universe [...]. The ten-year cycle of solar phenomena is reflected in the variations of Earth's magnetism and the electric display of the aurora borealis; this proves that another force, beyond gravity, begins at the Sun and is spread into space, pervading planets and determining the most complicated phenomena. We are aware of this force, but we ignore its behavior. Is it direct magnetic action, or is it a simple transformation of its caloric action?[157]

---

155 *La materia che compone* [*i corpi celesti*] *è però sempre la stessa. Gli elementi che il chimico studia in laboratorio sono gli stessi che lo spettroscopio ci svela nelle ultime Nebulose e nelle atmosfere stellari* [...]. Secchi, *Le stelle: Saggio di astronomia siderale*, 337.

156 *Le leggi che reggono la materia sono le stesse da noi e in quelle remote profondità.* Secchi, "La grandezza del Creato. Discorso secondo," 214.

157 *Il calore è la forza prima che anima l'Universo; e il calore non è che movimento. L'energia si trasmette da un corpo all'altro mediante un mezzo continuo, che chiamiamo etere, e noi siamo in contatto con le regioni più estreme dello spazio mediante questo mezzo misterioso, le cui vibrazioni costituiscono il calore radiante, la luce e l'attività chimica vitale, i cui squilibri di densità producono le attrazioni ed i fenomeni elettrici e magnetici* [...] *La gravità è una forza che regge tutto il creato, dal sassolino cadente sulla terra alla Nebulosa che si va condensando nella profondità dello spazio. Essa è la causa prima dell'incandescenza degli astri* [...] *Ma essa non è la sola che domina l'Universo* [...].
Le vicende decennali del Sole, sono riflesse nelle variazioni del magnetismo terrestre e nelle manifestazioni elettriche delle Aurore Boreali; ciò prova che un'altra forza oltre la gravità,

In Secchi's view, the universe is in evolution and strictly interconnected with life, which may germinate everywhere:

> Creation, contemplated by the astronomer, is not a simple mass of incandescent matter: it is a wonderful organism where, when the incandescence ceases, life starts. Even though it may be inaccessible to our telescopes, nonetheless, by analogy with our globe we are able to conclude that it exists in others. The atmospheric constitution of the other planets, [...] so similar to ours as well as that of the stars, so similar to that of the Sun, [...] convince us that those [celestial] bodies are in a stage similar to the current one of our system, or are running a stage that it has already passed, or is destined to pass. In the immense variety of creatures that had existed, and exist in our planet, we may argue the diversity of those that might exist there [...].[158]

Secchi was aware that "it would be a very restricted view, however, [...] to suppose that every celestial body is inhabited like ours":[159] other kind of beings, different from the terrestrial ones, could have developed in different atmospheric conditions. Ultimately, he was convinced that:

> Life fills the universe, and intelligence is associated to life and, since many beings inferior to us exist [on the Earth], in other conditions many others [beings] more capable than us might exist [...]. But this is a sphere inaccessible to the reign of the astronomer. His task is to study the material

---

parte dal Sole e si spande nello spazio la quale pervade i pianeti e ne determina le vicende più astruse. Noi siamo certi di questa forza ma ignoriamo il suo modo di azione. È dessa un'azione magnetica diretta, ovvero una semplice trasformazione della sua azione calorifica? Secchi, *Le stelle: Saggio di astronomia siderale*, 341–342.

158 *Il creato che contempla l'astronomo non è un semplice ammasso di materia incandescente: è un prodigioso organismo in cui dove cessa l'incandescenza della materia incomincia la vita. Benché questa non sia penetrabile ai suoi telescopii, tuttavia dall'analogia del nostro globo possiamo argomentarne la generale esistenza negli altri. La costituzione atmosferica degli altri pianeti che* [...] *è cotanto simile alla nostra, non che quelle delle stelle simili a quella del nostro sole, ci persuadono che que' corpi, o sono in uno stadio simile al presente del nostro sistema, o percorrono taluno di quei periodi che esso già percorse, o è destinato a percorrere.* [...] *Nell'immensa varietà delle creature che furono già, e che sono sul nostro pianeta, possiamo argomentare la diversità di quelle che possono esistere colà* [...]. Secchi, *Le stelle: Saggio di astronomia siderale*, 337.

159 *Sarebbe però ben angusta veduta quella di voler modellare l'Universo tutto sul tipo del nostro piccolo globo* [...] *né è filosofico pretendere che ogni astro sia abitato come il nostro* [...]. Secchi, *Le stelle: Saggio di astronomia siderale*, 338.

> and mechanical development of the world, tracking it back in space, and to come in aid to the geologist, who studies it in time […].[160]

Pluralism, however, was not something new in Secchi's thought since he had professed it since the beginning of his career; in 1856, in fact, he poetically wrote: "It is with a sweet sentiment that man thinks of these numberless worlds, where each star is a Sun that, as a minister of divine bounty, distributes life and goodness to other innumerable beings, who are blessed by the hand of the Omnipotent."[161]

Secchi also took a similar position on concordism, remarking on the risk of misinterpreting scripture in order to oppose the astonishing results of science:

> The vastness of the Creation is one of the ideas that frightens the small human mind. When it was announced for the first time, after the barriers of a material sphere enclosing the space were broken, that the stars were many suns, the mind was astonished by the vastness of the universe […], and the abundance of its bodies […]. It even tried to escape these consequences by entrenching itself behind misinterpreted scriptures! Do not be amazed about such in the past, because even today […] it can be hard to believe in the myriads of centuries that our globe has traversed […]. But the one helps the other, and we should be convinced that the work of the Creator is commensurate to the Creator himself […]; we shall never be unable to understand either entirely.[162]

---

160 *La vita empie l'Universo, e con la vita va associata l'intelligenza e come abbondano gli esseri a noi inferiori, così possono in altre condizioni esisterne di quelli immensamente più capaci di noi* […] *Ma questa è sfera ove l'astronomo non può estendere il suo regno. A lui è riservato lo sviluppo materiale e meccanico del mondo, rintracciandolo nello spazio e soccorrere il geologo che lo studia nel tempo.* Secchi, *Le stelle: Saggio di astronomia siderale*, 339.

161 *Inonda il cuore un dolce senso di gioia in pensare a que' mondi senza numero, nei quali ogni stella è un sole benefico che ministro della Divina Bontà sparge vita e giocondità su altri esseri innumerabili riempiti della benedizione della mano dell'Onnipotente.* Secchi "Conclusione," 158.

162 *La grandezza del Creato è una di quelle idee che spaventano la piccola mente umana. Quando si annunziò la prima volta, che rotte allo spazio etereo le barriere di una sfera materiale, le stelle erano tanti Soli, la mente restò tanto sbalordita dalla vastità dell'Universo che gli si veniva logicamente presentando, e dalla copia sterminata de' corpi che lo costituivano. Essa cercò quasi di sfuggire a queste conseguenze col trincerarsi dietro male interpretate sacre parole! Non ci meravigliamo del passato, perché anche oggidì* […] *si stenta a credere alle miriadi di secoli che deve avere attraversato il nostro globo* […] *Ma una cosa aiuterà l'altra, e saremo convinti che l'opera del Creatore è solo a lui commensurabile* […] *noi saremo sempre incapaci di comprenderla intieramente.* Secchi, *Le stelle*, 288.

Secchi was also open to the idea of evolution:

> The idea of consecutive transformations, taken with appropriate moderation, is not incompatible either with reason or religion. In effect, if we do not consider everything as carried out by pure inner forces of the brute matter, but we admit that these forces come from the first Reason who created matter and gave it the power to produce such effects, there is no intrinsic repugnance to believe that, until the intervention of a new force, some organisms could develop themselves in one way rather than in another, thus originating different beings. But when a series of these beings passes to another one containing a new principle, then things change. It is impossible to pass from the vegetable without sensitivity to the animal having sensations without a new power that cannot come from either organization or matter alone. You can say this even more when we pass from the brute animal to the man who thinks, reflects, and has a conscience. A new principle must be then associated to the physical forces of matter to produce these results [...].[163]

However, Secchi defended his position from those scientists who ridiculed those who "cannot do without the idea of a supreme Being who has made and rules everything":[164] "We are not so stupid as to believe that the Creator is the fluttering white-bearded old man, painted by Raphael in his Loggia."[165]

It seems appropriate to conclude here by reporting some excerpts from a prayer by Secchi to the Holy Spirit:[166]

---

163 *L'idea delle successive trasformazioni presa con debita moderazione non è punto inconciliabile colla ragione, né colla religione. Infatti, ove non si voglia tutto eseguito per pure forze innate e proprie della materia bruta, ma si ammetta che queste forze non d'altronde derivassero che dalla Cagione prima che creò la materia, e ad essa diede la potenza di produrre certi effetti, non vi è nessuna intrinseca repugnanza per credere che, fino a tanto che non interviene nessuna forza nuova, possano svilupparsi certi organismi in un modo piuttosto che in un altro, e dar origine così a differenti esseri. Ma quando da una serie di questi esseri si passa ad un'altra che contiene un nuovo principio, la cosa muta aspetto. Dal vegetale senza sensibilità non potrà passarsi all'animale che ha sensazioni, senza un nuovo potere che non può venire dalla sola organizzazione, né dalla sola materia. E molto più dovrà dirsi ciò quando si passa dal bruto animale all'uomo che ragiona, riflette ed ha coscienza. Un nuovo principio deve associarsi allora alle forze fisiche della materia per avere questi risultati* [...]. Secchi, *Lezioni elementari di fisica terrestre*, 199.

164 *Non* [*sanno*] *sbarazzar*[*si*] *dell'idea di un Ente supremo che tutto fece e tutto governa!* Secchi, *Lezioni elementari di fisica terrestre*, 191–192.

165 *Come se fossimo cosi' imbecilli da credere il Creatore quel vecchio dalla barba bianca e dallo svolazzo, dipinto da Raffaello nelle sue Logge.* Secchi, *Lezioni elementari di fisica terrestre*, 202.

166 It paraphrases, in the final lines, the known hymn *Veni, Creator*.

We therefore address ourselves to the Lord of all sciences [...] and especially to the very holy person of the very holy Trinity whose task is properly that to illuminate our intellect [...] we appeal to him to be enlightened, as we do not ask for this intelligence for our vanity [...] but to learn how to love and to serve him from the knowledge of him and his works.

May he [the Holy Spirit] visit our minds all opened to him [...]. May this gift be accompanied by the highest Grace that renders us friends of God and his dwellings [...] You, Paraclete Counselor, console our souls from the bitterness that is often the fruit of our struggles in studies [...]. May you, precious gift of the Highest Undivided Trinity [...], living source of every science [...], fire of true love of those whom always you link to your charity [...]. May you light up our mind with your light and our hearts with your ardor and give us fortitude and courage to overcome our rivals today and forever. [...] Let our studies be aimed at more and more knowing the divine Father, the Redemptor Son, and you.[167]

---

167 *Ricorriamo dunque al Signore di tutte le scienze* [...] *e in special modo alla SS. Persona della SS.a Trinità a cui è specialmente per appropriazione attribuito il rischiarare il nostro intelletto, ricorriamo a lui perché ci illumini, giacché noi non vogliamo questa intelligenza per nostra vanità* [...]. *Perché da questa cognizione di Lui e delle sue opere noi impariamo ad amarlo e a servirlo» «Egli visiti le nostre menti che sono tutte per lui non per la vanità* [...]. *Questi doni siano accompagnati da quelli della grazia suprema che ci fa amici di Dio e fa dell'uomo un suo abitacolo* [...]. *Né solo la mente ma anche il cuore, perché colla sola mente non potremo piacergli* [...]. *Voi paracleto consolatore, confortate le anime nostre nelle tristezze che spesso raccogliamo come frutto delle nostre fatiche negli studi* [...] *e dono prezioso della Triade altissima indivisa* [...] *fonte vivo di ogni scienza* [...] *fuoco di vero amore che ci leghi sempre alla vostra carità* [...] *e mozione spirituale colla quale voi potete lenire tutte le piaghe dell'anima e del corpo* [...] *illuminate col vostro chiarore la nostra mente, infiammate col vostro ardore i nostri cuori e dateci fortezza e coraggio per superare oggi e sempre i nostri nemici* [...]. *Fate che i nostri studi siano diretti a conoscere sempre meglio il divin Padre e il redentore Figlio e Voi.* APUG, FS 2.VII, 22.

CHAPTER 10

Secchi's Final Years: Uncertainty and Illness

After the annexation of Rome by the Kingdom of Italy, the eternal city was proclaimed its capital. For the politicians, most of whom were anti-clericals, it was time to demolish the temporal power of the pope and to establish a nation free from any religious influence. For this reason, hospitals, schools, and any other social institution held by religious orders were systematically confiscated in the territories that were annexed by the expanding Kingdom of Italy. However, once the unification was accomplished, the new Italian government had to manage the presence of the pope inside the capital city. In order to regulate their relationship, a law was promulgated (the Law of Papal Guarantees) in 1871 that fixed the properties and the rights of the pope. However, the law was unilateral and therefore considered unacceptable by Pius IX. The conflict between the papacy and the state thus continued on the diplomatic front, with a series of deliberations that provoked an escalation of tensions between the Italian government and the Holy See (see Introduction).

1 The Impact of Post-unity Policy

Naturally, the deterioration of the political relationship between the Italian government and the Vatican rendered Secchi's position increasingly difficult. The situation deteriorated in 1873 when a law was promulgated that extended the expropriation of the church's properties to the newly established capital. Consequently, the Collegio Romano had to be confiscated, as well as all other buildings belonging to religious orders. Secchi used all available legal means to defend the observatory, which had cost him so much of his energy and his money, and he wrote to the Ministry of Internal Affairs to argue that it ought to be considered a pontifical observatory, protected by the Law of Guarantees: "The observatory did not belong to the Jesuits, but to the holy father, who created and provided it with instruments, and established it for the use of the clergy, and therefore it was founded on a sacred and reserved place."[1] Secchi enclosed a list of instruments donated by the pope and purchased at Fr. Rosa's and his own expense, the value of which was not less than fifty thousand liras,

1 *L'osservatorio non era cosa dei gesuiti, ma del S. Padre da lui fondato, arricchito di strumenti e destinato all'uso del clero, e fondato perciò su luogo sacro e a se riservato.* APUG, FS 23.II.D.

 | DOI:10.1163/9789004387331_012

and concluded "that the observatory, rather than being subject to the suppression law, was defended by the other law, the one about Papal Guarantees."[2]

It is easy to imagine Secchi's mood at seeing the fruit of his labors and savings so seriously threatened. Despite several assurances, especially from Tacchini[3] and Schiaparelli,[4] Secchi thought seriously of leaving Italy and returning to the United States. Once more, Secchi became a pawn in the diplomatic war between Italy and the Vatican. Cardinal Antonelli, secretary of state of the Holy See, recommended that Secchi defend the pope's rights to the observatory and refuse any grant from the Italian government, but he did not assure any regular support from the Vatican if the observatory was not declared property of the pope. Not for the first time, Secchi was in a difficult position: "The cardinal works so that I am expelled, not by the government, but by the pope himself who denies my subsistence, because I am sure that the government will never make the declaration that the cardinal wants [...]."[5] Although his superiors could not openly align themselves with him, some of his confrères privately suggested that he accept financial aid from the Italian government if the Vatican did not provide the necessary support, and Secchi ultimately decided to take any money that was offered, given the absence of other forms of funding.

In that difficult moment, Secchi could count on the help of his friend Tacchini, who was called from Palermo to Rome "to make a deal and settle things amicably." Tacchini knew Secchi's position very well, since the Jesuit had already stated to him: "I would stay if the observatory remains pontifical, if not, I would not."[6] Thanks to the mediation of Tacchini, Secchi could continue as director of the observatory; all his conditions were accepted—the premises belonging to the observatory were to be isolated from the rest of the building of the Collegio Romano but would remain accessible to the staff of the observatory.

---

2 *Che l'Osservatorio anziché esser colpito dalla legge della soppressione, era difeso dall'altra delle Garanzie papali.* APUG, FS 23.II.D.

3 See Tacchini to Secchi, Palermo, April 14 and May 22, 1873, quoted in Chinnici and Gasperini, *Alle origini dell'astrofisica italiana*, 273 and 282.

4 See Schiaparelli a Secchi, Milano, April 21, 1875 [*sic*; 1873], quoted in Buffoni, Manara, and Tucci, *G.V. Schiaparelli, A. Secchi*, 233.

5 *Il cardinale lavora perché io sia cacciato dall'osservatorio non dal governo ma da papa in persona che mi nega la sussistenza; giacché son sicuro che il governo non darà mai la dichiarazione che vuole il cardinale.* APUG, FS 23.II.D.

6 *Tacchini astronomo di Palermo era stato chiamato a Roma evidentemente per trattare l'affare, e comporre le cose amichevolmente* [...] *Con Tacchini più volte avea detto che io sarei restato se l'osservatorio rimane pontificio, se no, no.* APUG, FS 23.II.D.

However, this compromise did not solve the financial problems facing the observatory. A climate of uncertainty and insecurity prevailed, which greatly influenced Secchi's health. The Collegio Romano Observatory become a sort of "no-man's land," unrecognized—and therefore unfunded—by any institution: neither by the resized papal government, which feared that it could be seized at some future date, nor by the Italian state. In short, it was neither Italian nor pontifical: it was simply Secchi's observatory. The Italian government was aware that to confiscate it would have been a very unpopular move, because it would have entailed the closure of the observatory. But it was impracticable—and prohibitive, given the kingdom's scarce financial resources—to maintain two observatories in Rome, where the Campidoglio Observatory was in full operation. As such, it made more sense to the Italian government to make no further moves to confiscate the observatory, simply abandoning it to its destiny and waiting for future events.

Another event that affected Secchi was his expulsion from the Academy of Lincei. In 1874, Minister Sella proposed reviving that ancient academy—which Pius IX, in 1847, had turned into a pontifical academy—in the name of secular science, renewing its statute and extending its classes. In this context, Secchi, being a Jesuit faithful to the pope, could no longer be included among its members. The astronomer was embittered by this event, which he viewed as a personal attack against him (see Chapter 8). The counteraction by the papacy was to keep the pontifical academy of New Lincei in existence and to appoint Secchi as president, in appreciation of his loyalty and in compensation for the abuse he had suffered.

2 The Congress of the Italian Scientists (1875)

In the meantime, the Italian government was facing the problem of maintaining twelve astronomical observatories—an expensive heritage resulting from the unification, since each of the Italian states pre-unification had its own astronomical observatory. Tacchini was aware that reform was necessary because their maintenance wasted a large amount of resources, and in 1874, he proposed a plan to reorganize the Italian astronomical observatories.[7] His proposal was discussed in the following year, in conjunction with the Congress of Italian Scientists in Palermo. Secchi took the opportunity to attend the meeting:

7 See Poppi, Bònoli, and Chinnici, "Il progetto Tacchini e la riforma degli Osservatori italiani."

> I had promised Tacchini to go to Palermo to do a comparative study of his spectroscope with mine and see if I could confirm the size of the solar diameter or explain the differences. In the end, I could not turn him down, recalling as well that if I am still at the observatory it is because he had managed and concluded the affair with the minister of education.[8]

Thus the main reason why Secchi decided to go to Palermo was to defend his observatory, as among the suggestions in the proposed reform was the closure of the Collegio Romano Observatory and the transfer of its instruments to the Florence Observatory.[9] Secchi saw this project, promoted by Sella and the mayor of Florence, Ubaldino Peruzzi (1822–91), as a trick "for eliminating," at the same time, "the expenditures for the observatory and the inconvenience of having a Jesuit around."[10]

As usual, the news of Secchi's participation at the Congress of Italian Scientists did not fail to give rise to controversy. He noted:

> Foolish voices were enthusiastically collected by clerical newspapers and they wrote all sorts of things about me [...]. Of course, they did not even neglect to inform the pope that I was embarked in that hubbub of the congress of Voltaire's disciples whose purpose was to conjure up the extermination of Christianity.[11]

This impression was not helped by the attendance of Renan, denier of the divinity of Jesus Christ (see Chapter 9), who had been invited to the same congress. Following the advice of the local Jesuit community, Secchi decided to stay with his confrères and to participate solely in the parts of the conference concerning

---

8 *Avea promesso a Tacchini di andare a Palermo a fare uno studio comparativo del suo spettroscopio col mio e vedere se si poteva provare riscontri pel diametro solare e spiegarne le divergenze. In fine non potei ricusarmi, anche ricordandomi che se sto all'osservatorio esso fu che maneggiò l'affare e lo concluse col Ministro dell'Istruzione* [...]. APUG, FS 23.II.D.

9 When Donati died, in 1873, the construction of the new observatory at Arcetri was almost finished, but its equipment remained incomplete. Several actions were undertaken by national and local authorities to bring the project to completion; see Simone Bianchi, Daniele Galli, and Antonella Gasperini, *Giovanni Virginio Schiaparelli e l'Osservatorio di Arcetri*, Fondazione Giorgio Ronchi CXI (Florence: Fondazione Giorgio Ronchi, 2011).

10 *Per levarsi la spesa dell'osservatorio e la incommodità del gesuita nel mezzo* [...]. APUG, FS 23.II.D.

11 *Stolide voci furono raccolte con entusiasmo dai fogli clericali e ne dissero di tutte le specie a mio carico* [...]. *Anche al papa naturalmente non mancarono di dire che io era imbarcato in quella baraonda del congresso di volteriani il cui scopo era congiurare lo sterminio del cristianesimo.* APUG, FS 23.II.D.

N. 220

XII. CONGRESSO DEGLI SCIENZIATI ITALIANI

E

I. CONGRESSO DELLA SOCIETÀ ITALIANA

PER IL PROGRESSO DELLE SCIENZE

CARTA DI AMMISSIONE

Il Sig. Prof. Angelo Secchi

è ammesso al Congresso ed ha dichiarato di ascriversi alla classe Matematiche

Palermo 28 Agosto 1875.

Per il Comitato

Questa carta vale per le adunanze del congresso e delle classi e per gli inviti ai convegni.

Tip. Lao.

FIGURE 10.1 Secchi's registration card for the Congress of Italian Scientists, held in Palermo in 1875. Courtesy of INAF-OAR.

the reform of the observatories, without attending any other meetings. Secchi wrote to Ferrari: Here there are rumors about the congress, but we have to distinguish congress from congress. Congress of demons with their patriarch Renan: I shall not attend it, but I cannot miss the astronomical and meteorological meetings to be held on demand of the Ministry at the observatory, they will end soon.[12] That was a decision, however, that Secchi regretted very much:

> I condemned myself to staying at home during all the hours of the public sessions. Thus I visited neither the national library, which is that of

12 *Qui molto si chiacchera per il congresso, ma vi è congresso e congresso. Congresso di demonii col loro Patriarca Renan. A questo non mi presentai, ma non posso mancare di andare alle conferenze che si terranno per domanda ministeriale dai meteorologisti e astronomi, le quali si faranno all'osservatorio, e si sbrigheranno presto.* Secchi to Ferrari, Palermo, August 25, 1875; APUG, FS 9.VII.

FIGURE 10.2 Satirical vignette about the Congress of Italian Scientists. Secchi is recognizable, playing with a kite; significantly, he is drawn close—though back to back—to Renan, who plays with a puppet, on the right. Private collection.

our ancient college, nor the University Hall, no exhibition, none other. On my word, I never stepped into any of those sites where I could have encountered Renan or Prince Umberto.[13] [...] In those days, my life in Palermo was worse than that of a friar, because I did nothing but go to the observatory and back home, and, once the congress was opened, to the university and then back home. Nor did I even have the time for walking [to and from the meeting] to myself, because M[arquis] Spedalotto[14] wanted me to accompany him in his carriage and this daily tour was

13 Son of Victor Emmanuel II, Umberto (1844–1900) was crown prince of the Kingdom of Italy.

14 Marquis Spedalotto and Secchi were in a friendly relationship from 1870 (see Chapter 5).

> a most cruel torment for me.[15] […] I could not have received a worse punishment.[16]

Despite his caution, Secchi was slandered in Rome, as he noted:

> "[Some agitators] said that I was being accommodated in the house of the archbishop who had been expelled, and I slept in his bed next to Renan! However, rather than retreat, I went ahead, having determined well my course of conduct […] [Nevertheless, some priests], animated not by zeal but envy, did not hesitate to slander me in Rome, and say that I had been with Renan, that talismanic word […]. So I was advised to pay a visit to the pope to undeceive him, with a pretext […]. The pope understood everything, patiently listened to me while I was telling him the story, and finally said: do not care about this chattering and go ahead, as I know you very well.[17]

However, it was worth going to Palermo, as Secchi's participation in the debate on the reform had a favorable outcome: in the reorganization of the observatories, provided for in a decree of 1876 by Minister Ruggiero Bonghi (1826–95), a special "status" was recognized for the Collegio Romano Observatory. The Bonghi decree essentially replicated the proposal discussed at Palermo and established that the observatories of Milan, Florence, Naples, and Palermo were classified as research institutions, and those of Turin, Padua, Bologna, and

15 The daily tour included the port of Palermo, and Secchi was generally disgusted by the odors and the chaos of the docks.

16 *Mi condannai a restare in casa durante tutte le ore che duravano le sedute pubbliche. Così non ho veduto nemmeno la biblioteca nazionale, che è quella del nostro antico collegio, né la sala universitaria, né nessuna esposizione o altro. In mia parola non misi mai piede in nessuno di que' siti ove poteva affrontarsi o il Renan o il Principe Umberto.* […] *La mia vita in questi giorni a Palermo fu quella di peggio che frate, perché non faceva altro che andare all'osservatorio e a casa, e cominciate le conferenze, alla università e poi a casa. Né anche aveva libere le ore del passeggio perché il M*[*arche*]*se Spedalotto mi voleva seco in carrozza e la passeggiata sua era per me il più crudo tormento.* […] *Pena maggiore non mi si poteva dare.* APUG, FS 23.II.D.

17 [*Alcuni agitatori*] *dicevano che io* […] *sarei stato alloggiato in Arcivescovado donde era stato espulso il vescovo, e che avrei dormito nel suo letto accanto a Renan!* […] *Ma io invece di ritirarmi credetti tirar di lungo avendo ben fissata la mia linea di condotta* […] [Ciononostante, *alcuni preti*] *non mancarono di calunniare in Roma, e dire che io era stato con Renan, parola talismanica, non per zelo ma per invidia* […] *Sicché mi fu consigliato di fare una visita al Papa per disingannarlo prendendo il pretesto di avere da lui delle indulgenze pel Marchese. Il Papa capì tutto, e mi sentì per tutta la storia con pazienza e mi soggiunse: non badate alle ciarle e tirate avanti che ben vi conosco.* APUG, FS 23.II.D.

FIGURE 10.3 Minister Ruggero Bonghi, who signed the Observatories' Reform decree in 1876; https://it.wikipedia.org/wiki/Ruggiero_Bonghi#/media/File:RuggeroBonghi.jpg (accessed June 8, 2018).

Rome (Campidoglio) as university observatories, while article 3 of the decree deferred any decision about the Collegio Romano Observatory to a future date, basically leaving its situation unchanged.[18] This legislative measure temporarily averted the confiscation of the observatory, although it did not secure its maintenance and financial support.

Secchi's vision of Italian astronomy can be seen in the minutes of the 1875 meeting discussing of the reform of the observatories proposed by Tacchini:

18 This reform, however, was never implemented (see Poppi, Bònoli, and Chinnici, "Il progetto Tacchini e la riforma degli Osservatori italiani").

> Fr. Secchi took the floor to express his ideas on the adjustment of Italian observatories. He says that the conditions of our astronomical observatories are unfortunate, and therefore there is much effort but little result. The topographic conditions are, according to him, the most fundamental, because an observatory placed on top of a very high building from the ground and in the middle of a big city can no longer satisfy all the current needs of science. Temperature variations in high buildings create too many sensitive movements, which must necessarily affect precision instruments. [...] He enumerates the more serious disadvantages in such observatories, such as the ground vibrations produced by the transit of vehicles through the streets, the sounds they produced, the irregularity of refractions, etc. [...] Among all the Italian observatories, only those of Naples and Florence are in impeccable conditions from the point of view of their location, although he realizes that, even for these two ones, there may be some irregularities from the surrounding soil configuration [...].[19]

Secchi's analysis was lucid and realistic. He was well informed about the state of the buildings and instruments, but he also addressed the question of the staff:

> Fr. Secchi says that to be charged with many tasks is an impediment to the work of the observers, and we should avoid it. He finds that there is much to complain about the lack of staff. For a meridian circle, if you want [...] the instrument to be operated without interruption, three astronomers must be involved: one will be the head of that section of the observatory; but the observers, including the head, must take turns using the instrument. This is the practice in large observatories such as in Greenwich and in Washington, where the astronomer has never two

19 *Il P. Secchi prende la parola per esporre le sue idee sul riordinamento degli osservatorii italiani. Dice che lo stato dei nostri osservatori astronomici è deplorabile, e perciò vi si fatica molto ma con poco frutto. Le condizioni topografiche sono, secondo lui, le più fondamentali, perché un osservatorio messo in cima ad un edificio molto elevato dal suolo, ed in mezzo ad una grande città non può più soddisfare a tutte le attuali esigenze della scienza. Le variazioni di temperatura inducono negli alti edifici dei movimenti troppo sensibili, che debbono necessariamente influire sopra gli strumenti di precisione.* [...] *Enumera ancora altri gravi inconvenienti in osservatorii di tal fatta, quali sono le oscillazioni del suolo prodotte dal transito dei veicoli per le vie adiacenti, i rumori che vi si producono, la irregolarità delle rifrazioni, ecc.* [...] *fra tutti gli osservatorii italiani solamente quelli di Napoli e di Firenze sono in condizioni inappuntabili sotto il punto di vista della ubicazione, quantunque egli dubiti che anche per questi due possa esservi qualche irregolarità proveniente dalla configurazione del suolo circostante* [...]. Poppi, Bònoli, and Chinnici, "Il progetto Tacchini e la riforma degli Osservatori italiani," 156.

consecutive nights of observations, and the same instrument never remains unused. The same goes for the necessary staff in the service of an equatorial [telescope].[20]

The conclusion was dramatic:

> Fr. Secchi says that there does not exist in Italy a complete observatory, one that does not need a major expenditure: we should make sure that the country has an astronomical observatory set up in good conditions with a good topographical location as well as these three main instruments: a large meridian circle, an equatorial of at least twelve-inch aperture, and a large vertical circle for observations of the moon.[21]

In this depressing landscape, according to Secchi it made sense to cultivate "physical astronomy" because it did not require the high precision instruments and stability that astrometry needed, yet nevertheless could provide important and original contributions to the understanding of celestial phenomena.

From these minutes, it clearly appears that Secchi's scientific authority was unquestioned. As a "leader," he easily took matters into his own hands: he presented the agenda, tried to synthesize and put to the vote the various resolutions. He respected his colleagues, but he did not hide the problems and showed that he sincerely cared about the fortunes of astronomy. In fact, there was no other prominent figure comparable to him in the Italian astronomical context of that time, as demonstrated by his unparalleled scientific productivity and the many awards he obtained. His belonging to a religious order did not seem to affect his relationship with fellow Italian astronomers, who seemed to

---

20 *Il p. Secchi dice che il cumulo degli impieghi è di impedimento all'attività degli osservatori, e si dovrebbe evitare. Trova inoltre che sia molto da lamentare la mancanza di personale. A un circolo meridiano quando si vuole* [...] *che l'instrumento sia adoperato senza interruzione, devono essere addetti tre astronomi: uno farà il Capo di quella sezione dell'osservatorio; ma gli osservatori, compreso il Capo, debbono alternarsi allo strumento. Così si pratica nei grandi osservatorii di Greenwich e di Washington, dove lo stesso astronomo non ha mai due notti di seguito di osservazioni, e pure gli stessi strumenti non restano mai inoperosi. Dicasi lo stesso per il personale necessario al servizio di una* [*macchina*] *Equatoriale.* Poppi, Bònoli, and Chinnici, "Il progetto Tacchini e la riforma degli Osservatori italiani," 157.

21 *Il P. Secchi dice che non esiste in Italia nessun osservatorio completo e che non abbia bisogno di grandi spese: che bisogna fare in modo che il paese abbia un osservatorio astronomico messo in buone condizioni topografiche, e fornito, oltre gli strumenti secondari, di questi tre principali strumenti cioè un grande Circolo meridiano, una Equatoriale di almeno dodici pollici di apertura, ed un grande circolo verticale per le osservazioni della Luna.* Poppi, Bònoli, and Chinnici, "Il progetto Tacchini e la riforma degli Osservatori italiani," 157.

take more account of his scientific achievements than his personal choices. When, during the discussion of the reform, they faced the question of the Collegio Romano Observatory, the members of the commission invited Secchi to prepare a report with a "desiderata [...] declaring now that his proposals will be held in the highest regard, and recommended to the government."[22]

3 Secchi's Journey to Cosenza

After recovering from a stomach disease, which appeared during his stay in Palermo and probably portended the malady that would ultimately lead Secchi to the end, the Jesuit astronomer kept a promise and went to Cosenza, in Calabria. In fact, the director of the local meteorological observatory, Domenico Conti, had asked him to help install a geomagnetic station there. Secchi traveled by train to Corigliano, where the welcome was "fitting for a prince": noblemen and authorities came to the railway station to escort him to the castle of Baron Francesco Compagna (1848–1925), where a reception was held in honor of the scientist.[23]

Secchi then traveled by cart to Cosenza, crossing the district of Sibari and the valley of the Crati River.[24] One hour before arriving at Cosenza, a committee headed by the mayor came to welcome him. Secchi was invited to leave the cart and to take the carriage of the mayor: "A minister would receive no better treatment [*non meno che per un ministro*]," Secchi commented. Once they arrived in the outskirts of the town, all the people came out to welcome him, in an atmosphere like a sort of popular festival.

Conti was most happy: "He is an excellent doctor and a good Christian," Secchi wrote in his diary; "he loves meteorology, but does not understand its difficulties very well. He wanted to erect a geomagnetic observatory without knowing what it was."[25]

22 Poppi, Bònoli, and Chinnici, "Il progetto Tacchini e la riforma degli Osservatori italiani," 168.

23 Secchi wrote: "The baron is a very polite young man, good, liberal in the good sense, religious, he has a friar as a chaplain, and is very rich but does not amass wealth and spends everything for the goodness of the local people [...]. May God preserve him; I shall always keep a good memory of him" (*Il Barone è un giovane compitissimo, buono, liberale in senso buono religioso, tiene un frate per cappellano, ricco assai che non accumula e spende tutto pel bene del paese che ne sono fanatici. Iddio lo conservi, io ne avrò sempre buona memoria*). APUG, FS 23.II.D.

24 Secchi was struck by the contrasts between the poor appearance of those huts and the silver sets that people used there to serve their guests.

25 *È un ottimo medico e buon cristiano. Ama la meteorologia, ma non capisce molto le difficoltà. Si era imbarcato nella erezione di un osservatorio magnetico senza sapere che cosa fosse* [...]. APUG, FS 23.II.D.

The choice of the site, as well as the adjustment of the instruments, was problematic; Secchi was quite unhappy with the final result, but recognized that it would have been difficult to do better in those conditions. Before Secchi's departure, the Cosentins organized what Secchi described as a "family party," with a "rich lunch": liberals and Catholics gathered together, "the first because of the arrogance of science, the second because of the memory of the Jesuits."[26] Secchi was then asked to give a public presentation on the connections between solar activity and geomagnetism and to draw a sundial on the wall of Compagna Palace, in memory of his sojourn.

Secchi left Cosenza with symptoms of fever. He preferred, however, to hide his illness and to travel by wagon to Eboli, where he had to take the train to Naples and Rome. Unfortunately, the trip was much longer than expected and troubled by many incidents:

> We spent the night in the Crati Valley with soldiers [...] because of the risk of bandits! When we came to the bridge crossing the river [...], the horses stopped in the middle of the bridge, and it was impossible to move them forward and they were stuck on the wooden bridge, in danger of capsizing into the river. [...] It was decided to detach the horses and to leave the wagon halfway across the bridge, and to send the postilion back to the previous postal station, in order to provide new horses. Meanwhile, the sky became dark and a frightful rain and storm began! I was totally soaked because of the effort to move the horses forward [...], and it was impossible for me to change my clothes [...], so that I felt iced and doomed. At that moment, the postilion returned and said that there were no available horses at the station, since the two ones there were ill. Of course, I became quite upset and ordered the wagon to get off the wooden bridge where we were in danger, turn around, and go back to the postal station, [...] threatening to give a flaming report to the prefect (he was from Reggio, of good stuff and we became friends, so I could act the blowhard). [...] While we were reversing the wagon, we heard a horn: a postilion and the mailman were coming, with two horses [...] to our aid. Really, they were angels, not men, and we calmed down and [...] got moving. I could not change my clothes before arriving at the postal station, where I did it almost in *propatulo* [i.e., publicly], oh well, it was night and you could not see anything. [...] Shortly after dawn, the horses balked again, and I [...] ruled the roost and ordered that two oxen be removed from a

26 *I primi per la boria della scienza, i secondi per la memoria dei Gesuiti.* APUG, FS 23.II.D.

> cart and attached before the horses, in order to move forward, shouting aloud, and warning that I would report everything to the prefect.[27]

After arriving at Eboli, Secchi then traveled by train to Naples, while his fever increased. Secchi thus concluded his tale:

> As God wished, I arrived at Naples and found shelter. After some rest, I went to the observatory to visit the astronomers [...]. The day after, I arrived in Rome, exhausted, and two days later I had a cough, spitting blood, and a pernicious fever that led me to death's door, and if I had not taken forty grains of quinine in four hours, I would be dead. This was the result of the expedition to Cosenza.[28]

4 The National Board for Meteorology (1877)

These various tensions and concerns wore Secchi out. As a consequence, in 1876 some of his scientific pursuits were regularly entrusted to his assistants

---

27 *Passammo la notte nella valle del Crati coi soldati [...] perché si temevano briganti! Ma arrivati al ponte che lo traversa [...] i cavalli non vollero fare la salita e si fermarono in mezzo al ponte e non volevano andare avanti e intraversarono il legno col pericolo di rovesciarci nel fiume. [...] si prese il partito di staccare i cavalli e di lasciare il legno così a traverso, e inviare il postiglione alla stazione di prima postale, per avere dei cavalli. Intanto si abbuia il cielo e comincia un'acqua e una bufera spaventosa! Io era tutto in acqua per la fatica fatta prima onde fare camminare i cavalli [...] e mi era impossibile mutarmi [...] onde mi sentii gelare e mi vidi perduto. Ritorna in quel punto il postiglione dicendo che alla posta non vi sono cavalli e che i due che vi sono sono malati. Come era naturale io stranii assai, e ordinai di rivoltare indietro col legno, andare alla posta, levarci da quel ponte che eravi estremo pericolo [...] minacciando al postiglione e al postiere di un rapporto fulminante al prefetto di Cosenza (questi era reggiano, di buona pasta, e avevamo fatto molto amicizia, onde io poteva fare il gradasso). [...] nel mentre che stiamo voltando il legno, ecco che si ode una cornetta, ed era un postiglione e con esso il postiere che venivano con due cavalli [...] per soccorrerci. Veramente erano angeli che venivano, e non uomini, onde calmatici alquanto [...] ci mettemmo in via. Non fu possibile cambiarmi se non alla posta vicina, e ciò anche dovetti fare quasi in propatulo, ma pazienza, era notte e non ci si vedeva davvero. [...] poco sopra al far del giorno i cavalli si impuntarono di nuovo, e [...]mi misi a fare io il padrone, e ordinai che si staccassero da un carro due bovi per attaccarli ai cavalli davanti onde andar avanti, e ciò feci sempre gridando forte, e facendomi forte col rapporto al prefetto.* APUG, FS 23.II.D.

28 *Come a Dio piacque, giunsi a Napoli e mi riparai. Riposatomi un poco andai all'osservatorio per visitare gli astronomi [...] Il dì appresso venni a Roma ma sfinito, e due giorni dopo mi si sviluppò una tosse con sputo sanguigno, e febbre che fu la famosa perniciosa che mi portò sull'orlo del sepolcro, e che se non erano 40 grani di chinino presi in 4 ore io era spacciato. Questo fu l'esito della spedizione a Cosenza.* APUG, FS 23.II.D.

Marchetti and Ferrari, and Secchi wrote to the father provincial begging him to exonerate them from some religious commitments:

> Sunday we have people coming in the morning from half past nine to midday. Being both occupied until ten or eleven, I would be alone, and if they are already tired when they get here, I cannot give them those instructions that are necessary for the distinguished people who visit us, and now I get tired so easily that I can no longer do this. [...] Among the scientific obligations [of Fr. Ferrari] at the observatory is the observation of the Sun [...], which I stopped carrying out because I get enormously tired.[29]

However, though he was no longer capable of regular observations, Secchi did not stop his scientific work. In fact, he intensified the production of books and treatises (see Chapter 6): an extended edition of *Le soleil* in two volumes, published in 1875 and 1877, and the treatise *Le stelle*, also published in 1877. He entrusted the editing of this book to Schiaparelli, who wrote that he felt honored "to act as a midwife of this new and beautiful work."[30] Secchi also prepared the manuscript of the *Lezioni di fisica terrestre* (Lessons on Earth physics), published posthumously in 1879. He was obliged to do all this work in order to find additional resources for the observatory.

In the meantime, Tacchini was organizing national scientific networks and proposed the creation of a Board for Meteorology, established by the three Ministries of Navy, Agriculture and Trade, and Public Education, with the purpose of standardizing the methods and the equipment to be adopted at the Italian meteorological observatories. Secchi was appointed chairman of the board, probably at Tacchini's suggestion. Secchi's achievements as a meteorologist (see Chapter 6) were so remarkable that his leadership was unquestioned in this field, and the government probably wanted to compensate for past conflicts.

---

29 *La domenica noi abbiamo la gente che viene alla mattina dalla 9½ a mezzodì. Essendo ambedue occupati fino alle 10 o alle undici io resterei solo, e venendo su già stanchi io non potrei servirmene per dare quelle istruzioni che occorrono alle persone di distinzione che capitano, e io ora sono così facile a stancarmi che non posso più fare io questo ufficio.* [...] ... *tra gli obblighi scientifici che* [*p. Ferrari*] *ha per appunto l'osservatorio vi è l'osservazione del sole ... avendo io cessato di farla perché mi stanco enormemente.* Secchi to Fr. Valeriano Cardella, S.J., Rome, February 4, 1876; ASPEM, Fondo Provincia Romana, "Fascicoli personali," 2604.

30 *Io mi tengo per assai onorato ch'ella voglia concedermi di far un poco da ostetrico a questa sua nuova e bella opera* [...]. Schiaparelli to Secchi, Milan, January 21, 1877, quoted in Buffoni, Manara, and Tucci, *G.V. Schiaparelli, A. Secchi*, 243.

But it was too late. The Jesuit's health deteriorated irreversibly. The board decided on the inspection of some selected meteorological observatories, and Secchi chose to visit those nearest to Rome, the travel being less tiring for him. He visited the observatories of Montecassino, Naples, Monte Cavo, Grottaferrata, and Potenza; for each observatory, he compared instruments, locations, and rooms, and edited complete and accurate reports, pointing out the various features of each observatory. "Even this modest work had sad consequences for him," wrote Fr. Ferrari later.[31]

5 Secchi's Death

In August 1877, Secchi showed symptoms of a stomach cancer that within a few months sapped his energy. In October, he wrote to Schiaparelli that he had not been able to observe Comet Tempel: "The weather was bad, and nothing could be seen. Now [...] I am plagued with excruciating stomach pains, and the doctor forbids me to get up at night, so I won't be able to deal with it."[32]

Brother Marchetti wrote:

> In August [1877], Father Secchi began to feel ill and said clearly that he felt his energy abandoning him and to be near to the end. He passed three months and more from bed to cot, but at the beginning of November, a lack of appetite [...], a general malaise [...], a great tiredness and nausea for everything, foretold the catastrophe [...] of his untimely end [...].[33]

He was examined by many noted doctors, all of whom gave the same verdict:

> His disease was declared life-threatening in December by Professors [Lorenzo] Poggiopollini, a close friend of the late father, [Alessandro] Ceccarelli [personal doctor of the pope], Targioni [possibly Adolfo

31 APUG, FS 24.I.

32 *Il tempo è stato pessimo, e non si è potuto veder nulla. Adesso poi è anche partito il P. Ferrari [...] e non so quando ritornerà, e io sono tormentato da atroci dolori di stomaco, e il medico mi vieta di alzarmi di notte, onde non me ne potrò occupare.* Secchi to Schiaparelli, Rome, October 8, 1877, quoted in Buffoni, Manara, and Tucci, *G.V. Schiaparelli, A. Secchi*, 251.

33 *Il p. Secchi fin dall'agosto [1877] cominciò a sentirsi male e diceva chiaro che egli sentiva venirgli meno le forze e che era in fin di vita. Passò fra letto e lettuccio buoni tre mesi, ma al principio di novembre una inappetenza caratteristica in un uomo piuttosto di cibo, un malessere in tutta la vita e [...] una grande stanchezza nelle membra e nausea di tutto ci fece presagire la catastrofe [...] dell'immatura sua morte.* APUG, FS 23.III.

> Targioni Tozzetti (1823–1902)] and Cheverie. However, their advice was not communicated to the patient. These gentlemen perceived that in the stomach of the patient, near the pylorus, there was a hard fibrous growth, which had necessarily to be a cancer or a *scirrhous*, or a perforated ulcer.[34]

Denza paid a visit to Secchi at the end of November 1877 and was alarmed by his condition:

> I was in Rome in the modest room of the excellent father, in friendly and ever pleasant conversations, when the news arrived that the astronomer von Littrow had died. Hence, with a sad premonition but without perturbation, he exclaimed: *Now, it is my turn!* I could not remove him from this dark thought, despite my comforting words, which, in truth, I tried to pronounce with a calm and tranquil voice, but without outspokenness, because I too saw that it was glimpsing the light of that torch, once so much shining. His lively look, which, though not singular, was pleasant and delighting, had become sad and weary. At the same time, his speaking was no longer enriched by those nice and sapid traits that he used with me as well as with people who enjoyed his company and that rendered his conversations so pleasant.[35]

---

34 *Il p. Secchi fin dall'agosto* [1877] *cominciò a sentirsi male e diceva chiaro che egli sentiva venirgli meno le forze e che era in fin di vita. Passò fra letto e lettuccio buoni tre mesi, ma al principio di novembre una inappetenza caratteristica in un uomo piuttosto di cibo, un malessere in tutta la vita e* [...] *una grande stanchezza nelle membra e nausea di tutto ci fece presagire la catastrofe* [...] *dell'immatura sua morte* [...] *Il suo male venne dichiarato mortale fin dal decembre dai sigg. Professori Poggiopollini, intimo amico del defunto Padre, Ceccarelli, Targioni e Cheverie. Ma non venne tale decisione comunicata all'infermo. Questi signori si avvidero che nello stomaco dell'infermo presso al piloro vi era una durezza ben provata, la quale doveva necessariamente risolversi o in un cancro o in uno scirro ovvero in un'ulcera perforante.* APUG, FS 23.III.

35 *Io era a Roma nella modesta stanza dell'ottimo Padre in amichevole ma sempre gratissimo conversare, quando ci arrivava l'annunzio della morte dell'astronomo De Littrow. Allora egli, con mesto presentimento, ma senza conturbarsi, esclamò:* Ora tocca a me! *Non valsero a distornarlo dal lugubre pensiero le mie parole di conforto, le quali, a dire il vero, mi studiavo sì di comporre con voce calma e serena, ma non già con animo franco, perché anch'io scorgevo assai fioca la luce di quella face, che altra volta splendeva cotanto. Il suo aspetto vivace, che, sebbene nulla avesse di singolare, piaceva tuttavia e rallegrava, era addivenuto triste ed oppresso. Né più il suo parlare era condito da quei modi lieti ed arguti, coi quali egli usava trattare con me, come con tutti coloro che godevano della sua confidenza, e che ne rendevano tanto cara e piacevole la conversazione.* Francesco Denza, *Il Padre Angelo Secchi* (Turin: Collegio Artigianelli -Tipografia e Litografia S. Giuseppe, 1878), 30–31.

In an effort to relieve his suffering, Secchi wanted to get out of Rome. Someone proposed that he to go to Nice, but Secchi declined: he suffered in the sea air,[36] and probably also felt he did not have enough energy for such a long journey. Moreover, he did not want to stray too far from the Collegio Romano, a place where he had spent most of his life and was home to his fondest memories. Superior General Beckx then invited him to spend some time in his residence Villa San Girolamo at Fiesole, where the headquarters of the Jesuit curia had been moved after the Kingdom of Italy's annexation of Rome. Fiesole, near Florence, was not too far from Rome. Secchi accepted the invitation, and on November 29, accompanied by Denza, he left for Fiesole, where his doctor friend Poggiopollini could continue to treat him from Rome. Denza wrote: I was […] much saddened and moved in seeing weary and exhausted, during the travel, that dearest man, whose dynamism and lust for life I have always previously admired; and with deep sadness and greatest pain, I made my farewell in the evening of that same day.[37]

But Secchi could not stray far from his observatory, and when he felt that the end was near, he asked to return to the Collegio Romano, to spend his last days in the place that was so special to him. At the very beginning of 1878, Ferrari published in the *Bullettino meteorologico* some studies by Secchi on electricity, in particular on Ohm's law.[38] It is unclear if this was done on his own initiative or at Secchi's request. This is, however, Secchi's final publication: by a strange coincidence, his first important work, published more than thirty years before, had been on the same subject—a sort of closing of a scientific loop.[39]

When news spread of the seriousness of Secchi's disease, the reactions were immediate. According to Marchetti:

> There was no nuns' convent or religious house where Fr. Secchi was known that did not raise fervent prayers to God for his recovery. They sent him objects used by the saints of their religious orders [a pious practice to ask for the intercession of those saints], but while accepting the pitiful offerings with gratitude, he would respond: "I just prayed for them

36 *Il p. Secchi non poteva sentire affatto l'aria di mare e me lo scrisse chiaramente in una lettera il 16 Xbre.* (*Fr. Secchi absolutely detested the sea air as he wrote to me in a letter dated December 16*) Marchetti in APUG, FS 23.III.

37 *Rimasi* […] *rattristato oltremodo e commosso nel vedere lungo il viaggio spossato ed affranto quell'uomo carissimo, che le altre volte tutte avevo ammirato pieno di vigore e di brio; e con profonda tristizia e con pena grandissima, presi da lui commiato la sera di quel giorno medesimo.* Denza, *Il Padre Angelo Secchi*, 31.

38 See Angelo Secchi, "Sulla legge di Ohm delle correnti elettriche," *Bullettino meteorologico dell'Osservatorio del Collegio Romano* 17 (1878): 9–10.

39 Posthumously, Ferrari also published a short text by Secchi about weather forecasting (see Angelo Secchi, "Sopra il prognostico del tempo," *Bullettino meteorologico dell'Osservatorio del Collegio Romano* 17 [1878]: 20; 25–26).

> to help me to die well. I've had enough of this world, nor do I care for the grace to live longer." The prayers of so many religious souls were not unsuccessful: if our Fr. Secchi, during that long and painful illness, had an admirable and prodigious calm and tranquility of mind, this was certainly due—he often repeated—to the incessant prayers of many pious souls and of those many nuns whom he patiently taught and instructed in the studies to let them pass the exams.[40]

Marchetti's report could of course be considered mere hagiography, an effort to portray Secchi in a positive light if not for apologetic reasons then perhaps simply out of his affection for the Jesuit. However, it is clear that Secchi was aware that the end was near. He had fought hard in the last years of his life, and the persisting uncertainty about his future and the straitened circumstances and difficulties he faced caused him great stress and undermined his health.

As a religious, Secchi prepared himself for death. His assistants reported that he obtained from the pope permission to receive Communion twice a week without fasting (due to his health condition) and that he wanted to keep holy water in his room for sprinkling. He also asked to receive the Viaticum publicly from the nearest parish, S. Maria in via Lata, in order to give a public attestation of his faith, but his superiors did not want to alert the people. Though the doctors said that it was not urgent, he asked to receive the Last Rites on January 27; those who attended described it as a very moving event: "[He said]: 'I can die at any moment with such excessive vomiting, and don't want to miss such a sacrament.'" He made an almost public confession and accused himself of being an unworthy son of the Holy Church and the Society of Jesus; he asked forgiveness from all the servants and bystanders "for the poor and scanty edification in his life [*cattiva e poca edificazione data nella sua vita*]," in the midst of a general commotion, and added: "Now I die quietly [...] and I look with calm for my departure from the world."[41]

---

40 *Non vi fu monastero di Vergini o casa religiosa la quale conobbe il p. Secchi che non alzasse fervide preghiere a Dio benedetto pel suo ristabilimento. Gli mandano oggetti usati dai Santi loro religiosi, ma egli accettando con viva gratitudine le pietose offerte, soleva rispondere così =Mi raccomando solo a loro affinchè mi aiutino a ben morire. Di questo mondo ne ho abbastanza, né curo la grazia di più sopravvivere= Le preghiere di tante anime religiose però non rimasero infruttuose: se il nostro p. Secchi, in quella lunga e penosissima infermità, ebbe una calma e tranquillità di animo ammirabile e prodigiosa si dovette certo, cone Egli pure si ripeteva sovente, alle incessanti preghiere di tante anime pie [e] di quelle tante religiose che egli pazientemente istruì e diresse negli studii per farle presentare agli esami.* APUG, FS 23.III. After the annexation of Rome, all nuns who were teachers in their own colleges were obliged to pass exams by the Italian Government, in order to be allowed to keep teaching.

41 *Posso morire da un momento all'altro fra questi eccessivi sbocchi di vomito, e non voglio rimanere privo di un tanto Sacramento [...] Ora muoio tranquillo [...] ed aspetto con calma la mia partenza dal mondo.* APUG, FS 23.III.

FIGURE 10.4 Portrait of Angelo Secchi. From *Tituli in supremis honoribus Angeli Secchii e Societate Jesu* (Rome: Tipografia delle Scienze Matematiche e Fisiche, 1878).

A notebook containing daily updates on Secchi's worsening health condition is still preserved,[42] with the signatures of the people that came to hear the latest news: from January 28 to February 25, 1878, it gave brief notes about his alternating condition up to its progressive deterioration. On January 29, the notebook bears fifty-eight signatures, including Princes Torlonia, Boncompagni, Lampedusa, Bonaparte; many other names appear in the following pages: nobles, reporters from the local newspapers, cardinals, priests, Jesuits, nuns, and so on. On February 1, a delegation came from Florence, including Tempel, Cantoni, Matteucci, and mathematician Costantino Pittei (1839–1912).[43] Secchi's friend Tacchini was also ill during this period, affected by the recurrent fevers he had contracted in India,[44] but Ferrari kept him informed: "Even this last night Father Secchi had his characteristic vomit and is quite depressed [...]. The mind is perfectly calm. You can imagine how we are."[45]

A few days before Secchi's death, Ferrari wrote:

> Unfortunately, you are right not to be quiet about our Fr. Secchi because, if there were several days of relative calm, from the day before yesterday the disease is exacerbated, and alarming symptoms appeared this morning, the first of which is some irregularity in his pulse. His energy is ever more worn out, and he suffers very much but quietly; you wouldn't recognize him anymore. We have long resigned ourselves to endure this immense misfortune.[46]

Marchetti extended his narration to the very last minutes of Secchi's life, and wishing to express how Secchi kept his sense of humor until the very end, he reported an anecdote:

---

42 See APUG, FS 23A.VII.

43 Secchi appreciated very much the kindness of the Florentine people and stated that "of strangers all I've seen, just the Tuscans are kind, other men, bears, bears" (*di forestieri tutti che ho visto, solo i toscani sono gentili, gli altri uomini, orsi orsi*). APUG, FS 23.II.D.

44 See Lorenzoni to Tacchini, Padua, February 21, 1878; ACREA, Fondo Tacchini.

45 *Anche questa notte scorsa il P. Secchi ha riavuto i vomiti caratteristici ed è alquanto abbattuto* [...] *La mente è perfettamente serena. Può immaginarsi come ci troviamo.* Ferrari to Tacchini, Rome, February 10, 1878; ACREA, Fondo Tacchini.

46 *Purtroppo ha ragione di non essere tranquillo sul conto del nostro P. Secchi poiché se vi furono alcuni giorni di relativa calma, da ieri l'altro però il male si è esacerbato e questa mattina sono comparsi dei sintomi allarmanti primo de' quali si è una certa intermittenza nel polso. Le forze sono più abbattute e soffre assai ma tranquillissimo; non si riconosce più. Noi ci siamo da gran tempo rassegnati a sopportare quest'immensa sventura.* Ferrari to Tacchini, Rome, February 10, 1878; ACREA, Fondo Tacchini.

> The day before his death, shortly before he lost consciousness [...], Father Provincial Valeriano Cardella asked him if he died contented in the lap of the Holy Catholic Church and in the bosom of the Society of Jesus, to which he added: "And at the Collegio Romano, next to the blessed St. Aloysius [Gonzaga], whom I knighted," alluding to the cross of Officer of the Légion d'honneur, which he had arranged to be attached to the urn that holds the mortal remains of that young angelic saint. These were the last words he uttered and, a few moments later, almost falling down flat in bed as one who would lie down to sleep, he no longer opened his lips and remained in this state until 7 p.m. the following day.[47]

On February 26, the *Bullettino meteorologico dell'Osservatorio del Collegio Romano* reported: "26th. Beautiful day, but saddest for our observatory, where at 7 p.m. the great soul of Father ANGELO SECCHI flew into the bosom of his Creator."[48]

Secchi ended his existence at the age of fifty-nine years. Pius IX—who had died but a few weeks earlier, on February 7—said this about him:

> Father Secchi has always been able to match religious virtues with science, but the most singular ones that stood out in him were humility and obedience. We know how many times, when he was invited to accept honors and other charges, he always came to Us to ask for advice and guidelines. He never took a step without first hearing Our opinion and viewpoint. He really was an excellent religious.[49]

---

47 *Il dì innanzi la sua morte, poco prima che perdesse i sensi* [...] *il p. Valeriano Cardella Provinciale gli domandò se moriva contento in grembo alla Santa Chiesa Cattolica e nel seno della Compagnia di Gesù, a cui Egli aggiunse: e in Collegio Romano, accanto a San Luigi benedetto, che ho fatto cavaliere" alludendo alla croce d'ufficiale della Legion d'Onore, che Egli legò che fosse attaccata all'urna che racchiude le spoglie mortali di quel santo angelico Giovane. Queste furono le ultime parole che proferì e pochi momenti dopo prostratosi quasi supino nel letto come uno che volesse adagiarsi per dormire, più non aprì il suo labbro e durò così fino alle ore sette pom*[*eridiane*] *del giorno seguente.* APUG, FS 23.III.

48 *26. Bella giornata, ma tristissima pel nostro osservatorio, in cui alle 7 pom. precise la grande anima del P. ANGELO SECCHI volava in seno del Suo Creatore* (*Bullettino meteorologico del Collegio Romano* 17 [1878]: 24).

49 *Il padre Secchi ha sempre saputo accoppiare colla scienza le virtù religiose, ma le due singolari che più spiccavano in lui furono l'umiltà e l'obbedienza. Sappiamo Noi quante volte venne stimolato ad accettare onori, cariche ed altro, Egli recavasi sempre da Noi a chieder direzione e consiglio. Mai fece un passo senza prima sentire quale era il Nostro parere e divisamento. Veramente egli è stato un ottimo religioso.* APUG, FS 23.III.25.

Secchi's funeral was probably held in the parish of S. Maria in via Lata, where many people were expected to attend: four hundred chairs were prepared for the general public, thirty-six for noblemen.[50] Denza described it as a "pious and modest ceremony" with Secchi's simple coffin covered by his "priestly emblems" and by "a candid wreath of camellias, jasmines, and lilies donated by an unspecified nobleman from northern Italy" (probably Count Almerico da Schio [1836–1930]) and surrounded by twelve lights.[51] His remains were brought by his pupils and friends to be buried at Campo Verano cemetery in Rome, in the mortuary chapel of the Jesuit fathers, located in the second rank of sarcophagi on the left of the visitors' entrance, and indicated by the number XXXVIII.[52]

6 Commemorations

Secchi's friends, however, probably considered that ceremony too modest for such an illustrious scientist. After his death, in fact, a public subscription was launched by the Accademia Tiberina to collect money for solemn tributes in memory of the Jesuit astronomer, former President.[53] The tributes were held on April 13 in St. Ignatius, the church having been sumptuously decorated for the occasion, especially thanks to the generosity of Prince Boncompagni: the church was entirely adorned with black and golden drapes of silk and velvet; a monumental catafalque, twelve-meters high, was set up with four statues in the basement and, in the upper part, a sort of temple with columns, containing the sarcophagus designed by the architect Enrico Gennari (1846-1915).[54] Many inscriptions in Latin by Angelini, Secchi's former professor, decorated the church and the basement of the catafalque, highlighting the life and achievements of the illustrious Jesuit from Reggio. Music by Wolfgang Amadeus Mozart (1756-1791) was performed, together with a eulogy by archeologist Enrico Fabiani (1815-1883).[55] The Academy of Nuovi Lincei went further and

50 See APUG, FS 23.VII.

51 See Denza, *Il p. Angelo Secchi*, 34.

52 See APUG, FS 23.XV.

53 Some of Secchi's friends privately took initiatives to remember the Jesuit astronomer. At Frascati, in the College Mondragone, the family Santovetti, friends of Secchi, hung a medallion made by Prinzi (see below) with the profile of the Jesuit astronomer. Angelini, Secchi's former professor, chose the words to be engraved on the medallion: *In astrorum scientia aetatis suae princeps* (First of his age in the knowledge of the stars).

54 See *Al p. Angelo Secchi nel XXV dalla morte* (Rome: Desclée, Lefebvre et C.ie, 1903), 77.

55 See *Tituli in supremis honoribus Angeli Secchii e Societate Jesu* (Rome: Tipografia delle Scienze Matematiche e Fisiche, 1878).

FIGURE 10.5 Sketch of the monumental meteorological station that was to have been built in Rome in memory of Secchi. From Cialdi Alessandro et al., "Progetto di un monumento meteorologico da erigersi in Roma alla memoria del p. Angelo Secchi," *Atti dell'Accademia Pontificia de' Nuovi Lincei* 35 (1882): 1–5, pl. 1.

FIGURE 10.6 Plan of a tower for a solar telescope to be erected at Reggio and dedicated to the compatriot Jesuit astronomer; this project, as with the one in fig. 10.5, was never realized. From Gaetano Tacchini, *Torre telescopio in cemento armato da erigersi in Reggio-Emilia all'insigne astronomo padre Angelo Secchi* (Modena: n.p., 1914), pl. [1].

promoted the erection of a monument to be financed with the public subscription. A special academic committee was appointed for this purpose;[56] in 1879, it proposed the erection of a public "meteorological monument," a work of art that included some meteorological instruments, in memory of the constant attention Secchi paid to a science in the service of the people.

The initiative was advertised in many Italian and foreign newspapers,[57] but the fundraising was not as successful as had been expected. In 1881, a sketch of the monument was presented to the academy; it was made by sculptor Giuseppe Prinzi (1825–95), who offered his collaboration for free, with the further intent of enhancing the beauty of the town.[58]

The funds, however, were insufficient; and although the committee repeatedly postponed the deadline, resized the original project, and removed some ornamental parts, the monument was never realized.[59] The reason for this failure was mainly due to the concomitant public subscription promoted by Tacchini, especially among liberals and scientists, to build another "scientific monument" in honor of Secchi: a big solar telescope, of seventy centimeters aperture, to be installed at Reggio—another project, however, that remained only on paper, despite the contributions of many Italian and foreign scientists.[60]

Other commemorations were made at Reggio, where, on February 26, 1879, an engraved stone was placed in the facade of the home where Secchi was

56 See Michele Stefano De Rossi, "Comunicazioni del Segretario," *Atti dell'Accademia Pontificia de' Nuovi Lincei* 31 (1878): 245–246, here 246. Members of the Committee were Alessandro Cialdi (President), prince Baldassarre Boncompagni, colonel Mattia Azzarelli, Augusto Statuti, fr. Stanislao Ferrari, fr. Giuseppe Lais, microscopist Francesco Castracane degli Antelminelli (1817–1899) (Treasurer), Michele Stefano de Rossi.

57 See, e.g., François-Napoléon Moigno, "Académie des Nuovi Lincei," *Les mondes*, n.s. 2 (1878): 295–296.

58 See Commissione Accademica, "Progetto di un monumento meteorologico da erigersi in Roma alla memoria del p. Angelo Secchi," *Atti dell'Accademia Pontificia de' Nuovi Lincei* 35 (1882), 1–5.

59 In 1891, thirteen years after Secchi's death, the Pontifical Academy of Nuovi Lincei inaugurated a simple marble bust in memory of the Jesuit scientist, which was placed in the Palace of Cancelleria Apostolica, where the headquarters of the academy was located. Another bust was placed in Pincio park (on the column of the mire of the Collegio Romano Observatory) because of the project—which was never realized—of erecting a popular observatory, dedicated to Secchi, on that hill (see Francesco Denza, "Appunti storici sulla Specola Vaticana," *Pubblicazioni della Specola Vaticana* 1 [1891], xi–xvii, here xiv).

60 See Riccardo Finzi, "Il mancato monumento scientifico al Padre Angelo Secchi," *Atti e memorie della deputazione di storia patria per le antiche province modenesi* 6 (1971): 137–147. To help with the fundraising, in 1888 Tacchini published a book on the total solar eclipses that he had observed and devolved the income of its sale to the subscription (see Tacchini, *Eclissi totali di Sole*). The plan of the building for the solar telescope was designed by one of Tacchini's brothers, Gaetano (1844–1928).

born (with the inscription: QUESTA SÌ POVERA CASA / DOVE NEL XXVIII GIUGNO MDCCCXVIII / NACQUE / ANGELO SECCHI / ASTRONOMO E FISICO INSIGNE AL MONDO / DICE QUANTO POSSA / CON FORTUNA D'INGEGNO VIRTÙ DI VOLERE (This is the humble house where on June 28, 1818 Angelo Secchi was born, astronomer and physicist, of worldwide renown; it speaks to how much one can accomplish with the virtues of willpower and the fortune of ingenuity). A bust was also installed at the local museum of Storia Patria and a medallion at the high school "L. Spallanzani." which was located in the old Jesuit college of Reggio, where Secchi had studied. In addition, the municipality gave his name to a district and, Ferrari donated two items—a manuscript by Secchi about atmospheric electricity and an aneroid barometer used by Secchi in his expeditions—to the City Library.[61]

61 The two objects were brought to Reggio by the lawyer, Deputy Giuseppe Fornaciari (1836–96); see APUG, FS 23.VII. Although Secchi's scientific legacy has not received the attention it deserves from Italian scholars and institutions, this is not necessarily the case elsewhere: in 1935, for instance, the International Astronomical Union named a crater, a mountain chain, and a crevasse on the moon's surface after Secchi; in 1973, a crater on Mars was named after Secchi, and in 1988 an asteroid (4705). A NASA instrument, launched in 2006, bears the acronym SECCHI (see foreword by Consolmagno).

CONCLUSION

Secchi's Scientific Legacy

By the end of the nineteenth century, the leadership in astrophysical research had passed from the Old World to the New. In the twentieth century, the United States became a leader in astrophysics research for a number of different reasons, including the lack of an entrenched tradition in celestial mechanics, the increasing technical expertise in the construction of large instruments, competition among local communities to have bigger and bigger telescopes, and the substantial contributions coming from private patronage, which were almost negligible in Europe. The United States was in a full economic boom, and the science of astronomy benefited from this. Indeed, the main US astronomical observatories of that time were built by wealthy businessmen and equipped with large telescopes thanks to substantial donations from several patrons.

The situation was quite different in Europe. At the end of the century, European astronomers preferred to invest large financial and human resources in the *Carte du ciel* (Sky chart) project,[1] which essentially continued the classical tradition in astrometry, though using the modern technology of photography. It is significant that, among the eighteen participating observatories in that project, there was no US observatory: in the United States, the course of astronomical research was very much oriented to the "new astronomy." Moreover, within a few years, Europe was overwhelmed by the First World War, with catastrophic socioeconomic consequences.

The use of large US telescopes accelerated the development of astrophysical research in the fields of both stellar and solar astronomy, starting from Secchi's results. His spectral types, for instance, were the starting point for the impressive work of spectral classification carried out at the Harvard College Observatory at the end of the 1880s. Secchi's classification scheme was extended and refined by adding subclasses, leading to the more sophisticated systems of spectral classes currently in use.[2]

---

1 The *Carte du ciel* was an international project, promoted by Paris Observatory in 1887, which aimed to produce a detailed photographic map and catalog of the entire celestial vault; see Ileana Chinnici, *La Carte du ciel: Correspondance inédite conservée dans les archives de l'Observatoire de Paris* (Palermo: Observatoire de Paris & Osservatorio Astronomico di Palermo "G.S. Vaiana," 1999), and Jérôme Lamy, ed., *La Carte du ciel* (Paris: EDP Science, 2008).

2 See Hearnshaw, *Analysis of Starlight*, 104–134.

 | DOI:10.1163/9789004387331_013

Similarly, Secchi's studies on solar physics were improved thanks to the development of solar towers (vertical telescopes for solar spectroscopy), which greatly increased the spectral resolution. Almost twenty years after Secchi's death, one of the most renowned American astrophysicists, George Ellery Hale (1868–1938), founder of *The Astrophysical Journal*, recognized Secchi, Tacchini, and the other Italian spectroscopists as forerunners in the solar spectroscopic research carried out at Mount Wilson and Yerkes Observatories. He wrote to Tacchini:

> No one appreciates more fully than I do how much of us who are engaged in solar investigations owe to the spectroscopic workers of Italy. The volumes of the *Memorie* [...] stand in a case near my table and are used almost every day. I have good reasons to know how much I am indebted to Tacchini, Secchi, Respighi, Lorenzoni [...], not to mention the other members of the Society.[3]

A few years later, on the occasion of the twenty-fifth anniversary of Secchi's death, Hale commemorated the Italian scientist with the following words:

> Few men contributed more than Secchi to the present widespread interest in Astro-physical research. His investigations are notable for the diversity of their objects, the ingenuity of the instrumental methods devised, and the tireless perseverance of the observer. In these respects they will stand as an inspiration to future investigators, and as a striking example to those who are called upon to obtain important results with modest instrumental means.[4]

In Italy, however, Secchi's legacy was largely forgotten. Italian astrophysics, which had been pioneering in the second half of the nineteenth century, lost its luster over the following years. Astronomers with a background in mathematics, who were numerically preponderant, tended to confine astronomical research to the traditional fields of astrometry and celestial mechanics, with some interest in photometry and celestial photography. The golden age of Italian spectroscopy quickly waned; in 1920, the Italian Spectroscopic Society

3 Hale to Tacchini, Chicago, March 9, 1896, quoted in Ileana Chinnici, "La Società degli Spettroscopisti Italiani e la fondazione di *The Astrophysical Journal* nelle lettere di G.E. Hale a P. Tacchini," in *Atti del XVI Congresso di Storia della Fisica e dell'Astronomia, Como, 1996*, ed. Pasquale Tucci (Como: Consiglio Nazionale delle Ricerche, 1997), 299–321, here 306.

4 AA. VV. *Al p. Angelo Secchi nel XXV dalla morte* (Rome: Desclée, Lefebvre et C.ie, 1903), 83.

became the Italian Astronomical Society—a change in the name that reflected a change in the nature of the society.

Even after his death, Secchi remained a highly controversial figure: a monument-telescope for solar observations, proposed by Tacchini and to be dedicated to Secchi, which was to be built in Reggio Emilia with funds provided by a public petition, was never constructed despite the support of the international astronomical community (see Chapter 10).

It is also significant that, on the occasion of the twenty-fifth anniversary of his death, almost all the directors of the Italian observatories declined an invitation to attend the memorial event sponsored by the Italian government in Rome, merely sending a telegram instead. The trend in Italian astronomy was to retreat more and more from the new astronomy that had been inaugurated by Secchi; Italy, which once led the way in the field of spectral analysis,[5] soon became the tail light.

The crisis in Italian astrophysics was situated in the context of a crisis of European astrophysics, which was unable to keep up with the powerful tools available in those years in the United States, where the new discipline experienced a period of tremendous growth. Unhampered by the legacy of the classical tradition, US astronomy moved decisively toward the unexplored realms opened up by the new astronomy, acquiring increasingly powerful instruments in spectroscopic analysis. Astrophysics research in the United States made great progress thanks to the contribution of private initiatives, which led to the creation of important institutions such as Lick and Yerkes Observatories, where world-famous astrophysicists worked such as the aforementioned Hale, James E. Keeler (1857–1900), and Henry N. Russell (1877–1957). The country that welcomed Secchi as a refugee during the exile of 1848 ultimately inherited his scientific legacy.

After more than thirty years of decline, astrophysical studies were finally relaunched in Italy by Giorgio Abetti, who had carried out research on solar spectroscopy in the United States under the supervision of Hale; and Italy had its first solar tower in 1925 at Arcetri Observatory, in Florence, thanks to the crucial technical and financial support of the Californian astrophysicist.

The reasons for the prolonged historiographical silence on Secchi and his scientific contributions are probably attributable to the complexity of his personality and the political implications of his being a Jesuit in an anti-clerical age. His role in the history of science has therefore been poorly understood and insufficiently appreciated, despite his important contributions to a whole host of scientific disciplines.

---

5 See Chapter 8, note 34.

However, where history has been negligent, the arts have attempted to compensate. Though no monument was ever erected at Reggio or in Rome (see Chapter 10), many poems and sonnets were devoted to Secchi both while he was alive and to commemorate his death (a list, though almost certainly incomplete, is given in the footnote).[6] A few paintings, a couple of busts, and some medallions are still preserved, mainly in Rome. The many sundials he drew on the walls of palaces and monasteries, largely forgotten and in disrepair, are today his most significant memorials. These old and fading instruments contain Secchi's scientific spirit, even though his name does not always appear on them.

As we have seen throughout this book, Secchi had a surprising impact on the collective imagination, even in chronologically and geographically distant contexts. Indeed, one unexpected homage to the famous scientist has come from one of the best known poets of the Soviet Union, Arseny Tarkovsky (1907–89), who was censored for years on charges of mysticism. It therefore seems fitting to close this account of Secchi's life with one of Tarkovsky's short poems[7] recalling the figure of the famous Italian astronomer:

> "I'm sorry, my dear,
> Merz equatorial!"
> [...]
> Here, in Rome, after a long exile,
> Gray-haired, half-blind, half-dead,
> Alone among the heavenly lights,
> He stands with his head uncovered.
> The breath of Rome—like dry grass.

6 See (1) Latin Hexameters by Luigi Tripepi (1836–1906), vice-president of the Accademia Tiberina (located at Altemps Palace), with dedication to Leo XIII, *Alla memoria del sommo astronomo/P. Angelo Secchi d.C.d.G./la cui morte fu perdita dolorosissima alla/Scienza ed alla religione*, s.d.; APUG, FS 23.VI.C; (2) verses by Andrea Leonetti (1838–87), rector of the Collegio Nazareno, *In funere Clarissimi Viri Angeli Secchii e Societate Iesus*, December 12, 1878; APUG, FS 23.VI.C; (3) Italian poem by Giovanni Battista De Dominicis-Tosti, undated and untitled; APUG, FS 23.VI.C; (4) song by Bernardo Mattiauda (printed, with portrait, Rome: Tipografia delle Scienze matematiche e fisiche, via Lata n. 3, 1878) *In morte del Padre Angelo Secchi*, dedicated to the Prince Boncompagni; APUG, FS 23.XV; (5) poem about the Sun (1858), dedicated to Secchi, by Giovanni Bacci, professor of rhetoric at the Seminary of Prato (near Florence) (another poem, about the moon, was dedicated to Ferrari); APUG, FS 23.VII.

7 The initial verses of Tarkovsky's poem are remarkably similar to an obituary by Victor Van Tricht, S.J. (1842–97), which is reported by Moigno (see Van Tricht, "Les écrits et la mort de Secchi," 202, in Moigno, *Le révérend Père Secchi*, 188–205), but it is unclear if the poet explicitly took inspiration from this text.

Greetings to you, last step!
The fate is crafty, and kings were wrong,
And yet that day has come true.

From the Merz equatorial,
He is unable to tear his old hands;
Urania will not, as before,
feast in this lonely tower.

Swallowing the bitter air, Secchi strokes
The copper, uncleaned for a long time.
"Lovely friend, we will be separated forever,
Let me die in peace now."

He goes down the decrepit steps
To nothingness, to dust, to the last judgment,
And the swallows over the equatorial,
As messengers of oblivion, scurrying.

Since I was a child, I wept for this
High, kin shadow,
in order, following it passing through the world,
To bless the last step.[8]

8 Arseny Tarkovsky, *Poems from Different Years* (Moscow: "Sovremennik," 1983).

Bibliography

AA. VV. *Al p. Angelo Secchi nel XXV dalla morte*. Rome: Desclée, Lefebvre et C.ie, 1903.

Abetti, Giorgio. "Le nebulose oscure: Una scoperta del p. Angelo Secchi." *L'Universo* 9 (1928): 315–319.

Aebischer, Tullio. "Le misure geodetiche." In *Angelo Secchi: L'avventura scientifica del Collegio Romano*, edited by Aldo Altamore and Sabino Maffeo, 227–240. Foligno: Edizioni Quater, 2012.

Aebischer, Tullio. ed. *La scoperta del Caposaldo B della Base Geodetica di A. Secchi lungo la via Appia Antica*. Rome: Edizioni Nuova Prhomos, 2013.

Alatri, Paolo. *Lotte politiche in Sicilia sotto il governo della Destra: 1866–74*. Rome: Einaudi Editore, 1954.

Alder, Ken. *The Measure of All Things*. London: Simon & Schuster, 2002.

Altamore, Aldo. "La nascita della Nuova Astronomia." In *Angelo Secchi: L'avventura scientifica del Collegio Romano*, edited by Aldo Altamore and Sabino Maffeo, 109–144. Foligno: Edizioni Quater, 2012.

Altamore, Aldo, and Natalia Ptitsyna. "Il primo osservatorio geomagnetico d'Italia." In *Angelo Secchi: L'avventura scientifica del Collegio Romano*, edited by Aldo Altamore and Sabino Maffeo, 177–196. Foligno: Edizioni Quater, 2012.

Angeletti, Angelo. "Padre Francesco De Vico, astronomo maceratese un ricordo nel bicentenario della nascita." *Astronomia* 1 (2006): 4–10.

Alberi Auber, Paolo, and Maria Luisa Tuscano. "L'obelisco di Augusto alla luce di due manoscritti di padre Angelo Secchi e dei testi di Giuseppe Vasi e Cornelio Meyer." In *Atti del XIX Seminario nazionale di Gnomonica*, edited by Maria Luisa Toscano, 25–33. Palermo: Visiva Marketing Tools, 2014.

Aubin, David. "Orchestrating Observatory, Laboratory, and Field: Jules Janssen, the Spectroscope, and Travel." *Nuncius* 17, no. 2 (2002): 615–633.

Balley, François. *Endemo-epidémie et météorologie de Rome: Etudes sur les maladies dans leur rapports avec les divers agents météorologiques*. Paris: V. Rozier, 1863a.

Balley, François. *Météorologie et météorographie, pathogénie et nosographie, ou: Éléments de recherches sur la connexion entre les divers agents météorologiques et la pathogénie civile et militaire à Rome de 1850 à 1861; Atlas*. Paris: V. Rozier, 1863b.

Bayart, Pierre. "La méridienne de France et sa prolongation jusqu'aux Baléares." *Revue XYZ* 116 (2008): 58–62.

Becchetti, Piero, and Maria Francesca Bonetti. "La dagherrotipia a Roma." In *L'Italia d'argento: 1839–1859. Storia del dagherrotipo in Italia*, edited by Maria Francesca Bonetti and Monica Maffioli, 238–250. Florence: Edizioni Alinari, 2003.

Becker, Barbara J. *Unravelling Starlight: William and Margaret Huggins and the Rise of the New Astronomy*. Cambridge: Cambridge University Press, 2011.

Bertone, Gianfranco, and Dan Hooper. "A History of Dark Matter." US Department of Energy, High Energy Physics Division, 2016.

Bianchi, Simone. "Un imperatore ad Arcetri." *Giornale di astronomia* 38, no. 2 (2012): 2–12.

Bianchi, Simone, and Daniele Galli. "Les observatoires astronomiques en Italie: An 1863 Report by Otto Wilhelm Struve." *Nuncius* 30, no. 1 (2015): 195–227.

Bianchi Simone, Daniele Galli, and Antonella Gasperini "Le due inaugurazioni dell'Osservatorio di Arcetri," *Giornale di Astronomia* 3, (2013): 19–30.

Bianchi, Simone, Daniele Galli, and Antonella Gasperini. *Giovanni Virginio Schiaparelli e l'Osservatorio di Arcetri.* Fondazione Giorgio Ronchi CXI. Florence: Fondazione Giorgio Ronchi, 2011.

Bigg, Charlotte, and Klaus Staubermann, eds. "Spectroscope Histories." *Nuncius* 17, no. 2 (2002): 583–689; 18, no. 2 (2003): 735–852.

Biographical Encyclopedia of Astronomers. New York: Springer, 2014.

Bovio, Giovanni. "Discorso." In *Atti del Parlamento Italiano, Camera dei Deputati, Sessione del 1876–77: Discorsi 1*, 690–697. Rome: Tipografia Eredi Botta, 1877.

Branchetti, Maria Grazia, and Daniela Sinisi, eds. *"La meravigliosa invenzione": Strade ferrate nel Lazio 1846–1930*. Rome: Gangemi Editore, 2003.

Brenni, Paolo. "Il meteorografo di Padre Angelo Secchi." *Nuncius: Annali di storia della scienza* 1 (1993): 197–247.

Brenni, Paolo. "Prometheus's Tools: Instruments and Apparatus Used in Atmospheric Electricity Research and Experiments." In *Playing with Fire: Histories of the Lightning Rod,* edited by Peter Heering, Oliver Hochadel, and David J. Rhees, 230–255. Transactions of the American Philosophical Society 99, part 5. Philadelphia: American Philosophical Society, 2009.

Brevi memorie intorno al Collegio Romano. Rome: Stamperia della S.C. De Propaganda Fide, 1870.

Bud, Robert, and Deborah Jean Warner, eds. *Instruments of Science: An Historical Encyclopedia*. New York: Garland, 1998.

Buffoni, Letizia, Alessandro Manara, and Pasquale Tucci, eds. *G. V. Schiaparelli, A. Secchi: Corrispondenza (1861–1878)*. Milan: Edi. Artes, 1991.

Buonanno, Roberto. *Il cielo sopra Roma: I luoghi dell'astronomia*. Milan: Springer-Verlag, 2008.

Calandrelli Giuseppe and Andrea Conti, Opuscoli astronomici. Rome: Stamperia Salomoni, 1808.

Carrara, Benedetto. *Il Padre Angelo Secchi*. Padua: Tipografia del Seminario, 1903.

Casanovas Juan, "Secchi, Angelo (Pietro Angelo)" in: *Diccionario Histórico de la Compañia de Jesús* 4, edited by Charles E. O'Neill and Joaquín M. Domínguez, 3542–43. Rome: Institutum Historicum, and Madrid: Universidad Pontificia Comillas, 2001.

Casella Mario, "Cornoldi, Giovanni Maria," in Dizionario biografico 29. Rome: Treccani, 1983; http://www.treccani.it/enciclopedia/giovanni-maria-cornoldi_(Dizionario-Biografico)/ (accessed June 3, 2018).

Castellani, Giuseppe. "Il P. Angelo Secchi S.I. e le sue missioni scientifiche." *La civiltà cattolica* 4 (1933): 606–620.

Castellani, Giuseppe. "Storia di una spedizione del P. Angelo Secchi (agosto–ottobre 1872)." *La civiltà cattolica* 11 (1944): 159–162.

Chinnici, Ileana. "19th-Century Spectroscopic Instruments in Italian Astronomical Observatories." *Nuncius* 15, no. 2 (2000): 671–680

Chinnici, Ileana. "An 'Italian' Observatory in India: The History of the Calcutta Observatory." *Studies in the History of Medicine and Science* 14 (1996): 91–115.

Chinnici, Ileana. "Il profilo scientifico e umano di Angelo Secchi." In *Angelo Secchi: L'avventura scientifica del Collegio Romano*, edited by Aldo Altamore and Sabino Maffeo, 43–64. Foligno: Edizioni Quater, 2012.

Chinnici, Ileana. "La Società degli Spettroscopisti Italiani e la fondazione di *The Astrophysical Journal* nelle lettere di G. E. Hale a P. Tacchini." In *Atti del XVI Congresso di Storia della Fisica e dell'Astronomia, Como, 1996*, edited by Pasquale Tucci, 299–321. Como: Consiglio Nazionale delle Ricerche, 1997.

Chinnici, Ileana. "Le voyage scientifique de Georges Rayet en Italie et la fondation de l'Observatoire de Bordeaux." In *Sur les traces de Cassini: Astronomes et observatoires du Sud de la France; 121ème Congrès national des sociétés historiques et scientifiques, Nice, 1996*, edited by Paul Brouzeng and Suzanne Débarbat, 259–269. Paris: Editions du C.T.H.S., 2001.

Chinnici, Ileana. *La Carte du ciel: Correspondance inédite conservée dans les archives de l'Observatoire de Paris.* Palermo: Observatoire de Paris & Osservatorio Astronomico di Palermo "G. S. Vaiana," 1999.

Chinnici, Ileana. *L'eclisse totale di sole del 1870 in Sicilia.* Palermo: INAF-Osservatorio di Palermo, 2008a.

Chinnici, Ileana. "The Maker and the Scientist: The Merz–Secchi Connection." In *Merz Telescopes: A Heritage Worth Preserving*, edited by Ileana Chinnici, 39–68. New York: Springer, 2017.

Chinnici, Ileana. "Per una storia istituzionale degli Osservatori Astronomici in Italia." *Giornale di astronomia* 41 (2015): 11–21.

Chinnici, Ileana. "Personaggi e istituzioni della Chiesa Cattolica nella fondazione degli osservatori astronomici in Italia." Documentazione interdisciplinare di scienza e fede (2009). http://www.disf.org/altriTesti/Chinnici.asp (accessed April 5, 2018).

Chinnici, Ileana. "Pietro Tacchini (1838–1905): A Key Figure in the Post-unitarian Italian Astronomy." *Memorie della Società Astronomica Italiana: Supplementi* 9 (2006): 28–34.

Chinnici, Ileana. "The Società degli Spettroscopisti Italiani: Birth and Evolution." *Annals of Science* 65, no. 3 (2008b): 393–438.

Chinnici, Ileana, and Carlo Blanco. "L'Etna e le stelle: La fondazione dell'Osservatorio Astrofisico di Catania." In *Atti del XXXIII Convegno annuale della Società Italiana degli Storici della Fisica e dell'Astronomia*, edited by Lucio Fregonese and Ivana Gambaro, 67–83. Pavia: Pavia University Press, 2016.

Chinnici, Ileana, and Paolo Brenni. "The Palermo Merz Equatorial Telescope: An Instrument, a Manuscript, Some Drawings." *Nuncius* 30 (2015): 228–279.

Chinnici, Ileana, and Antonella Gasperini. *Alle origini dell'astrofisica italiana: Il carteggio Secchi–Tacchini*. Florence: Fondazione Giorgio Ronchi, 2013.

Chinnici, Ileana, and Wiktor Gramatowski. "Le carte di Angelo Secchi conservate presso l'archivio della P. Università Gregoriana: Un inventario inedito rivisitato." *Nuncius* 16, no. 2 (2001): 571–627.

Chinnici, Ileana, and Sabino Maffeo, "Angelo Secchi e gli astronomi del suo tempo." In *Angelo Secchi: L'avventura scientifica del Collegio Romano*, edited by Aldo Altamore and Sabino Maffeo, 65–88. Foligno: Edizioni Quater, 2012.

Cialdi, Alessandro. *Cenni storici dei fari antichi più famosi e di alcuni moderni*. Rome: Tipografia delle Scienze Matematiche e Fisiche, 1877.

Cialdi, Alessandro. *Sul moto ondoso del mare e su le correnti di esso*. Rome: Tipografia delle Belle Arti, 1866.

Cialdi, Alessandro, and Angelo Secchi. "Sur la transparence de la mer." *Comptes rendus de l'Académie des Sciences* 61 (1865): 100–104.

Cialdi, Alessandro [et al.], "Progetto di un monumento meteorologico da erigersi in Roma alla memoria del p. Angelo Secchi," *Atti dell'Accademia Pontificia de' Nuovi Lincei* 35 (1882): 1–5.

Clerke, Agnes Mary. *The System of the Stars*. London: Longmans, Green and Co., 1890.

Commission Internationale du Mètre. "Rapport de la première commission." In *Réunions générales de 1872: Procès-verbaux*, 122–131. Paris: Imprimerie nationale, 1872.

Comte, Auguste. *Cours de philosophie positive*. Tome 2. Paris: Bachelier, 1835.

Conlin, Michael F. "Popular and Scientific Reception of the Foucault Pendulum in the United States." *Isis* 90 (1999): 181–204.

Cornoldi, Giovanni Maria. *I sistemi maccanico e dinamico circa la costituzione delle sostanze corporee considerati rispetto alle scienze fisiche*. Verona: Tipografia Vicentini e Franchini, 1864.

Crookes, William. "On the Photography of the Moon." *Proceedings of the Royal Society* 8 (1857): 363–371.

Curran, Robert Emmett. *The Bicentennial History of Georgetown University: From Academy to University, 1789–1889*. Washington, DC: Georgetown University Press, 1993.

De la Rue, Warren. "Comparison of Mr. De la Rue's and Padre Secchi's Eclipse Photographs." *Proceedings of the Royal Society* 13 (1863): 442–444.

Deleschamps, Albert. *Etude physique du son de la parole*. Paris: F. Savy, 1869.

Denza, Francesco. *Il Padre Angelo Secchi*. Turin: Collegio Artigianelli -Tipografia e Litografia S. Giuseppe, 1878.

De Rossi, Michele Stefano De Rossi."Comunicazioni del Segretario." *Atti dell'Accademia Pontificia de' Nuovi Lincei* 31 (1878): 245–246.

Di Legge, Alfonso, Francesco Giacomelli, and Antonio Prosperi. "Lorenzo Respighi: Una pagina di storia dell'astronomia romana." *Rivista di astronomia e scienze affini* 6, no. 9 (1912): 625–640.

"Discussioni: Tornata del 25 novembre 1872." In *Atti Parlamentari, Camera dei Deputati, 1872–73*, 3311–32. Rome: Eredi Botta, 1874.

Documenti diplomatici italiani, seconda serie: 1870–1896. Rome: Istituto Poligrafico dello Stato, 1976.

Donati, Giovan Battista. "Intorno alle strie degli spettri stellari." *Nuovo cimento* 15 (1862): 292–304; 366–376.

Donati, Giovan Battista. "Osservazioni spettroscopiche di macchie solari." *Memorie della Società degli Spettroscopisti Italiani* 1 (1872): 52–56.

Donati, Giovan Battista. "Schreiben des Herrn Professors Donati, Directors der Sternwarte in Florenza, an den Herausgeber." *Astronomische Nachrichten* 62 (1864): 375–378.

Elenco dei membri dell'Accademia Filosofico–Medica di San Tommaso d'Aquino. Bologna: Istituto Tipografico, 1876.

Engels, Friedrich. *Dialettica della natura*. Rome: Editori Riuniti, 1978.

Ericsson, John. "The Difference of Thermal Energy Transmitted to the Earth by Radiation from Different Parts of the Solar Surface." *Nature* 12 (1875): 517–520.

Fabri Scarpellini, Erasmo. *La scienza contemporanea nello Stato Pontificio*. Rome: Tipografia della Reverenda Camera Apostolica, 1857.

Fargis, George A. "Georgetown College Observatory, 1843–1893." *Publications of the Astronomical Society of the Pacific* 7, no. 41 (1895): 89–97.

Faye, Hervé A. "Rapport de la Commission nommée le 17 août pour préparer une réponse à la lettre adressée par M. Le Ministre de l'Instruction Publique, au sujet de l'opportunité de la création d'un Observatoire d'Astronomie Physique aux environs de Paris." Académie des Sciences, séance du 2 novembre 1874, Paris Observatory Archives, Meudon Section, Dossier Janssen.

Fergola, Emanuele, and Angelo Secchi. *Sulla differenza di longitudine fra Napoli e Roma determinata per mezzo della trasmissione telegrafica delle osservazioni*. Napoli: Accademia delle Scienze, 1871.

Ferrari, Gaspare Stanislao. "Terza serie delle misure micrometriche delle stelle doppie fatte all'equatoriale del Collegio Romano dal 22 giugno 1872 a tutto il 1874." *Atti della Pontificia Accademia dei Nuovi Lincei* 28 (1875): 207–228.

Filograssi, Giuseppe. "Teologia e filosofia nel Collegio Romano dal 1824 ad oggi." *Gregorianum* 35, no. 1954 (1965): 512–540.

Finzi, Riccardo. "Il mancato monumento scientifico al Padre Angelo Secchi." *Atti e memorie della deputazione di storia patria per le antiche province modenesi* 6 (1971): 137–147.

Fioravanti, Rita. "*Intorno alla soluzione di un problema fisico-cosmologico*: Il domenicano Vincenzo Nardini e il gesuita Angelo Secchi." In Angelo Secchi, astronomo e fisico: Attualità scientifica e luoghi storici a Roma, edited by Tullio Aebischer and Rita Fioravanti, 23–48. Rome: Edizioni Nuova Prhomos, 2014.

Fioravanti, Rita. "Nardini, Vincenzo." In *Dizionario biografico degli Italiani* 77. Rome: Treccani, 2012a. http://www.treccani.it/enciclopedia/vincenzo-nardini_(Dizionario-Biografico)/ (accessed June 3, 2018).

Fioravanti, Rita. "Per una bibliografia secchiana." In *Angelo Secchi: L'avventura scientifica del Collegio Romano*, edited by Aldo Altamore and Sabino Maffeo, 277–290. Foligno: Edizioni Quater, 2012b.

Fizeau, Hyppolyte. "Acoustique et optique." In *Société Philomathique de Paris, Année 1848*, 81–83. Paris: Imprimerie de Cosson, 1848.

Forel, François-Alphonse. *Le Léman, monographie limnologique*. Lousanne: F. Rouge Editeur, 1895.

Gallon, Feliciano. "La stampa romana e la vita religiosa." In *La vita religiosa a Roma intorno al 1870: Ricerche di storia e sociologia*, edited by Paul Droulers, Giacomo Martina, and Paolo Tufari, 47–96. Rome: Università Gregoriana Editore, 1971.

Gasparini, Laura. "Padre Angelo Secchi e l'applicazione della fotografia nelle osservazioni astronomiche." *Fotologia* 12 (1990): 34–47.

Gasperini, Antonella, Massimo Mazzoni, and Alberto Righini. "La costruzione della Torre Solare di Arcetri nel carteggio Hale-Abetti." *Giornale di astronomia* 3 (2004): 23–30.

Grandi, Taddeo. *Ciro Menotti e i suoi compagni*. Modena: Tipografia Azzoguidi, 1880.

Green, Nathaniel Everett. "Mars and Schiaparelli's Canals." *Observatory* 3 (1879): 252.

Green, Nathaniel Everett. "On Some Changes in the Markings of Mars since the Opposition of 1877," *Monthly Notices of the Royal Astronomical Society* 40 (1880): 331–332.

Guillemin, Amédée. *Le ciel: Notions élémentaires d'astronomie physique*. Paris: Hachette, 1877.

Haifley, Julie Link. "Capital Images: The Photography of Titian Ramsay Peale, 1855–1885." *Records of the Columbia Historical Society* 50 (1980): 229–244.

Hall, David. *Fred Dibnah's Victorian Heroes*. London: Bantam Press, 2011.

Hearnshaw, John B. *The Analysis of Starlight*. Cambridge: Cambridge University Press, 1986.

Hearnshaw, John B. *The Measurement of Starlight*. Cambridge: Cambridge University Press, 1996.

Henry, Joseph. "On the Application of the Thermo-Galvanometer to Meteorology, etc." In *Scientific Writings of Joseph Henry*, edited by the Smithsonian Institution, 1:215. Washington, DC: Smithsonian Institution Press, 1886.

Hentschel, Klaus. *Mapping the Spectrum*. Oxford: Oxford University Press, 2002.

Herschel, William. "Catalogue of One Thousand New Nebulae and Clusters of Stars." *Philosophical Transactions of the Royal Society of London* 76 (1786): 457–499.

Hufbauer, Karl. *Exploring the Sun*. Baltimore: Johns Hopkins University Press, 1993.

Huggins, William. "On the Spectrum of Comet I, 1866." *Proceedings of the Royal Society* 15 (1866): 5–7.

Huggins, William. "On the Spectrum of Comet II, 1868." *Proceedings of the Royal Society* 16 (1868): 481–482.

Huggins, William, and William Allen Miller. "On the Spectra of some Nebulae." *Philosophical Transactions of the Royal Society* 154 (1864): 437–444.

Hughes, Stefan. *Catchers of the Light. The Forgotten Lives of the Men and Women Who First Photographed the Heavens*. n.p.: ArtDeCiel Publishing, 2012.

Husillos Tamarit, Ignacio. "El Desierto de Las Palmas, centro de espiritualidad en la diócesis de Tortosa." In *Teresa y Enrique: Cara a cara; Actas del Congreso internacional teresiano realizado en Tortosa del 16 al 18 de Enero de 2015*, edited by Gemma Bel and Pilar Liso, 35–134. Ávila: CITeS. Universidad de la Mística, 2015.

Indice generale delle Memorie della Società degli Spettroscopisti Italiani, vol. I–XL. Catania: Stabilimento Tipografico S. Di Mattei e C., 1912.

Istruzione sui parafulmini. Naples: Reale Tipografia della Guerra, 1842.

James, Frank A. J. L. "'The civil engineer's talent': Michael Faraday, Science, Engineering and the English Lighthouse Service, 1836–1865." *Transactions of the Newcomen Society* 70 (1998–99): 153–160.

Janssen, Jules C. "Note sur de nouveaux spectroscopes." *Atti della Pontificia Accademia dei Nuovi Lincei* 16 (1862–63): 73–75.

Janssen, Jules C. "Note sur la présence du sodium dans l'étoile alpha Orion." *Bullettino meteorologico dell'Osservatorio del Collegio Romano* 2 [1863]: 4).

Kelland, Philip M. A. "Father Secchi." *Proceedings of the Royal Society of Edinburgh* 10 (1880): 11–12.

Kirchhoff, Gustav R. "Über den Zusammenhang zwischen Emission und Absorption von Licht und Wärme." *Monatsberichte Berliner Akademie* (1859): 783–787.

Kronk, Gary. *Cometography*. Cambridge: Cambridge University Press, 2003.

Lais, Giuseppe. "Le comete," in AA. VV. *Al p. Angelo Secchi nel XXV dalla morte*. Rome: Descl.e, Lefebvre et C.ie, 1903.

L'Esposizione Universale di Vienna del 1873 illustrata. Milan: Edoardo Sonzogno, 1873.

Lamy, Jérôme, ed. *La Carte du ciel*. Paris: EDP Science, 2008.

Launay, Françoise. *Un globe-trotter de la physique céleste: L'astronome Jules Janssen*. Paris: L'Observatoire de Paris-Vuibert, 2008.

Lay, Renzo. "Angelo Secchi e il triplice osservatorio del Collegio Romano." *Giornale di fisica* 74, no. 3 (2013): 217–239.

Lay, Renzo. "Il nuovo Osservatorio del Collegio Romano." In *Angelo Secchi: L'avventura scientifica del Collegio Romano*, edited by Aldo Altamore and Sabino Maffeo, 91–107. Foligno: Edizioni Quater, 2012.

Leverington, David. *Babylon to Voyager and Beyond: A History of Planetary Astronomy*. Cambridge: Cambridge University Press, 2003.

Leverington, David. *A History of Astronomy: From 1890 to the Present*. London: Springer-Verlag, 1995.

Lockyer, Norman J. *Contributions to Solar Physics*. London: MacMillan, 1874.

Lockyer, Norman J. "Father Secchi." *Nature* 17 (1878): 370.

Lockyer, Norman J. "Father Secchi on Solar Discoveries." *The Academy* 3, no. 61 (1872): 451–452.

Lockyer, Norman J. "Father Secchi's Work on the Sun." *Nature* 10 (1874): 3–4.

Lockyer, Norman J. "Notes." *Nature* 9 (1874): 389–391.

Lockyer, Norman J. "Observations of the Sun: No. V." *Proceedings of the Royal Society* 18 (1869): 74–79.

Lockyer, Norman J. "Physical Sciences, etc." *The Academy* 3, no. 53 (1872): 288–291.

Lugli, Mario Umberto. *Pietro Tacchini*. Modena: Edizioni Il Fiorino, 2001.

Maffeo, Sabino. "Il Collegio Romano e l'insegnamento delle scienze." In *Angelo Secchi: L'avventura scientifica del Collegio Romano*, edited by Aldo Altamore and Sabino Maffeo, 16–22. Foligno: Edizioni Quater, 2012.

Maffeo, Sabino. "Padre Angelo Secchi e la meteorologia." In AA. vv., *Presenze scientifiche illustri al Collegio Romano*, 21–31. Rome: Ufficio Centrale di Ecologia Agraria, 2001a.

Maffeo, Sabino. *The Vatican Observatory: In the Service of Nine Popes*. Rome: Libreria Editrice Vaticana, 2001b.

Malusa, Luciano. *Neotomismo e intransigentismo cattolico: Il contributo di Giovanni Maria Cornoldi per la rinascita del tomismo*. Milan: IPL, 1986.

Malusa, Luciano. "Palmieri, Domenico." In *Dizionario biografico degli Italiani* 80. Rome: Treccani, 2014. http://www.treccani.it/enciclopedia/domenico-palmieri_(Dizionario-Biografico)/ (accessed June 3, 2018).

Mangianti, Franca. "L'Ufficio Centrale di Ecologia Agraria e la sua sede nel Palazzo del Collegio Romano." *Agricoltura: Speciale "120° Anniversario dell'UCEA"* 277 (1996): 15–26.

Mantovani, Roberto. "Un fisico dimenticato: Il gesuita G. B. Pianciani." http://www.uniurb.it/PhysLab/immagini/mantovani_sisfa2002.pdf (accessed May 29, 2018).

Marchetti, Francesco. "Cenni necrologici del p. Nazareno Mancini d.C.d.G.." *Bullettino di bibliografia e di storia delle scienze matematiche e fisiche* 3 (1870): 429–438.

Marchi, Giuseppe. *Monumenti delle arti cristiane primitive nella metropoli del Cristianesimo*. Rome: Tipografia di C. Puccinelli, 1844.

Marini, Heber. *Trente visites à l'Exposition Universelle de 1867*. Paris: A. Lacroix et C.ie Editeurs, 1868.

Martina, Giacomo. "Al Collegio Romano il 20 settembre 1870: Dalla relazione del p. Pietro Ragazzini." *Archivium historiae pontificiae* 8 (1970): 332–347.

Maury, Antonia. "The Spectral Changes of Beta Lyrae." *Annals of Harvard College Observatory* 84, no. 8 (1933): 207–255.

Mazzotti, Massimo. "The Jesuit on the Roof: Observatory Sciences, Metaphysics, and Nation Building." In *The Heavens on Earth: Observatories and Astronomy in Nineteenth-Century Science and Culture*, edited by David Aubin, Charlotte Bigg, and H. Otto Sibum, 58–85. Durham, NC: Duke University Press, 2010.

McKevitt, Gerald. *Brokers of Culture: Italian Jesuits in the American West 1848–1919*. Stanford: Stanford University Press, 2007.

Meadows, Arthur J. *Early Solar Physics*. Oxford: Pergamon Press, 1970.

Meadows, Arthur J. *Science and Controversy: A Biography of Sir Norman Lockyer*. Cambridge, MA: MIT Press, 1972.

Meadows, Arthur J. "The New Astronomy." In *The General History of Astronomy, 4A: Astrophysics and Twentieth-Century Astronomy to 1950*, edited by Owen Gingerich, part A, 59–72. Cambridge: Cambridge University Press, 1984.

Meadows, Arthur J. "The Origin of Astrophysics." In *The General History of Astronomy, 4A: Astrophysics and Twentieth-Century Astronomy to 1950*, edited by Owen Gingerich, part A, 3–15. Cambridge: Cambridge University Press, 1984.

Merz, Georg. "Objectiv-Spektralapparat." *Repertorium für Experimental-Physik* 6 (1870): 164–165.

Merz, Georg, and Sigmund Merz. "Aus einem Schreiben der Herren G. & S.Merz an den Herausgeber." *Astronomische Nachrichten* 65, no. 1537 (1865): 13–16.

Middleton, W. E. Knowles. *The History of the Barometer*. Baltimore: Johns Hopkins University Press, 1964.

Miotto, Enrico, Giuseppe Tagliaferri, and Pasquale Tucci. *La strumentazione nella storia dell'Osservatorio di Brera*. Milan: Unicopli Editore, 1989.

Moigno, François-Napoléon. "Académie des Nuovi Lincei." *Les mondes*, n.s. 2 (1878): 295–296.

Moigno, François-Napoléon. "Météorograph du R. P. Secchi (S.J.)." *Les mondes* 14 (1867): 61–66.

Moigno, François-Napoléon ed. *Le révérend Père Secchi.* Paris: Gauthier-Villars, 1879.

Mori, Attilio. *Cenni storici sui lavori geodetici e topografici e sulle principali produzioni cartografiche eseguite in Italia dalla metà del secolo XVIII ai nostri giorni.* Florence: Istituto Geografico Militare, 1903.

Nardini, Vincenzo. *Risposta* [...] *alla lettera del p. Angelo Secchi.* Rome: Tipografia di Giovanni Cesaretti, 1862.

Newcomb, Simon. "The Sun." *Nature* 4 (1871): 41–43.

Newcomb, Simon. *Popular Astronomy*. London: MacMillan, 1878.

O'Reilly, Fintan. "Stonyhurst and Angelo Secchi." *Stonyhurst Magazine* (1999): 28–34.

Paladini, Leone. *Petizione ai molto reverendi Padri del Concilio per la soppressione dell'Osservatorio Romano*. Turin: Tipografia del giornale "Il Conte di Cavour," 1869.

Parisi, Flora. "Lettere di p. Angelo Secchi." In *Angelo Secchi astronomo e fisico: Attualità scientifica e luoghi storici a Roma*, edited by Tullio Aebischer and Rita Fioravanti, 45–48. Città di Castello: Edizioni Nuova Prhomos, 2014.

Pianciani, Giovanni Battista. *Istituzioni fisico-chimiche* 1–4. Rome: Crispino Puccinelli, 1833–35.

Pieralisi, Sante. *Sopra la nuova edizione del processo originale di Galileo Galilei*. Rome: Tipografia Poliglotta della Propaganda, 1879.

Piggott, Jan R. *Palace of the People: The Crystal Palace at Sydenham 1854–1836*. London: Hurst, 2004.

Poitevin, Prosper. "V. Les États pontificaux." *L'Exposition Universelle de 1867 illustrée* 44 (1867): 218–219.

Ponzi, Giuseppe. "Sui manufatti in focaja rinvenuti all'Inviolatella nella campagna romana e sull'uomo all'epoca della pietra." *Atti della Pontificia Accademia dei Nuovi Lincei* 20 (1867): 41–52.

Poppi, Francesco, Fabrizio Bònoli, and Ileana Chinnici. "Il progetto Tacchini e la riforma degli Osservatori italiani." In *Cento anni di astronomia in Italia 1860–1960*. Atti dei Convegni Lincei 217, 123–171. Rome: Bardi Editore, 2005.

Ptitsyna, Natalia. "Il disco di Secchi." *Angelo Secchi: L'avventura scientifica del Collegio Romano*, edited by Aldo Altamore and Sabino Maffeo, 241–253. Foligno: Edizioni Quater, 2012.

Radau, Rudolph. "Académie des Sciences." *Le moniteur scientifique* 9 (1867): 888–891.

Rayet, Georges. "Rapport sur une mission astronomique en Italie." *Archives des missions scientifiques et littéraires: Choix de rapports et instructions*, ser. 3, t. 3–2 (1876): 529–544.

Respighi, Lorenzo. "Réponse à une note précédente du P. Secchi sur quelques particularité de la constitution du soleil." *Comptes rendus de l'Académie des Sciences* 74 (1872a): 1387–90.

Respighi, Lorenzo. "Sulle osservazioni spettroscopiche del bordo e delle protuberanze solari, fatte all'Osservatorio dell'Università Romana sul Campidoglio." *Atti della Reale accademia dei Lincei* 25 (1872b): 247–316.

Rigge, William F. "Father Angelo Secchi," *Popular Astronomy* 26 (1918): 589–598.

Royal Astronomical Society Council. "Report […] to the Forty-First Annual General Meeting of the Society." *Monthly Notices of the RAS* 12 (1861): 73–123.

Sawyer Hogg, Helen. "Variable Stars." In *The General History of Astronomy, 4A: Astrophysics and Twentieth-Century Astronomy to 1950*, edited by Owen Gingerich, part A, 73–89. Cambridge: Cambridge University Press, 1984.

Scalera, Giancarlo. *I moti e la forma della terra*. Rome: Istituto Nazionale di Geofisica, 1999.

Schjellerup, Hans C. F. C. "Catalog der roten, isolierten Sterne," *Astronomische Nachrichten* 67 (1866): 97–112.

Schmidinger, Heinrich. "Centros tomistas en Roma, Nápoles, Perugia, etc.: S. Sordi, D. Sordi, L. Taparelli D'Azeglio, M. Liberatore, C. M. Curci, G. M. Cornoldi y otros." In *Filosofía cristiana en el pensamiento católico de los siglos XIX y XX*, edited by Emerich

Coreth, S.J., Walter M. Neidl, and Georg Pfligersdorffer, 2:103–122. Madrid: Ediciones Encuentro, 1994.

Secchi, Angelo. "Sopra alcune esperienze di reometria elettrica", *Gazzetta scientifica* 3 (1847c): 385–394.

Secchi, Angelo. "II. Considerazioni sulla vera maniera di valutare il raggiamento solare, e ricerche sulla forza assorbente dell'atmosfera terrestre." *Memorie dell'Osservatorio del Collegio Romano: Anno 1851; Appendice alle osservazioni del 1851* (1852): xvi–xxvi.

Secchi, Angelo. "III. Sull'intensità del calore nelle varie parti del disco solare." *Memorie dell'Osservatorio del Collegio Romano: Anno 1851; Appendice alle osservazioni del 1851* (1852): xxvii–xxxvii.

Secchi, Angelo. "IV. Esperimenti sulla deviazione del pendolo diretti a provare la rotazione della terra." *Memorie dell'Osservatorio del Collegio Romano: Anno 1851; Appendice alle osservazioni del 1851* (1852): xxxviii–xxxx.

Secchi, Angelo. "Alcune considerazioni su le tre ultime grandi comete," *Memorie dell'Osservatorio del Collegio Romano 1860–62* (1863): 2:18–32.

Secchi, Angelo. "Ristauri fatti all'Osservatorio." *Memorie dell'Osservatorio del Collegio Romano: Anno 1850*, 5–8. Rome: Tipografia delle Belle Arti, 1851.

Secchi, Angelo. "Astrometria." *Memorie dell'Osservatorio del Collegio Romano: Anno 1850*, 35–45. Rome: Tipografia delle Belle Arti, 1851.

Secchi, Angelo. "L'astronomia in Roma nel pontificato di Pio IX." In *Triplice omaggio alla Santità di Papa Pio IX nel suo giubileo episcopale*, 29–77. Rome: Tipografia della Pace, 1877a.

Secchi, Angelo. "Atlas photographique lunaire." *Comptes rendus de l'Académie des Sciences* 47 (1858a): 362–364.

Secchi, Angelo. "Aurora borealis." *Proceedings of the American Association for the Advancement of Science* 11 (1849): 2–9.

Secchi, Angelo. *Catalogo di 1321 stelle doppie misurate col grande equatoriale di Merz all'Osservatorio del Collegio Romano*. Rome: Tipografia delle Belle Arti, 1860a.

Secchi, Angelo. "Catalogo delle stelle di cui si è determinato lo spettro luminoso all'Osservatorio del Collegio Romano." *Memorie della Società Italiana delle Scienze*, serie 3, t. I-I (1867a): 105–152.

Secchi, Angelo. "Catalogue d'étoiles double." *Comptes rendus de l'Académie des Sciences* 47 (1858b): 366.

Secchi, Angelo. "Cometa del 6 marzo 1853 scoperta all'Osservatorio del Collegio Romano." *Memorie dell'Osservatorio del Collegio Romano 1852–55* (1856f): 149.

Secchi, Angelo "Cometa di Biela 1852." *Memorie dell'Osservatorio del Collegio Romano 1852–55* (1860c): 147–149.

Secchi, Angelo. "Conclusione." *Memorie dell'Osservatorio del Collegio Romano 1852–55* (1856a): 157–158.

Secchi, Angelo. "Correspondance." *Comptes rendus de l'Académie des Sciences* 36 (1853a): 543.

Secchi, Angelo. "Descrizione del nuovo Osservatorio del Collegio Romano." *Memorie dell'Osservatorio del Collegio Romano 1852–55*, (1856b): 1–24.

Secchi, Angelo. "Descrizione di un meteorografo ossia registratore meteorologico universale eretto all'osservatorio." In *Memorie dell'Osservatorio del Collegio Romano 1857–1859*, 1:1–8. Rome: Tipografia delle Belle Arti, 1859a.

Secchi, Angelo. "Descrizione di un nuovo apparato per trasmettere i segni nei telegrafi elettrici." *Raccolta scientifica di Roma*, anno terzo (February 1, 1847a): 40–45.

Secchi, Angelo. "Eclipse solaire du 18 juillet 1860." *Comptes rendus de l'Académie des Sciences* 51 (1860d): 749–751.

Secchi, Angelo. "Elettricità atmosferica." *Bullettino meteorologico dell'Osservatorio del Collegio Romano* 1 (1862a): 158–160.

Secchi, Angelo. "Études sur la planète Mars." *Comptes rendus de l'Académie des Sciences* 48 (1858c): 364–366.

Secchi, Angelo. "Extrait d'une lettre à M. Arago." *Comptes rendus de l'Académie des Sciences* 34 (1852): 949–950.

Secchi, Angelo. "Fotografie lunari e degli altri corpi celesti." *Memorie dell'Osservatorio del Collegio Romano N.S.I* (1859b): 158–160.

Secchi, Angelo. "Gruppi di stelle." *Memorie dell'Osservatorio del Collegio Romano 1852–55* (1856e): 71–79.

Secchi, Angelo. "Guida dei naviganti di lungo corso di Vincenzo Gallo." *Annali di scienze matematiche e fisiche* 4 (1853b): 245–260.

Secchi, Angelo. *Intorno ad alcuni avanzi di opere idrauliche antiche rinvenuti nella città di Alatri*. Rome: Tipografia delle Belle Arti, 1865a.

Secchi, Angelo. "Intorno alla corrispondenza che passa fra le variazioni meteorologiche e quelle del magnetismo terrestre." *Memorie dell'Osservatorio del Collegio Romano*, n.s. 2, nos. 11–15 (1863): 81–120.

Secchi, Angelo. *Intorno alla soluzione di un problema fisico-cosmologico. Lettera al R. P. Nardini*. Rome: Tipografia Marini, 1862b.

Secchi, Angelo. "Intorno alla vita e alle opere del P. Giambattista Pianciani." *Giornale Arcadico di Scienze: Lettere ed arti*, n.s., 62 (1862c): 1–48.

Secchi, Angelo. "Intorno all'influenza solare sull'atmosfera terrestre: Parte III." *Bullettino meteorologico dell'Osservatorio del Collegio Romano* 2 (1863c): 105–108.

Secchi, Angelo. "Introduzione: Sullo stato dell'osservatorio e sui lavori intrapresi nell'anno 1851." *Memorie dell'Osservatorio del Collegio Romano: Anno 1851* (1852): 1–16.

Secchi, Angelo. "Lettera a Salvatore Betti intorno alla condanna del falsario degli autografi di Galileo Galilei ec." *Giornale arcadico di scienze, lettere ed arti* 207, no. 62 (1869a [*sic*; 1870]): 170–171.

Secchi, Angelo. *Lezioni elementari di fisica terrestre, con l'aggiunta di due discorsi sulla grandezza del creato*. Turin: Ermanno Loescher, 1879.

Secchi, Angelo. *Memorie dell'Osservatorio del Collegio Romano: Anno 1850*. Rome: Tipografia delle Belle Arti, 1851.

Secchi, Angelo. *La météorologie et le météorographe du P. Secchi S.J. à l'Exposition Universelle*. Paris: Gauthier-Villars, 1867b.

Secchi, Angelo. "Nébuleuses découvertes avec l'équatorial de Merz au Collège Romain de 11 novembre 1865 au 18 janvier 1866." *Astronomische Nachrichten* 66, no. 1571 (1866a): 161.

Secchi, Angelo. "Note sur les spectres prismatiques des corps célestes." *Comptes rendus de l'Académie des Sciences* 57 (1863): 71–75.

Secchi, Angelo. "Nouvelle mesure de la base de Boscovich." *Comptes rendus de l'Académie des Sciences* 50 (1860f): 377–378.

Secchi, Angelo. "La nuova stella del Cigno," *Bullettino meteorologico dell'Osservatorio del Collegio Romano* 16 (1877b): 1–3.

Secchi, Angelo. "Nuove apparenze dell'anello di Saturno." *Annali di scienze matematiche e fisiche* 2 (1851): 39–44.

Secchi, Angelo. "Observations de la double Comète de Biela." *Astronomische Nachrichten* 35 (1853c): 251–252.

Secchi, Angelo. "Observations faites pendant l'éclipse totale du 18 juillet 1860 sur le sommet du mont Saint-Michel au Desierto de Las Palmas en Espagne." *Comptes rendus de l'Académie des Sciences* 51 (1860g): 156–162; 276–279; 386–388.

Secchi, Angelo. "On Foucault's Experiment." *Monthly Notices of the Royal Astronomical Society* 11, no. 8 (1851d): 199.

Secchi, Angelo. "On Stellar Spectrometry." In *Report of the Thirty-Eighth Meeting of the British Association of Science, Held at Norwich in August 1868*, 165–170. London: John Murray, 1869b.

Secchi, Angelo. "Osservazioni di nebulose." *Memorie dell'Osservatorio del Collegio Romano 1852–55* (1856g): 80–96.

Secchi, Angelo. "Osservazioni di Saturno in occasione della disparizione dell'anello negli anni 1861 e 62." *Memorie dell'Osservatorio del Collegio Romano*, n.s. 2, no. 10 (1863): 73–96.

Secchi, Angelo. "Pendolo Foucault." *Annali di scienze matematiche e fisiche* 2 (1851e): 325–345.

Secchi, Angelo. *Quadro fisico del sistema solare secondo le più recenti osservazioni*. Rome: Tipografia delle Belle Arti, 1859c.

Secchi, Angelo. "Questioni spettroscopiche." *Bullettino meteorologico dell'Osservatorio del Collegio Romano* 11 (1872a): 53–57.

Secchi, Angelo. "Rapporto del prof. P. Angelo Secchi." In *Rapporti sulle osservazioni dell'ecclisse totale di sole del 22 dicembre 1870 eseguite in Sicilia dalla Commissione italiana*, edited by Gaetano Cacciatore, 15–30. Palermo: Stabilimento Tipografico Lao, 1872b.

Secchi, Angelo. "Le recenti scoperte astronomiche." *Giornale Arcadico di Scienze: Lettere ed arti* 54 (1868a): 168–196.

Secchi, Angelo. "Relazione delle osservazioni fatte in Spagna durante l'eclisse totale del 18 luglio 1860." *Giornale Arcadico di Scienze: Lettere ed arti*, n.s. 20 (1860i): 53–121.

Secchi, Angelo. "Researches on Electrical Rheometry." *Smithsonian Contributions to Knowledge* 3, no. 2 (1852): 1–59.

Secchi, Angelo. "Résultats fournis par l'analyse spectrale de la lumière d'Uranus, de l'étoile R des Gémeaux, et des taches solaires." *Comptes rendus de l'Académie des Sciences* 68 (1869c): 761–765.

Secchi, Angelo. "Ricerche nelle protuberanze solari." *Nuovo cimento* 3 (1870a): 217–220.

Secchi, Angelo. "Schreiben des des Herrn Professors Secchi [...] an den Herausgeber." *Astronomische Nachrichten* 67, no. 1608 (1866b): 381–382.

Secchi, Angelo. "Serie seconda delle misure micrometriche fatte all'equatoriale di Merz del collegio romano dal 1863 al 1866 inclusive: Stelle doppie e nebulosi." *Atti della Accademia Pontificia de' Nuovi Lincei* 21 (1868b): 159–173.

Secchi, Angelo. *Le soleil*. 2 Vols. Paris: Gauthier-Villars, 1875–77.

Secchi, Angelo. "Sopra alcune osservazioni fatte nell'Osservatorio del Collegio Romano durante l'eclisse del 28 luglio 1851." *Annali di scienze matematiche e fisiche* 2 (1851f): 383–393.

Secchi, Angelo. "Sopra il prognostico del tempo." *Bullettino meteorologico dell'Osservatorio del Collegio Romano* 17 (1878a): 20; 25–26.

Secchi, Angelo. *Le stelle: Saggio di astronomia siderale*. Milan: Fratelli Dumolard, 1877c.

Secchi, Angelo. "Studi fisici sulle Comete del 1874." *Memorie della Società degli Spettroscopisti Italiani* 3 (1874a): 117–120.

Secchi, Angelo. "Studi spettrali." *Bullettino meteorologico dell'Osservatorio del Collegio Romano* 8 (1869d): 81–83.

Secchi, Angelo. "Sugli orologi elettromagnetici." *Raccolta scientifica di Roma*, anno secondo (November 15, 1846): 353–358.

Secchi, Angelo. "Sugli spettri prismatici della luce de' corpi celesti." *Bullettino meteorologico dell'Osservatorio del Collegio Romano* 2 (1863): 108–110; 114–116; 124–126.

Secchi, Angelo. "Sugli spettri prismatici della luce de' corpi celesti." *Memorie dell'Osservatorio del Collegio Romano*, n.s. 2 (1863): 121–128.

Secchi, Angelo. "Sugli spettri prismatici delle stelle fisse." *Memorie della Società Italiana delle Scienze*, serie 3, t. I-I (1867c): 67–104.

Secchi, Angelo. "Sugli spettri prismatici delle stelle fisse: Memoria II^{a}." *Memorie della Società Italiana delle Scienze*, serie 3, t. 2 (1869e): 73–133.

Secchi, Angelo. "Sul clima di Roma." *Bullettino meteorologico dell'Osservatorio del Collegio Romano* 5 (1866c): 44–48; 57–58.

Secchi, Angelo. "Sul modo di valutare la forza del raggiamento solare." *Annali di scienze matematiche e fisiche* 2 (1851g): 508–520.

Secchi, Angelo. "Sulla correlazione delle forze fisiche e sulla sua influenza nel concetto dell'universo." *Giornale arcadico*, n.s. 155, no. 9 (1858d): 1–27.

Secchi, Angelo. "Sulla legge di Ohm delle correnti elettriche." *Bullettino meteorologico dell'Osservatorio del Collegio Romano* 17 (1878b): 9–10.

Secchi, Angelo. "Sulla struttura della fotosfera solare: Lettera del Rev. Sig. Dawes." *Bullettino meteorologico dell'Osservatorio del Collegio Romano* 3 (1865b): 93–99.

Secchi, Angelo. "Sulla teoria delle macchie solari proposta dal sig. Kirchhoff." *Bullettino meteorologico dell'Osservatorio del Collegio Romano* 3 (1864a): 1–4.

Secchi, Angelo. "Sulle macchie e sulla temperatura del Sole." *Memorie dell'Osservatorio del Collegio Romano 1852–55* (1856c): 130–133.

Secchi, Angelo. "Sulle macchie solari e sul modo di determinarne la profondità." *Memorie dell'Osservatorio del Collegio Romano*, n.s. 1 (1859d): 9–12.

Secchi, Angelo. "Sulle oscillazioni del pendolo avuto riguardo alla rotazione della terra." *Annali di scienze matematiche e fisiche* 2 (1851h): 238–242.

Secchi, Angelo. "Sullo spettro della Cometa Tempel." *Memorie della Società degli Spettroscopisti Italiani* 3 (1874b): 29–30.

Secchi, Angelo. "Sullo spettro delle protuberanze solari osservato in pieno sole." *Bullettino meteorologico dell'Osservatorio del Collegio Romano* 7 (1868c): 89–91.

Secchi, Angelo. "Sull'ecclisse solare totale osservato in Spagna nel 18 luglio 1860." *Memorie dell'Osservatorio del Collegio Romano 1860–62* (1863): 2:33–52.

Secchi, Angelo. *Sull'ecclisse totale del Sole che avrà luogo ai 22 dicembre 1870: Notizie ed istruzioni*. Milan: Vallardi Editore, 1870b.

Secchi, Angelo. "Sull'ultima pubblicazione del prof. Volpicelli intorno alla cecità del Galileo: Lettera del p. Secchi al ch.mo prof. cav. Betti uno dei compilatori del Giornale Arcadico." *Giornale arcadico di scienze, lettere ed arti* 57 (1868d): 233–246.

Secchi, Angelo. *Sull'unità delle forze fisiche*. Rome: Tipografia Forense, 1864b.

Secchi, Angelo. "The Sun." *Nature* 4 (1871b): 82–83.

Secchi, Angelo. "Sur les expériences du pendule qui vont être entreprises dans le tunnel des Alpes occidentales." *Comptes rendus de l'Académie des Sciences* 73 (1871c): 1192–93.

Secchi, Angelo "Sur un nouveau baromètre à balance." *Comptes rendus de l'Académie des Sciences* 44 (1857): 31–34.

Secchi, Angelo. "Sur les perturbations magnétiques observées à Rome le 2 septembre 1859." *Comptes rendus de l'Académie des Sciences* 49 (1859e): 458–460.

Secchi, Angelo. "La teoria delle macchie solari proposta da Galileo." *Bullettino meteorologico dell'Osservatorio del Collegio Romano* 13 (1874b), 45–48; 53–55.

Secchi, Angelo. *Il traforo delle Alpi nella Catena del Moncenisio al Colle di Fréjus*. Rome: Tipografia e Libreria Alessandro Befani, 1872c.

Secchi, Angelo. *L'unité des forces physiques*. Paris: F. Savy, 1869.

Secchi, Angelo. "Selenografia." *Memorie dell'Osservatorio del Collegio Romano 1852–55* (1856d): 134–135.

Secchi, Angelo. "Entdeckung und Beobachtungen eines kleinen Cometen", *Astronomische Nachrichten* 35, no. 822 (1853d): 89–90.

Secchi, Angelo. "Beobachtungen des Biela'schen Cometen und Bemerkungen über densellen und seinen aufgefunde Begleiter", *Astronomische Nachrichten* 35, no. 828 (1853e): 191–192.

Secchi, Angelo. *Misure della base trigonometrica eseguita sulla Via Appia*. Rome: Tipografia della Rev. Camera Apostolica, 1858.

Secchi, Angelo. "Spectre de la comète de Tempel," *Comptes rendus de l'Académie des Sciences* 62 (1866): 210.

Secchi, Angelo. *Sulla struttura delle macchie solari*. Rome: Tipografia delle Belle Arti, 1866.

Secchi, Angelo. "Rapporto della Commissione per la Misura del Meridiano Centrale Europeo." *Atti della R. Accademia dei Lincei* 24 (1871a): 232–258.

Secchi, Angelo. *Le soleil*. Paris: Gauthier-Villars, 1870.

Secchi, Angelo. "Studi fisici fatti all'Osservatorio del Collegio Romano sulle comete di Tempel II e Coggia III nel 1874." *Memorie della Società degli Spettroscopisti Italiani* 3 (1874): 121–26.

Secchi, Angelo. "Un omaggio alla memoria del Commodoro Maury." *Bullettino Meteorologico dell'Osservatorio del Collegio Romano* 12 (1873): 41–42.

Serpieri, Alessandro. *Sulla riduzione delle osservazioni meteorologiche e specialmente delle osservazioni triorarie*. Pesaro: Tipografia Nobili, 1854.

Serra-Carpi, Giuseppe. *Sulle linee isotermiche dell'Italia de' suoi mari ed isole adiacenti*. Rome: Tipografia delle Belle Arti, 1865.

Shea, John Gilmary. *Memorial of the First Centenary of Georgetown College, D.C.*. New York: Collier, 1891.

Simon, Josep. *Communicating Physics: The Production, Circulation and Appropriation of Ganot's Textbooks in France and England*. London: Routledge, 2011.

Soret, Jacques-Louis. "On the Intensity of the Solar Radiation." *Philosophical Magazine* 34, no. 231 (1867): 404–407.

Struve (von), Friedrich Wilhelm. *Stellarum duplicium et multiplicium mensurae micrometricae per magnum Fraunhoferi tubum annis a 1824 ad 1837 in Specula dorpatensi institutae*. Petropoli: Ex Typographia academica, 1837.

Swan, William. "On the Prismatic Spectra of the Flames of Compounds of Carbon and Hydrogen" *Transactions of the Royal Society of Edinburgh* 21 (1857): 411–430.

Swings, Polydore F. F. "Cometary Spectra." *Quarterly Journal of the Royal Astronomical Society* 6 (1965): 28–69.

Tacchini, Gaetano. *Torre telescopio in cemento armato da erigersi in Reggio-Emilia all'insigne astronomo padre Angelo Secchi*. Modena: n.p., 1914.

Tacchini, Pietro. "Della convenienza ed utilità di erigere sull'Etna una stazione astronomico-meteorologica." *Atti Accademia Gioenia* 3 (1878): 7–26.

Tacchini, Pietro. "Un disegno del Sole nell'opera di Kircher Mundus Subterraneus." *Memorie della Società degli Spettroscopisti Italiani* 5 (1876): 135–136.

Tacchini, Pietro. *Eclissi totali di Sole*. Rome: Tipografia Eredi Botta, 1888.

Tacchini, Pietro. "Fisica solare." *Bullettino meteorologico del R. Osservatorio di Palermo* 7 (1871): 93–95.

Tacchini, Pietro. "Nuova Società di Spettroscopisti Italiani." *Memorie della Società degli Spettroscopisti Italiani* 1 (1872): 3–6.

Tacchini, Pietro. "Protuberanze solari osservate contemporaneamente a Palermo, Roma e Padova nel luglio ed agosto 1871." *Memorie della Società degli Spettroscopisti Italiani* 1 (1872): 25–32.

Tacchini, Pietro. "Il Sole e le aurore boreali osservate a Palermo nell'aprile 1871." *Rivista Sicula* 5 (1871): 415–427.

Tarkovsky, Arseny. *Poems from Different Years*. Moscow: "Sovremennik," 1983.

Terby, François. "The Markings on Mars." *Observatory* 3 (1880): 416.

Tomassini, Stefano. *Storia avventurosa della rivoluzione romana: Repubblicani, liberali e papalini nella Roma del '48*. Milan: Il Saggiatore, 2008.

Tricht (van), Victor. "Les écrits et la mort de Secchi." In *Le révérend Père Secchi*, edited by Francois Moigno, 188–205. Paris: Gauthier-Villars, 1879.

Tuscano, Maria Luisa. "Angelo Secchi e la gnomonica." In *Angelo Secchi: L'avventura scientifica del Collegio Romano*, edited by Aldo Altamore and Sabino Maffeo, 255–73. Foligno: Edizioni Quater, 2012.

Tyler, John E. "The Secchi Disc." *Limnology and Oceanography* 13, no. 1 (1968): 1–6.

Udías, Agustín. *Jesuit Contribution to Science: A History*. New York: Springer, 2015.

Udías, Agustín. *Searching the Heavens and the Earth: The History of Jesuit Observatories*. Dordrecht: Kluwer Academic, 2003.

Vasi, Giuseppe. *Delle magnificenze di Roma antica e moderna. Libro nono*. Rome: Nella Stamperia di Niccolò e Marco Pagliarini, 1759.

Violle, Jules. "Sur la température du soleil." *Annales scientifiques de l'ENS* 4 (1875): 363–370.

Volpicelli, Paolo. *Sull'epoca della completa cecità di Galileo: Risposta al Ch. e Rev. P. Angelo Secchi*. Rome: Tipografia delle Belle Arti, 1868.

Von Lamont, Johann. "Ueber die Anordnung und physische Beschaffenheit der Planeten." *Jahrbuch der Königliche Sternwarte bei München für 1838* (1838): 175–202.

Waterston, John J. "An Account of Observations on Solar Radiation." *Monthly Notices of the Royal Astronomical Society* 22, no. 2 (1862): 60–68.

Weaver, Harold F. "The Development of Astronomical Photometry-I." *Popular Astronomy* 54 (1946): 211–230.

Webb, William Henry. *Descriptive Particulars of the* Great Eastern *Steamship*. London: Marshall & Sons, 1854.

Wernand, Marcel R. "On the History of the Secchi Disc." *Journal of the European Optical Society: Rapid Publications* 5 (2010): 1–6.

Young, Charles A. *The Sun*. New York: Dr. Appleton & Company, 1896.

[Zantedeschi, Francesco]. *Delle dottrine del R.P. A. Secchi intorno ai presagi delle meteore e delle burrasche e di quelle del prof. Zantedeschi con documenti storici*. Padova: Tipografia di A. Bianchi, 1866.

Zantedeschi, Francesco. *Risposta documentata del professore Francesco Zantedeschi all'articolo del p. A. Secchi inserito nel Bullettino meteorologico dell'Osservatorio del Collegio Romano: Vol. pag. 107, n° di Ottobre 1866 intorno ai presagi delle meteore e delle burrasche con documenti storici*. Padua: Tipografia di A. Bianchi, 1866.

Zanut, Stefano. "L'osservatorio meteorologico di Pordenone e l'ambiente scientifico di fine Ottocento." In AA. vv., "Pordenone, città tra passato, presente e future." *La loggia* 10 (2013): 19–25.

Index

www.ingramcontent.com/pod-product-compliance
Lightning Source LLC
La Vergne TN
LVHW010639300926
849161LV00005B/988